ISBN 978-1-333-25857-3
PIBN 10479386

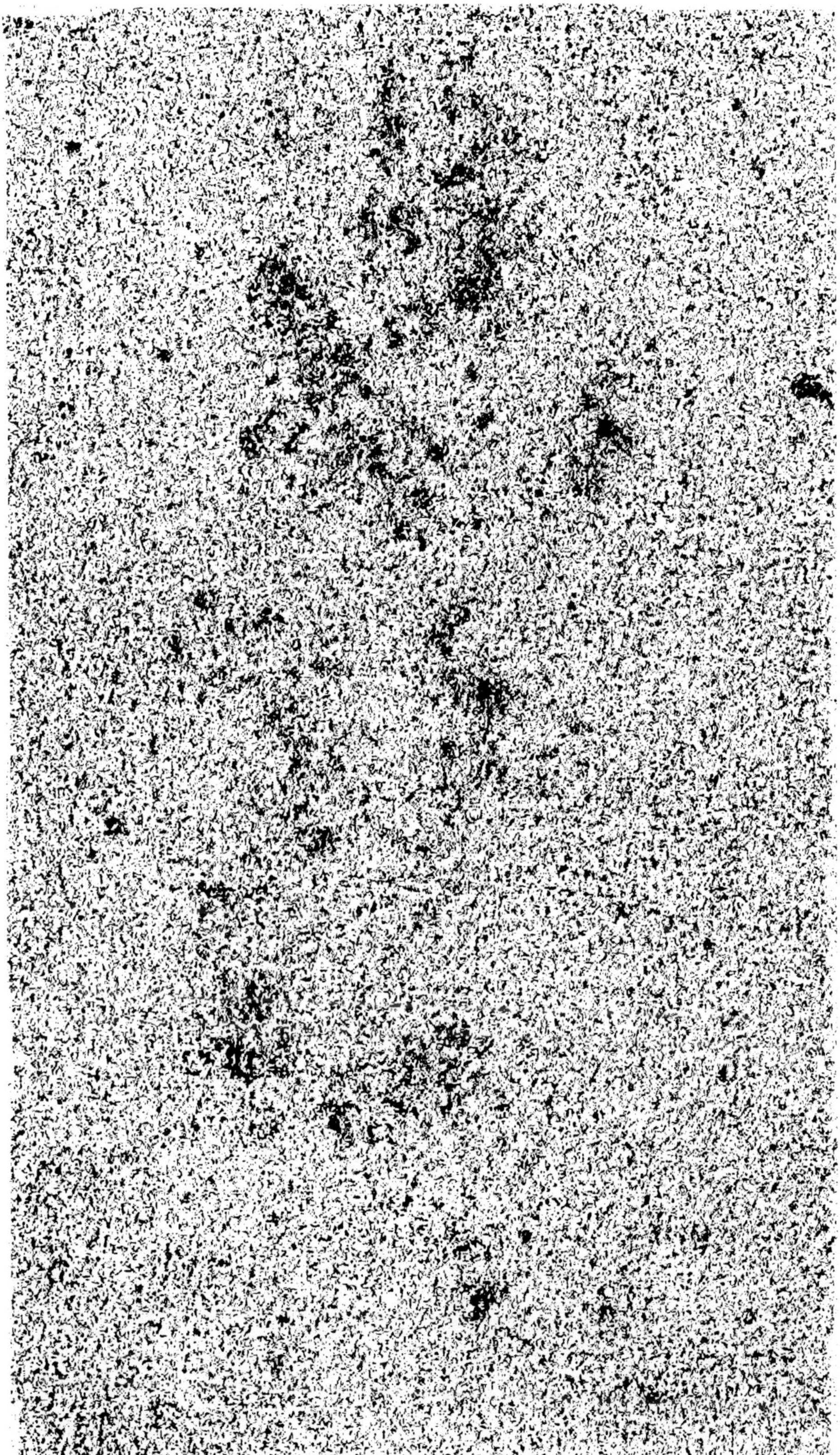

ELEMENTARY ILLUSTRATIONS

OF

THE CELESTIAL MECHANICS

OF

LAPLACE.

ELEMENTARY ILLUSTRATIONS

OF THE

ELESTIAL MECHANIC

OF

LAPLACE.

PART THE FIRST,

COMPREHENDING THE FIRST BOOK.

LONDON:

PRINTED FOR JOHN MURRAY,

BOOKSELLER TO THE ADMIRALTY, AND TO THE BOARD OF LONGITUDE,

ALBEMARLE STREET.

1821.

Howlett and Brimmer, Printers
10, Frith Street, Soho.

TO

THE RIGHT HONOURABLE

ROBERT VISCOUNT MELVILLE,

IRST LORD COMMISSIONER OF THE BOARD OF ADMIRALT

&c. &c. &c.

AS AN ACKNOWLEDGMENT

OF THE ENLIGHTENED LIBERALITY

WHICH HE HAS DISPLAYED

ON MANY PUBLIC OCCASIONS

IN THE ADVANCEMENT OF

ASTRONOMICAL AND GEOGRAPHICAL KNOWLEDGE

THESE ILLUSTRATIONS

ARE GRATEFULLY AND RESPECTFULLY DEDICATED

BY THE AUTHOR.

PREFACE.

ALTHOUGH it is intended that these Illustrations, if they be found useful to the British student, should be extended not only to the whole of the Celestial Mechanics of Laplace, but possibly to some other works relating to astronomy and the higher mathematics; yet they may be considered as forming, even in their present state, a work completely independent of all others: and the separate publication of each part has been considered as possessing the advantage of dividing a long journey, into stages of a less formidable appearance, for the convenience both of the traveller and of his conductor, so that if either party should discontinue the undertaking, before the whole tour is completed, the part actually travelled over may be considered as making a whole within itself, and affording sufficient information and improvement to repay the labour of the journey, even without any ulterior view to the completion of the remaining part.

The translator having been accustomed to consider the elementary doctrines of motion, and some other parts of the subjects discussed, in a point of view, which has from habit become more familiar to him, and which he is, perhaps on that account, involuntarily disposed to think more natural and satisfactory, he has extracted, from his own former publications, such parts as he has felt himself compelled to substitute for Mr. Laplace's introductory investigations, but without omitting, as collateral illustrations, such of Mr. Laplace's demonstrations as appear to be the most ingenious and satisfactory. In these earlier parts, he has found it most convenient to adopt the order and arrangement of his own elementary works, inserting any of Mr. Laplace's remarks in the form of Scholia or otherwise: but in the principal part of the book he has followed the order of the original sections, introducing any additions of his own in the form of Lemmas or Scholia, besides the explanatory remarks, and details of demonstration, which are distinguished by being included in crotchets. The text is, however, throughout the whole, divided into distinct propositions, enunciated at the beginning of each investigation, which is perhaps a departure from a strict analytical order, but which affords the memory, as well as the apprehension of the student, a very material advantage. The steps required for each demonstration are filled up by a recurrence to the fundamental principles of mathematics and mechanics, without reference to any other introductory work, which indeed would have been insufficient for the information of the mere English reader: but these summary demonstrations must not be understood as intended to be fully comprehended by a mere beginner, or as calculated to supersede the necessity of the study of many other works, on the different branches of mathematical science. The translator flatters himself, however, that he has not expressed the author's meaning in English words alone, but that he has rendered it perfectly intelligible to any person, who is conversant with the English mathematicians of the old school only, and that his book will serve as a connecting link between the geometrical and algebraical modes of representation. A Mosaic work of this kind may perhaps possess less of perfect harmony, than if it had been more regularly modelled into a continuous system: but the want of strict method is in part compensated, by the greater interest, which naturally arises from a mixture of the direct application to the phenomena of nature, with the abstract investigation of purely mathematical truths. To the illustrious author of the work, however, some apology is certainly due, for having ventured to depart from the original symmetry of his design; and the best excuse, that can be assigned, will perhaps be the universal acquaintance of all judges of the higher mathematics, with the Mécanique Céleste in its original form, which will enable them at once to attribute to the translator any want of analytical refinement, that may have been admitted by the alterations.

To those who are desirous of confining their attention to whatever is absolutely new and original, or placed in a decidedly new light, it may be proper to point out the extreme simplicity which is given, at the end of the book, to the theory of waves and of sounds, and the still greater novelty of that of the cohesion of fluids, which, it is presumed, will be allowed to be deduced in a most unexceptionable manner from the general principle of virtual velocities. There are, also, some remarks on the application of Taylor's theorem, which may be found of considerable utility in computing the forms of the surfaces of fluids, and which are still more important on account of the great assistance, which may be derived from them, in calculations respecting the figure of the earth, as connected with its compressibility.

It is almost superfluous to add, that any corrections, which may occur to the mathematical reader, whether of errors of the press, or of more serious mistakes, will be gratefully received, and candidly acknowledged, by the author of these illustrations.

London, 28 *Feb*. 1821.

CONTENTS.

INTRODUCTION.

RUDIMENTS OF THE MATHEMATICS.

SECTION I. OF QUANTITY AND NUMBER.

1. **Definition.** The letters of the alphabet are employed at pleasure for denoting any quantities, as algebraical symbols or abbreviations. But, in general, the first letters in order are used to denote known quantities, and the last to denote unknown quantities; and constant quantities are often distinguished from variable quantities in the same manner.

2. **Definition.** Quantities are equal when they are of the same magnitude.

Scholium. The abbreviation $a=b$ implies that a is equal to b; $a>b$ that a is greater than b; and $a<b$ that a is less than b.

3. **Definition.** Addition is the joining of magnitudes into one sum.

Scholium. The symbol of addition is an erect cross: $a+b$ implies the sum of a and b, and is called a more b.

4. **Definition.** Subtraction is the taking as much from one quantity as is equal to another.

Scholium. Subtraction is denoted by a single line, as $a-b$, or a less b, which is the part of a remaining, when a part equal to b has been taken from it.

5. **Definition.** A negative quantity is of an opposite nature to a positive one, with respect to addition or subtraction; the condition of its determination being such, that it must be subtracted where a positive quantity would be added, and the reverse.

Scholium. A negative quantity is denoted by the sign of subtraction: thus if $a+b=a-c$, $b=-c$ and $c=-b$. A debt is a negative kind of property, a loss a negative gain, and a gain a negative loss.

6. **Definition.** A unit is a magnitude considered as a whole complete within itself.

Scholium. When any quantities are enclosed in a parenthesis, or have a line drawn over them, they are considered as one quantity with respect to other symbols; thus $a-(b+c)$, or $a-\overline{b+c}$, implies the excess of a above the sum of b and c.

7. **Definition.** A whole number is a number composed of units by continued addition.

Thus one and one compose two, $2+1=3$, $3+1=4$, or $2+2=4$. Such numbers are also called multiples of unity.

8. **Definition.** A simple fraction is a number which by continual addition composes a unit, and the number of such fractions, contained in a unit, is denoted by the denominator, or number below the line.

Thus $\frac{1}{3}+\frac{1}{3}+\frac{1}{3}=1$.

9. **Definition.** A number composed of such simple fractions, by continual addition, may properly be termed a multiple fraction; the number of simple fractions composing it is denoted by the upper figure or numerator.

In this sense $\frac{2}{3}$, $\frac{3}{3}$, $\frac{4}{3}$, are multiple fractions, and $\frac{3}{3}=1$, $\frac{4}{3}=\frac{3}{3}+\frac{1}{3}=1+\frac{1}{3}$, or $1\frac{1}{3}$.

10. **Definition.** Such quantities as are expressible by the relations denoted by whole numbers, or fractions, are called commensurable quantities.

SCHOLIUM. All quantities may, in practice, be considered as commensurable, since all quantities are expressible by numbers, either accurately, or with an error less than any assignable quantity.

11. DEFINITION. Multiplication is the adding together so many numbers equal to the multiplicand as there are units in the multiplier, into one sum, called the product.

SCHOLIUM. Multiplication is expressed by an oblique cross, by a point, or by simple apposition; $a \times b = a.b = ab$.

12. DEFINITION. Division is the subtraction of a number from another as often as it is contained in it; or the finding of that quotient, which, when multiplied by a given divisor, produces a given dividend.

SCHOLIUM. Division is denoted by placing the dividend before the sign $\div$ or $:$, and the divisor after it; as $a \div b = a : b$.

13. AXIOM. When no difference can be shown or imagined between two quantities, they are equal.

14. AXIOM. Quantities, equal to the same quantity, are equal to each other.

If $a = b$ and $c = b$, then $a = c$.

15. AXIOM. If to equal quantities equal quantities be added, the wholes will be equal.

If $a = b$, then $a + c = b + c$; if $a - b = c$, then adding b, $a = b + c$; if $a + b - c = d$, then adding c, $a + b = c + d$.

16. AXIOM. If from equal quantities equal quantities be subtracted, the remainders will be equal.

If $a = b$, $a - c = b - c$, if $a + b = b + c$, $a = c$.

17. AXIOM. If equal quantities be multiplied by equal numbers, the products will be equal.

If $a = b$, $3a = 3b$; if $a = b : 3$, $3a = b$; and if $a = b$, $ca = cb$.

18. AXIOM. If equal quantities be divided by equal numbers, the quotients will be equal.

If $5a = 10b$, $a = 2b$; and if $ca = cb$, $a = b$.

SCHOLIUM. Articles 16, 17, 18, might have been deduced from art. 15, but they are all easily admitted as axioms. We must however observe that this proposition does not extend to the case of 0 for a divisor.

19. THEOREM. A multiple fraction is equal to the quotient of the numerator divided by the denominator.

Or, $\frac{a}{b}=a:b$, for $\frac{a}{b}=\frac{1}{b}.a$ (9); and $b.\frac{a}{b}=b.\frac{1}{b}a$ (17); but $b.\frac{1}{b}=1$ (8); and $b.\frac{1}{b}.a=1.a=a$, therefore $b.\frac{a}{b}=a$ (14), and $a:b=\frac{a}{b}$ (12).

SCHOLIUM. Hence $\frac{a}{b}$ is a common symbol for $a:b$.

20. THEOREM. A quantity, multiplied by a simple fraction, is equal to the same quantity divided by its denominator.

Or $a.\frac{1}{b}=a:b$; for $a.\frac{1}{b}=\frac{a}{b}$(9), and $\frac{a}{b}=a:b$ (19), therefore $a.\frac{1}{b}=a:b$ (14).

21. THEOREM. A quantity, divided by a simple fraction, is equal to the same quantity multiplied by its denominator.

Or $a:\frac{1}{b}=ab$, for if $a:\frac{1}{b}=c, a=c.\frac{1}{b}(12)=\frac{c}{b}=c:b$ (20), and multiplying by b, $ab=c=a:\frac{1}{b}$.

22. THEOREM. A quantity multiplied by a multiple fraction is equal to the same quantity multiplied by the numerator, and then divided by the denominator.

Or $a.\frac{b}{c}=ab:c$; for $a.\frac{b}{c}=a.b.\frac{1}{c}=ab.\frac{1}{c}=ab:c$ (20).

23. THEOREM. A quantity divided by a multiple fraction is equal to the same quantity multiplied by the denominator, and divided by the numerator.

Or $a : \frac{b}{c} = ac : b$; for $a : \frac{b}{c} = a : \left(b . \frac{1}{c}\right) = (a : b) : \frac{1}{c} = (a : b) . c$ (21), $= ac : b$.

SCHOLIUM. A beginner may perhaps render these demonstrations more intelligible, by substituting any numbers at pleasure for the characters. For example, the demonstration of the first theorem may be written thus. Twelve fourths, $\frac{12}{4}$, are equal to 12 divided by 4; for, by the definition of a multiple fraction, $\frac{12}{4} = 12 . \frac{1}{4}$, and multiplying these equals by 4, $4 . \frac{12}{4} = 4 . 12 . \frac{1}{4}$; but by the definition of a simple fraction $4 . \frac{1}{4} = 1$, therefore $4 . 12 . \frac{1}{4} = 12$, whence $4 . \frac{12}{4} = 12$, and by the definition of division, $12 : 4 = \frac{12}{4}$. But, in fact, the proposition is too evident to admit much demonstrative confirmation.

24. THEOREM. A positive number or quantity being multiplied by a positive one, the product is positive.

For the repeated addition of a positive quantity must make the result actually greater. What is true of numbers may practically be affirmed of quantities in general (10).

25. THEOREM. A negative number or quantity being multiplied by a positive one, the product is negative.

For since adding a negative quantity is equivalent to subtracting a positive one, the more of such quantities that are added, the greater will the whole diminution be, and the sum of the whole, or the product, must be negative.

26. THEOREM. A negative number or quantity being multiplied by a negative one, the product is positive.

Or $-a . -b = ab$. For $a . -b = -ab$(25): that is, when the positive quantity a is multiplied by the negative b, the product indicates that a must be subtracted as often as there are units in b: but when a is negative, its subtraction is equivalent to the addition of an equal positive number; therefore in this case an equal positive number must be added as often as there are units in b.

27. DEFINITION. If the quotients of two pairs of numbers are equal, the numbers are proportional, and the first is to the second, as the third to the fourth; and any quantities, expressed by such numbers, are also proportional.

If $a : b = c : d$, a is to b as c to d, or $a : b :: c : d$.

28. **Theorem.** Of four proportionals, the product of the extremes is equal to that of the means.

Since $a : b = c : d$, $a : b.\ bd = c : d.\ bd$. (17), or $ad = cb$.

29. **Theorem.** If the product of the extremes of four numbers is equal to that of the means, the numbers are proportional.

If $ad = cb$, $ad : bd = cb : bd$ (18), and $a : b = c : d$; also $ad : cd = cb : cd$, and $a : c = b : d$.

30. **Theorem.** Four proportionals are proportional alternately.

If $a : b :: c : d$, $ad = bc$ (28), therefore $a : c :: b : d$ (29).

31. **Theorem.** Four proportionals are proportional by inversion.

If $a : b :: c : d$, $ad = bc$, $ad : ac = bc : ac$, and $d : c = b : a$.

32. **Theorem.** Four proportionals are proportional by composition.

If $a : b :: c : d$, $a+b : b :: c+d : d$; for since $ad = bc$, $ad + bd = bc + bd$ (15), or $(a+b).\ d = (c+d).\ b$, therefore $a+b : b :: c+d : d$ (29).

33. **Theorem.** Four proportionals are proportional by division.

If $a : b :: c : d$, $a-b : b :: c-d : d$; for since $ad = bc$, $ad - bd = bc - bd$ (16), $(a-b).\ d = (c-d).\ b$, and $a-b : b :: c-d : d$ (29).

34. **Theorem.** If any number of quantities are proportional, the sum of the antecedents is in the same ratio to the sum of the consequents.

If $a : b :: c : d$, $a : b :: a+c : b+d$; for since $ad = bc$, $ab + ad = ab + bc$, $a.\ (b+d) = b.\ (a+c)$, and $a : b :: a+c : b+d$ (29).

35. **Theorem.** If any number of antecedents and any number of consequents be added together, the ratio of the sums will be less than the greatest of the single ratios, when those ratios are unequal.

Let $\frac{a}{b} > \frac{c}{d}$, then $\frac{a+c}{b+d} < \frac{a}{b}$; for if $\frac{a}{b} = \frac{e}{d}$, $e > c$, and $\frac{a+e}{b+d} > \frac{a+c}{b+d}$ (34); consequently $\frac{a}{b} > \frac{a+c}{b+d}$. The same demonstration may be extended to any number of ratios.

36. **Definition.** A series of numbers, formed by the continual addition of the same number to any given number, is called an arithmetical progression.

2, 5, 8, 11, 14, 17, 20, by adding 3.
20, 17, 14, 11, 8, 5, 2, by adding—3.
a, $a+b$, $a+2b$, $a+3b$, ... $a+(n-1).b$, in general.

Scholium. It may be observed that the sum of each pair of the numbers of these equal progressions is $22=2+20=a+a+(n-1).b=2a+(n-1).b$; the whole sum $22\times7=(2a+(n-1).b).n$, and the sum of each, $na+\frac{nn-n}{2}.b$, a being the first term, b the difference, and n the number of terms.

37. **Definition.** A series of numbers, formed by continual multiplication by a given number, is called a geometrical progression.

As 2, 6, 18, 54; multiplying 2 continually by 3.
a, ab, abb, $abbb$; multiplying a by b.

38. **Definition.** If one of the terms of a geometrical progression is unity, the other terms are called powers of the common multiplier.

As $\frac{1}{32}$, $\frac{1}{16}$, $\frac{1}{8}$, $\frac{1}{4}$, $\frac{1}{2}$, 1, 2, 4, 8, 16, 32. Each term is denoted by placing obliquely over the common multiplier a number expressive of its distance from unity, as $8=2^3$: negative numbers, implying a contrary situation to positive ones, denote that the term precedes instead of following the unit, as $\frac{1}{8}=2^{-3}$.

By reversing the series it is obvious that $\frac{1}{8}=(\frac{1}{2})^3$, and $8=(\frac{1}{2})^{-3}$

It appears that the addition of the indices denoting the places of any terms will point out a term which is their product, as $2^3\times2^2=2^5$, or $8\times4=32$; and that the subtraction of the index is equivalent to division by the term. Hence if $a^2=b=b^1$, a^1 must be equal to $b^{\frac{1}{2}}$ in order that $b^{\frac{1}{2}} \times b^{\frac{1}{2}}$ may make $b^1=a^2$. So that simple fractional numbers serve as indices of the number of times that the quantity must be multiplied together, in order that the product may be the common multiplier of the series, or the simple number b.

SCHOLIUM. Fractional powers are sometimes denoted by the mark $\sqrt{}$, meaning root: thus $\sqrt{a}=a^{\frac{1}{2}}$, $\sqrt[3]{a}=a^{\frac{1}{3}}$. The second power of a number a being called its square, and the third its cube, the fractional powers are called square and cube roots.

The sums of geometrical progressions may be thus computed, if $a+ab+ab^2 \ldots +ab^{n-1}=x$, $ab+ab^2+ab^3 \ldots +ab^n=bx$, and subtracting the former equation from the latter $ab^n-a=bx-x$, therefore $x=\frac{ab^n-a}{b-1}$: which, when $b<1$ and $n=\infty$, or infinite, becomes $\frac{}{1-b}$.

The binomial theorem, for involution, is $(a+b)^n=a^n+n.a^{n-1}b+n.\frac{n-1}{2}.a^{n-2}b^2+n.\frac{n-1}{2}.\frac{n-2}{3}.a^{n-3}b^3+\ldots$ In simple cases, its truth may be shown by induction. See 244.

POWERS OF NUMBERS.

1st	2d	3d.	4th.	5th.	6th.	7th.	8th.
2	4	8	16	32	64	128	256
3	9	27	81	243	729	2187	6561
4	16	64	256	1024	4096	16384	65536
5	25	125	625	3125	15625	78125	390625
6	36	216	1296	7776	46656	279936	1679616
7	49	343	2401	16807	117649	823543	5764801
8	64	512	4096	32768	262144	2097152	16777216
9	81	729	6561	59049	531441	4782969	43046721

$2^{\frac{1}{2}}=1.414213$; $3^{\frac{1}{2}}$,1.732; $5^{\frac{1}{2}}$, 2.236; $6^{\frac{1}{2}}$,2.449; $7^{\frac{1}{2}}$, 2.646; $8^{\frac{1}{2}}$, 2.828; $10^{\frac{1}{2}}$, 3.162.

$2^{\frac{1}{3}}=1.26$; $3^{\frac{1}{3}}$, 1.442: $4^{\frac{1}{3}}$, 1.587; $5^{\frac{1}{3}}$, 1.71; $6^{\frac{1}{3}}$, 1.817; $7^{\frac{1}{3}}$, 1.913; $9^{\frac{1}{3}}$, 2.08; $10^{\frac{1}{3}}$, 2.154.

39. Definition. In decimal arithmetic, each figure is supposed to be multiplied by that power of 10, positive or negative, which is expressed by its distance from the figure before the point.

Thus 672.53 means $6\times10^2+7\times10^1+2\times10^0$, or 2×1, $+5\times10^{-1}$, or $\frac{5}{10}$ or $\frac{50}{100}+3\times10^{-2}$, or $\frac{3}{100}$, together $672\frac{53}{100}$.

Scholium. On some occasions other numbers are substituted for 10 in calculations: particularly 12, which has many advantages, and is used in operations respecting carpenter's work; and sometimes the number 2 facilitates computations; and it may be employed where it is inconvenient to multiply characters; since two different marks, or a mark and a vacant place, are sufficient, when continually repeated, to express all numbers. The powers of 60 are also used in the subdivisions of time, and of angles.

40. Definition. The reciprocal of a number is the quotient of a given unit divided by that number.

Scholium. Mr. Barlow has inserted an ample table of reciprocals in his very useful collection of Tables.

41. Definition. The harmonic mean of two quantities is the quantity of which the reciprocal is the half sum of their reciprocals.

Thus, the harmonic mean of 3 and 6 is 4; for $\frac{1}{2}\left(\frac{1}{3}+\frac{1}{6}\right)=\frac{1}{4}$. And the harmonic mean is equal to the product divided by the half sum. Thus $\frac{36}{9}=4$.

42. Definition. The common logarithm of a number is that power of 10 which expresses it.

For instance, $l\ 1000=3$, since $10^3=1000$. $l\ 2=.30103$, for $10^{\frac{30103}{100000}}=2$. The principal use of logarithms is derived from that property of indices, by which their addition and subtraction is equivalent to the multiplication and division of the respective numbers.

43. Problem. To solve a quadratic equation.

Reduce the equation to the form $xx\pm ax=b$, add the square of half a; then $xx\pm ax+\frac{aa}{4}=b+\frac{aa}{4}$, whence $x\pm\frac{a}{2}=\pm\surd\left(b+\frac{aa}{4}\right)$ and $x=+\surd\left(b+\frac{aa}{4}\right)\mp\frac{a}{2}$.

SECT. II. OF THE COMPARISON OF VARIABLE QUANTITIES.

44. Definition. The quantities by which two variable magnitudes are increased or decreased, in the same time, are called their increments or decrements, or their increments positive or negative.

Scholium. They are sometimes denoted by an accent placed over the variable quantity; thus x' and $\acute{y}$ are the simultaneous increments of x and y.

45. Definition. The ratio, which is the limit of the ratios of the increments of two connected quantities, as they are taken smaller and smaller, is called the ratio of the velocities of their increase or decrease.

Scholium. It would be difficult to give any other sufficient definition of velocity than this. If both the quantities vary in the same proportion, the ratio of x' and $\acute{y}$ will be constant (18), and may be determined without considering them as evanescent; but if they vary according to different laws, that ratio must vary, accordingly as the time of comparison is longer or shorter: and since the degree of variation, at any instant of time, does not depend on the change produced at a finite interval before or after that instant, it is necessary, for the comparison of this variation, that the increments should be considered as diminished without limit, and their ultimate ratio determined; and it is indifferent whether these evanescent increments be taken before, or after the given instant, or whether the mean between both results be employed.

46. Definition. Any finite quantities, in the ratio of the velocities of increase or decrease of two or more magnitudes, are the fluxions of those magnitudes.

Scholium. They are denoted by placing a point over the variable quantity, thus, $\dot{x}, \dot{y}$. And $\frac{x'}{\acute{y}}$ is always ultimately equal to $\frac{\dot{x}}{\dot{y}}$. The variable quantity is called a fluent with respect to its fluxion, as x is the fluent of $\dot{x}$, or $x = \int \dot{x}$. On the continent the term fluxion is not used, but the evanescent increment is called a difference, and denoted by *d* or δ, and the variable quantity is conceived to consist of the entire sum or integral of such differences, and marked $\int$, as $x = \int dx$, or $\int \delta x$. This mark has the advantage of differing in form from the short *s*, which is used as a literal character. See 229.

47. THEOREM. **When the fluxions of two quantities are in a constant ratio, their finite increments are in the same ratio.**

For if it be denied, let the ratios have a finite difference; then if the time, in which the increments are produced, be continually divided, the ratio of the parts may approach nearer to the ratio of the fluxions than any assignable difference, for that ratio is their limit (46), and this is true, by the supposition, in each part; therefore the sums of all the increments will be to each other in a ratio nearer to that of the fluxions than the assigned difference (35).

48. THEOREM. **The fluxion of the product of two quantities is equal to the sum of the products of the fluxion of each into the other quantity.**

Or $(xy)^{\cdot} = y\dot{x} + x\dot{y}$. Let the quantities increase from x and y to $x + x'$ and $y + y'$, then their product will be first xy and afterwards $xy + yx' + xy' + x'y'$, of which the difference is $yx' + xy' + x'y'$, and the ratio of the increments of x and xy is that of x' to $yx' + xy' + x'y'$; or, when the increments vanish, to $yx' + xy'$, since in this case $x'y'$ vanishes in comparison with xy'. But $x' : (yx' + xy') :: \dot{x} : (y\dot{x} + x\dot{y})$, and the fluxion is rightly determined (46); for since $\frac{y'}{x'} = \frac{\dot{y}}{\dot{x}}$, $\frac{xy'}{x'} = \frac{x\dot{y}}{\dot{x}}$ (18); but $\frac{yx'}{x'} = \frac{y\dot{x}}{\dot{x}}$ (18), and $\frac{yx' + xy'}{x'} = \frac{y\dot{x} + x\dot{y}}{\dot{x}}$ (15).

SCHOLIUM. It is also obvious, that the fluxion of any quantity xy is equal to the sum of the results obtained by multiplying it by the fluxion of each variable quantity, and dividing it by that quantity: thus, $(xy)^{\cdot} = xy\left(\frac{\dot{x}}{x} + \frac{\dot{y}}{y}\right)$; $(xx)^{\cdot} = xx\left(\frac{\dot{x}}{x} + \frac{\dot{x}}{x}\right) = 2x\dot{x}$.

49. **Theorem.** The fluxion of any power of a variable quantity is equal to the fluxion of that quantity multiplied by the index of the power, and by the quantity raised to the same power diminished by unity.

Or $(x^n)^{\cdot} = nx^{n-1}\dot{x}$. Let $n=2$, then $(xx)^{\cdot} = x\dot{x} + x\dot{x}$ (48) $= 2x\dot{x} = nx^{n-1}x$. If $n=3$, $x^n = (xx).x$, and its fluxion is $x\,(xx)^{\cdot} + (xx)\dot{x} = 2xx\dot{x} + xx\dot{x} = 3x^2\dot{x} = nx^{n-1}\dot{x}^n$. And the same may be proved of any whole number. If n is a fraction, as $\frac{1}{p}$, put $y = x^n$, then $x = y^p$, and $\dot{x} = py^{p-1}\dot{y}$, $\dot{y} - \frac{\dot{x}}{py^{p-1}} = \frac{1}{p}.\ y^{1-p}\dot{x}$ (38) $= \frac{1}{p}y.\ y^{-p}\dot{x} = nx^{n-1}\dot{x}$, as before; and in the same manner the proof may be extended to all possible cases.

50. **Theorem.** When the logarithm of a quantity varies equably, the quantity varies proportionally.

Or if l $x = y$, $\frac{\dot{y}}{a} = \frac{\dot{x}}{x}$. For $x = b^y$ (42), and when y becomes $y + \acute{y}$, $x + x' = b^{y+\acute{y}} = b^y.b^{y'} = x.b^{y'}$, and $x' = x.b^{y'} - x = x.\left(b^{y'} - 1\right)$; but $\acute{y}$ being constant, by the supposition, $b^{y'} - 1$ is constant, and may be called $\frac{\acute{y}}{a}$, and $x' = \frac{x\acute{y}}{a}$; therefore $\dot{x} = \frac{x\dot{y}}{a}$, and $\frac{\dot{x}}{x} = \frac{\dot{y}}{a}$.

Scholium. Numerical logarithms do not, strictly speaking, vary by evanescent increments; but other quantities may flow continually, and be always proportional to logarithms: in either case the proposition is true. In Briggs's logarithms, commonly used, b is 10, and a, the modulus, is .4342944819; dividing all the system by a, or multiplying by 2.302585093, we have Napier's original hyperbolical logarithms, where $\dot{y}$ becomes $= \frac{\dot{x}}{x}$, and $a = 1$.

51. **Theorem.** The fluxion of any power of a quantity, of which the exponent is variable, is equal to the fluxion of the same power considered as constant, together with the fluxion of the exponent multiplied by the power and by the hyperbolical logarithm of the quantity.

If $x^y = z$, $\dot{z} = yx^{y-1}\dot{x} + (\text{hl } x).\ x^y\dot{y}$; for hl $z = y.$ (hl x), (42); now

$(\text{hl } z)^{\cdot} = \frac{\dot{z}}{z}$, (50); and $\dot{z} = z.(\text{hl } z)^{\cdot} = z.(y.\ \text{hl } x)^{\cdot} = z.(\frac{y\dot{x}}{x} + (\text{hl } x).\dot{y})$, (48, 50) $= yx^{y-1}\dot{x} + (\text{hl } x)\, z\dot{y}$.

52. THEOREM. When a variable quantity is greatest or least, its fluxion vanishes.

For a quantity is greatest when it ceases to increase, and before it begins to decrease; that is, when it has neither increment nor decrement; and it is least when it has ceased to have a decrement and has not yet an increment.

53. PROBLEM. To solve a numerical equation by approximation.

The most general and useful mode of solving all numerical equations is by approximation. Substitute for the unknown quantity a number, found by trial, which nearly answers to the conditions; then the error will be a finite difference of the whole equation; which, when small, will be to the error of the quantity substituted, nearly in the ratio of the evanescent differences, or of the fluxions; and this ratio may be easily determined.

Thus, if $x^3 - 6x^2 + 4x = 6699$, call 6699, y, then $3x^2\dot{x} - 12x\dot{x} + 4\dot{x} = \dot{y}$, and $\dot{x} = \frac{\dot{y}}{3x^2 - 12x + 4}$, and $x' = \frac{y'}{3x^2 - 12x + 4}$ nearly; now assume $x = 20$, then $y = 5680$, and $y' = 1019$, whence x' 1.05, and x corrected is 21.05; by repeating the operation we may approach still nearer to the true value 21.

If $x^n y$, $\dot{x} = \frac{\dot{y}}{nx^{n-1}}$, whence the common rule for the extraction of roots is derived. In order to find the nearest integer root, the digits must be divided, beginning with the units, into parcels of as many as there are units in the index, and the nearest root of the last or highest parcel being found, and its power subtracted, the remainder must be divided by its next inferior power multiplied by the given index, in order to find the next figure, adding the next parcel to the remainder before the division. There are also, in particular cases, other more compendious methods.

It is, however, often more convenient to solve an equation by the rule of double position, taking two approximate values of the root, and finding a third which differs from one of them by a quantity bearing the same proportion to their difference as the error of that one bears to the difference of the two errors.

SECTION III. OF SPACE.

54. **Definition.** A solid is a portion of space limited in magnitude on all sides.

Scholium. Space is a mode of existence incapable of definition, and supposed to be understood by tradition.

55. **Definition.** A surface is the limit of a solid.

56. **Definition.** A line is the limit of a surface.

57. **Definition.** A point is the limit of a line.

Scholium. The paper, of which this figure covers a part, is an example of a solid, the shaded portion represents a portion of surface: the boundaries of that surface are lines, and the three terminations or intersections of those lines are points. In conformity with this more correct conception, these definitions are illustrated by representations of the respective portions of space of which the limits are considered; and also by the more usual method of denoting a line by a narrow surface, and a surface by such a line surrounding it.

58. **Definition.** A line joining two points is called their distance.

59. **Definition.** When the distance of any two or more points remains unchanged, they are said to be at rest; and a space of which all the points are at rest, is a quiescent space.

60. **Definition.** A line which must be wholly at rest, with respect to any quiescent space, when two of its points are at rest in that space, is a straight line.

61. **Definition.** A line which is neither a straight line, nor composed of straight lines, is a curve line.

62. DEFINITION. A plane is a surface, in which if any two points be joined by a straight line, the whole of the straight line will be in the surface.

63. DEFINITION. An angle is the inclination of two lines to each other.

SCHOLIUM. An angle is sometimes denoted by this mark ∠, and is described by three letters placed near the lines, the middle letter at the angular point.

64. DEFINITION. When a straight line standing on another straight line makes the adjacent angles equal, they are called right angles.

65. DEFINITION. A straight line between two right angles is called a perpendicular to the line on which it stands.

66. DEFINITION. When a plane surface is contained by a circumference, such that all straight lines drawn to it from a certain point in the plane are equal, the surface is a circle.

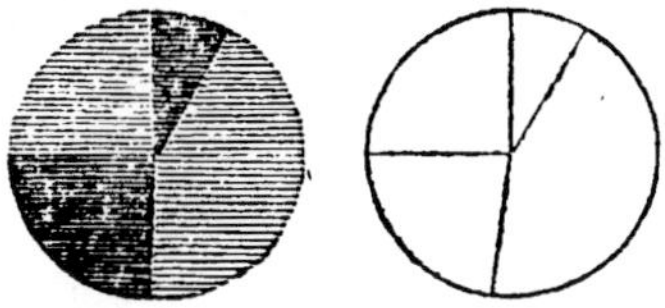

67. DEFINITION. The point, equally distant from the circumference, is called the centre.

68. DEFINITION. Any straight line, drawn from the centre to the circumference, is called a radius.

69. DEFINITION. The term circle also often implies the circumference, and not the circular surface.

70. DEFINITION. A portion of the circumference of a circle is called an arc.

71. DEFINITION. A straight line, joining the extremities of an arc, is its chord.

72. **Definition.** The surface, contained between an arc and its chord, is called a segment of a circle.

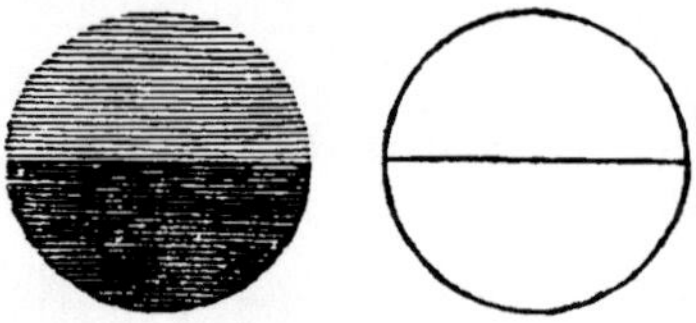

73. **Definition.** A chord passing through the centre is a diameter.

74. **Definition.** A triangle is a surface contained between three lines; and these lines are understood to be straight, unless the contrary is expressed.

75. **Definition.** When two straight lines, lying in the same plane, may be produced both ways indefinitely, without meeting, they are parallel.

Scholium. The parallelism of lines is sometimes denoted by this mark ||.

76. **Postulate.** It is required that the length of a straight line be capable of being identified, whether by the effect of any object on the senses, or merely in imagination, so that it may remain invariable.

Scholium. This is practically performed by making visible marks on a material surface; although, strictly speaking, no such marks remain at distances absolutely invariable, on account of changes of temperature, and of other circumstances.

77. **Postulate.** That a straight line of indefinite length may be drawn through any two given points.

78. **Postulate.** That a circle may be described on any given centre with a radius equal to any given straight line.

79. **Axiom.** A straight line joining two points is the shortest distance between them.

SCHOLIUM. With respect to all straight lines, this axiom is a demonstrable proposition; but since the demonstration does not extend to curve lines, it becomes necessary to assume it as an axiom.

80. AXIOM. Of any two figures meeting in the ends of a straight line, that which is nearer the line has the shorter circumference, provided there be no contrary flexure.

81. AXIOM. Two straight lines, coinciding in two points, coincide in all points.

SCHOLIUM. If they did not coincide in all points, the two points of coincidence being at rest, and one of the lines being made the axis of motion, the other must revolve round it, contrarily to the definition of a straight line. Although this is sufficiently obvious, it can scarcely be called a formal demonstration.

82. AXIOM. All right angles are equal.

83. AXIOM. A straight line, cutting one of two parallel lines, may be produced till it cut the other.

84. PROBLEM. From the greater of two right lines, AB, to cut off a part equal to the less, CD.

On the centre A describe a circle with a radius equal to CD (78), and it will cut off AB=CD (66).

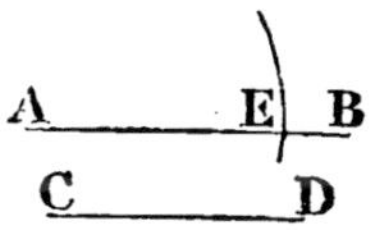

85. PROBLEM. On a given right line, AB, to describe an equilateral triangle.

On the centres A and B draw two circles, with radii equal to AB, and to their intersection C, draw AC and BC; then AB=AC=BC (66), and the triangle ABC is equilateral.

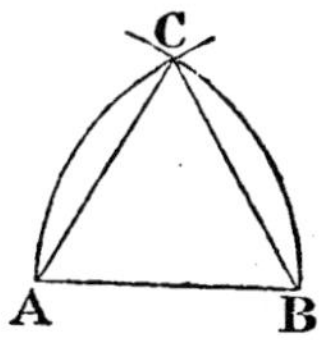

86. THEOREM. Two triangles, having two sides and the angle included, respectively equal, have also the base and the other angles equal.

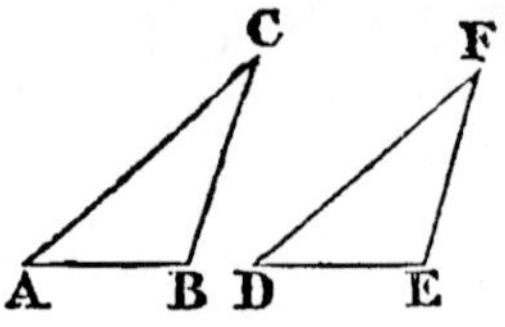

In the triangles ABC, DEF, let AC=DF, BC=EF, and ∠ ACB=DFE. Now supposing a triangle equal to DEF to be constructed on AC, the side equal to FE must coincide in position with CB, because ∠ ACB=DFE, and also in magnitude, for they are equal, therefore the point B will be an angular point of the supposed triangle; and since the base of both triangles must be a right line, it must be the same line AB (81), and the supposed triangle will coincide every where with ABC; therefore ABC=DEF, and the angles at A and B are equal to the angles at D and E.

87. **Theorem.** If two sides of a triangle are equal, the angles opposite to them are equal.

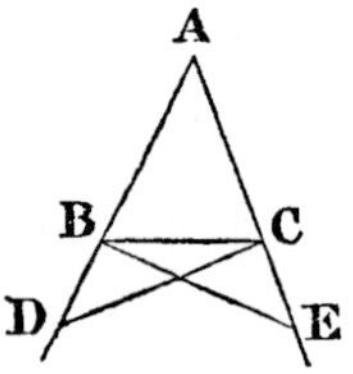

In the sides AB and AC produced, take at pleasure AD=AE, and join BE, CD; then since AD=AE, and AC=AB, and the angle at A is common to the triangles ADC, AEB, those triangles are equal (86), and ∠ ACD=ABE, ∠ ADC=AEB, and CD=BE; but BD=CE (16), therefore ∠ BCD=CBE (86), and ∠ ACD—BCD=ABE—CBE (16), or ∠ ACB=ABC.

88. **Theorem.** If two angles of a triangle are equal, the sides opposite to them are equal.

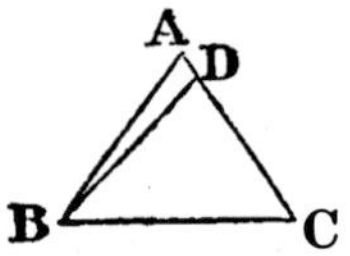

Let ∠ ABC=BCD; then AC=AB. If it be denied, take, in the greater AC, CD equal to the less AB; then, since ∠ ABC=DCB, AB=DC, and BC is common, the triangle ABC=DCB (86), the whole to a part, which is impossible.

89. **Theorem.** If two triangles have their bases equal, and their sides respectively equal, their angles are also respectively equal.

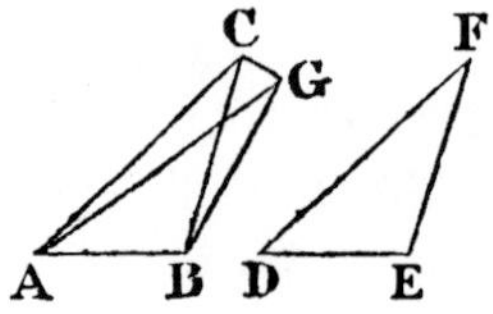

If a triangle be supposed to be constructed on AB, the base of ABC, equal to DEF, the vertex of the triangle must coincide with C, and the whole triangle with ABC. For if it be denied, let G be the vertex of the triangle so constructed; join CG; then since AC=AG, ∠ ACG=AGC (87), and in the same manner ∠ BGC=BCG; but BGC > AGC, therefore BGC > ACG; and ACG > BCG, therefore much more BGC > BCG, to which it was shown to be equal. And the same may be proved in any other position of the point G; therefore the triangle equal to DEF, supposed to be described on AB, coincides with ABC.

90. PROBLEM. To bisect a given angle.

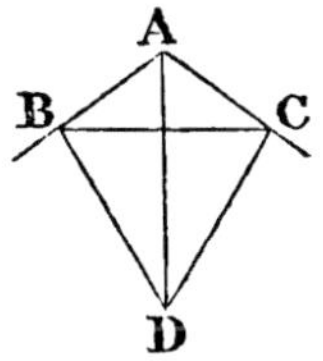

In the right lines forming the angle, take at pleasure AB=AC; on BC describe an equilateral triangle BCD, and AD will bisect the angle BAC. For AB=AC, BD=CD, and the base AD is common, therefore the triangle ABD=ACD (89), and ∠ BAD=CAD.

91. PROBLEM. To bisect a given right line, AB.

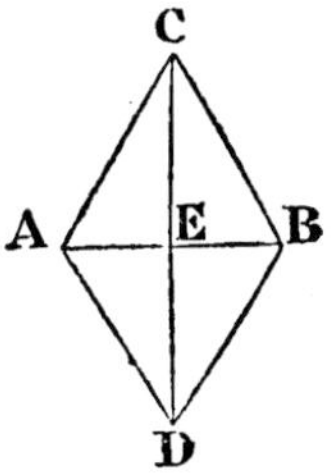

Describe on it two equilateral triangles, ABC, ABD; and CD, joining their vertices, will bisect AB in E. For since AC=CB, AD=BD, and CD is common to the triangles ACD, BCD, ∠ ACD=BCD (89); but CE is common to the triangles ACE and BCE, therefore AE=EB (86).

92. PROBLEM. To erect a perpendicular to a given right line at a given point.

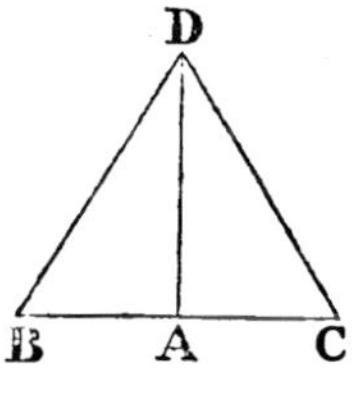

On each side of the point A, take at pleasure AB=AC, and on BC make an equilateral triangle, BCD. Then AD shall be perpendicular to BC. For the sides of BAD and CAD are respectively equal, therefore the angle BAD=CAD (89), and both are right angles (64), and AD is perpendicular to BC (65).

93. PROBLEM. From a point, A, without a right line, BC, to let fall a perpendicular on it.

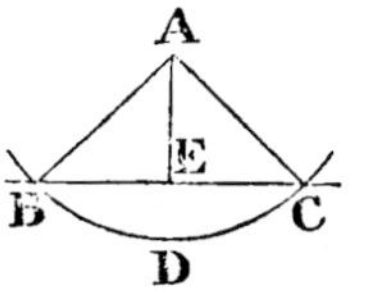

On the centre A, through any point D, beyond BC, describe a circle, which must obviously cut BC join AB and AC, and bisect the angle BAC by the line AE; AE will be perpendicular to BG. For ∠BAE=CAE, AB=AC, and AE is common to the triangles BAE, CAE; therefore ∠AEB=AEC (86), and both are right angles (64).

94. **Theorem.** The angles, which any right line makes on one side of another, are, together, equal to two right angles.

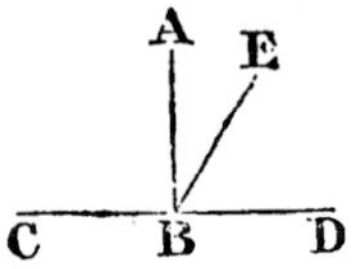

Let AB be perpendicular to CD, and EB oblique to it, then $CBE+EBD=CBA+ABE+EBD=CBA+ABD$ (14).

95. **Theorem.** If two right lines make with a third, at the same point, but on opposite sides, angles together equal to two right angles, they are in the same right line.

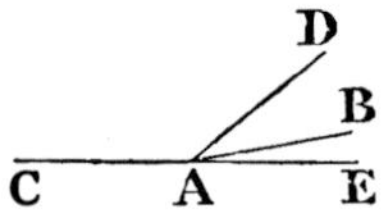

If it be denied, let AB, which together with AC, makes with AD, the angles BAD, DAC equal to two right angles, be not in the right line CAE. Then $BAD+DAC$, being equal to two right angles, is equal to $EAD+DAC$ (94), and $BAD=EAD$, the less to the greater, which is impossible.

96. **Theorem.** If two right lines intersect each other, the opposite angles are equal.

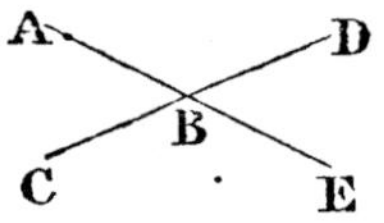

From the equals, $ABC+ABD$ and $ABD+DBE$ (94, 82), subtract ABD, and the remainders, ABC, DBE, are equal. In the same manner $ABD=CBE$.

97. **Theorem.** If one side of a triangle be produced, the exterior angle will be greater than either of the interior opposite angles.

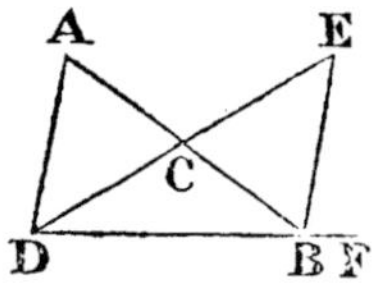

Bisect AB in C, draw DCE; take $CE=CD$ and join BE, then the triangle $ACD=BCE$ (96, 86), and $\angle CBE=CAD$; but $ABF>CBE$, therefore $ABF>CAD$. And in the same manner it may be proved, by producing AB, that ABF is greater than ADB.

98. **Theorem.** The greater side of any triangle is opposite to the greater angle.

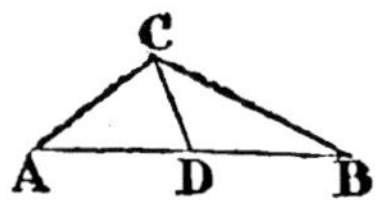

Let $AB > AC$, then $\angle ACB > ABC$. For taking $AD = AC$, and joining CD, $\angle ACD = ADC$ (87). But $\angle ADC > CBD$ (97), and $ACB > ACD$, therefore much more $\angle ACB > CBD$, or ABC.

99. **Theorem.** Of two triangles on the same base, the sides of the interior contain the greater angle.

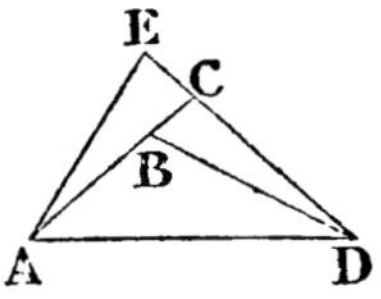

Froduce AB to C, then $\angle ABD > ACD$ (97), and $\angle ACD > AEC$, therefore much more $ABD > AED$.

100. **Problem.** To make a triangle, having its sides equal to three given right lines, every one of them being less than the sum of the other two.

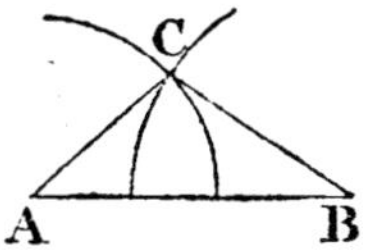

Take AB equal to one of the lines, and on the centres A and B describe two circles with radii equal to the other two lines; draw AC and BC to the intersection C, and ABC will be the triangle required.

101. **Problem.** At a given point in a right line, to make an angle equal to a given angle.

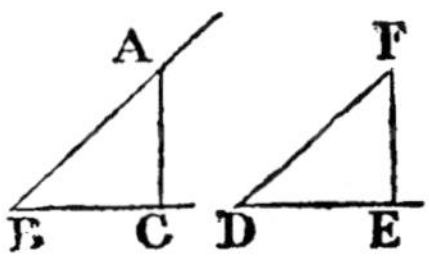

In the lines forming the given angle ABC, take any two points, A and C, join AC, and taking $DE = BC$, make the triangle DEF, having $DF = BA$ and $FE = AC$ (100), then $\angle FDE = ABC$ (89).

102. **Theorem.** If two triangles have two angles and a side respectively equal, the whole triangles are equal.

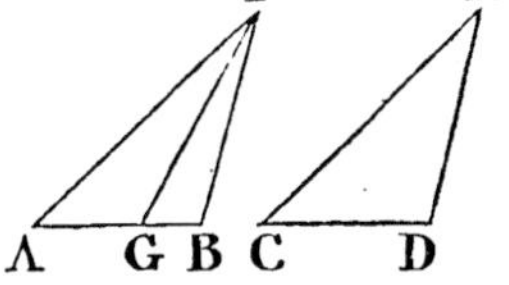

Let the equal sides be AB and CD, intervening between the equal angles, then if on AB a triangle equal to CDE be supposed to be constructed, the points A and B, and the angles at A and B being the same in this triangle and in ABF, the sides must coincide both in position and in length; therefore $ABF = CDE$.

If the equal sides are AF and CE, opposite to equal angles, then $AB = CD$, and the whole triangles are equal. For if AB is not equal to CD, let it be the greater, and let $AG = CD$; then, by what has been demonstrated, the triangle $AFG = CED$, and $\angle AGF = CDE - ABF$, by the supposition; but $AGF > ABF$ (97), which is impossible.

103. THEOREM. The shortest of all right lines, that can be drawn from a given point to a given right line, is that which is perpendicular to the line, and others are shorter as they are nearer to it.

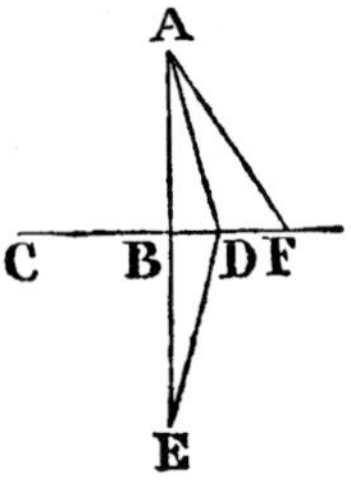

Let AB be perpendicular to CD, then AB is shorter than AD. Froduce AB, take $BE = AB$, and join DE; then the triangle $ABD = EBD$ (86), and $AD = DE$. But $AB + BE$ or $2AB$ is less than $AD + DE$ or $2AD$ (79), therefore $AB < AD$ (18). In a similar manner $2AD < 2AF$ (80), and $AD < AF$.

104. THEOREM. If a right line, cutting two others, makes the alternate angles equal, the two lines are parallel.

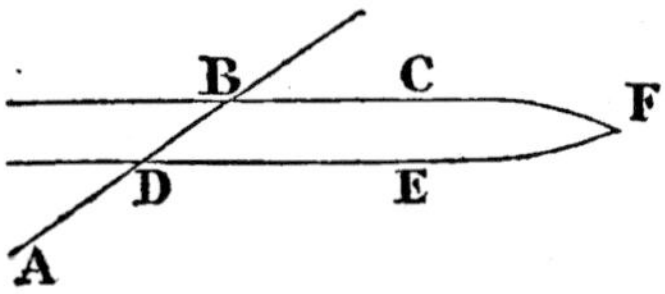

If $\angle ABC = ADE$; BC and DE are parallel; for if they meet, as in F, they will form a triangle BDF, and $\angle ADE > ABC$ (97).

105. THEOREM. A right line, cutting two parallel lines, makes equal angles with them.

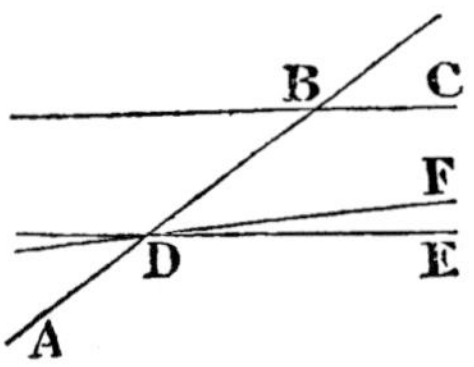

Let AB cut the parallels BC, DE; then if $\angle ABC$ is not equal to ADE, let it be equal to ADF, then BC and DF are parallel (104), and DE, which cuts DF, will also, if produced, cut BC (83), contrarily to the supposition.

106. THEOREM. Right lines, parallel to the same line, are parallel to each other.

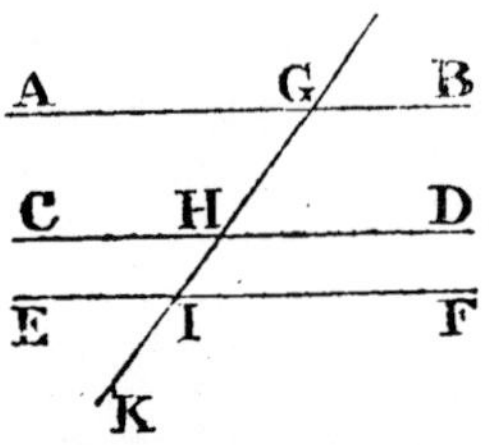

Let AB and CD be parallel to EF; draw GHI cutting them all, then $\angle KGB = KIF$ (105), and $\angle KHD = KIF$, therefore $\angle KGB = KHD$, and $AB \parallel CD$ (104).

107. **Problem.** Through a given point to draw a right line parallel to a given right line.

From A draw, at pleasure, AB, meeting BC in B, and make ∠BAD=ABC (101), then AD || CB (104).

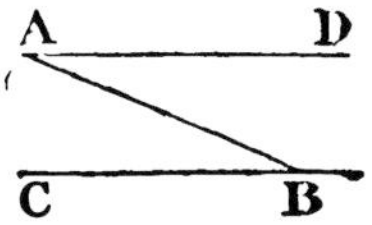

108. **Theorem.** The angles of any triangle, taken together, are equal to two right angles.

Produce AB to C, and draw BD parallel to AE. Then ∠EBD=AEB (105), and ∠DBC=EAB; therefore the external angle EBC is equal to the sum of the internal opposite angles, AEB, EAB, and adding ABE, the sum of all three is equal to ABE+EBC, or to two right angles (94).

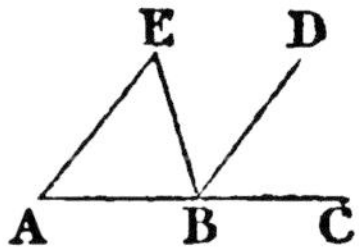

109. **Theorem.** Right lines, joining the extremities of equal and parallel right lines, are also equal and parallel.

Let AB and CD be equal, and parallel. Then AC will be equal and parallel to BD. For, joining BC, ∠ABC=BCD (105), and the triangles ABC, DCB, are equal (86), and AC=DB; also ∠ACB =DBC, therefore AC || BD (104).

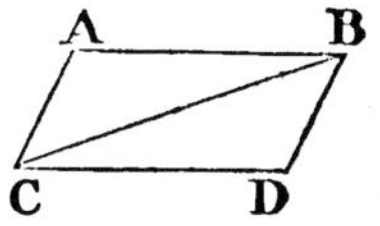

110. **Definition.** A figure, of which the opposite sides are parallel, is called a parallelogram.

111. **Definition.** A straight line, joining the opposite angles of a parallelogram, is called its diagonal.

112. **Definition.** A parallelogram, of which the angles are right angles, is a rectangle.

113. **Definition.** An equilateral rectangle is a square.

114. **Theorem.** The diagonal of a parallelogram divides it into two equal triangles, and its opposite sides are equal.

For ABC is equiangular with DCB (105), and BC is common, therefore they are equal (102), and AB=CD, and AC=BD.

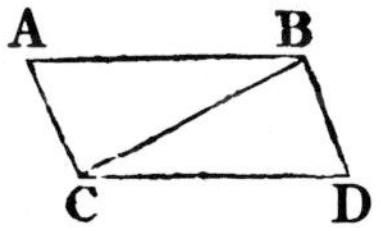

115. THEOREM. *Parallelograms on the same base, and between the same parallels, are equal.*

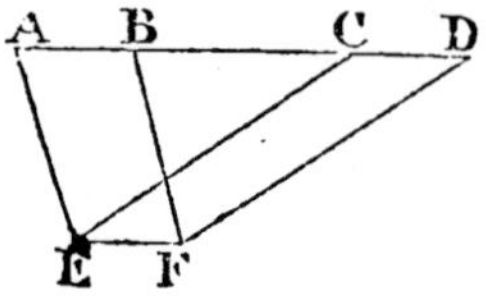

Since AB=CD, both being eqnal to EF, AC=BD (15, or 16), and the triangle AEC is equiangular (105) and equal (102) to BFD; therefore deducting each of them from the figure ABFD, the remainder ED is equal to the remainder AF.

116. THEOREM. *Parallelograms on equal bases, and between the same parallels, are equal.*

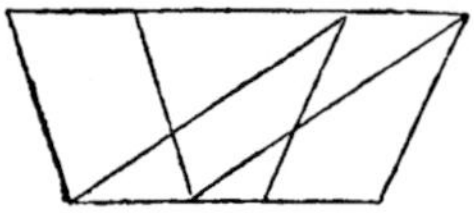

For each is equal to the parallelogram formed by joining the extremities of the base of the one, and of the side opposite to the base of the other (115).

117. THEOREM. Triangles on equal bases, and between the same parallels, are equal.

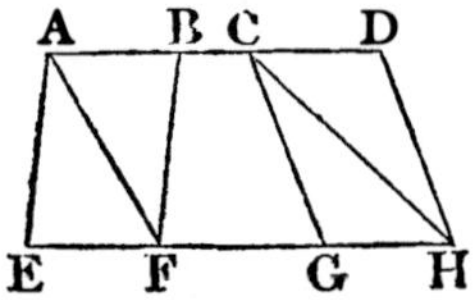

Take AB and CD equal to the base EF or CH, and join BF and DH. Then EB and CD are parallelograms between the same parallels (109), and on equal bases, therefore they are equal (116), and their halves, the triangles AEF, CGH (114), are also equal (18).

118. THEOREM. In any right angled triangle, the square described on the hypotenuse is equal to the sum of the squares described on the two other sides.

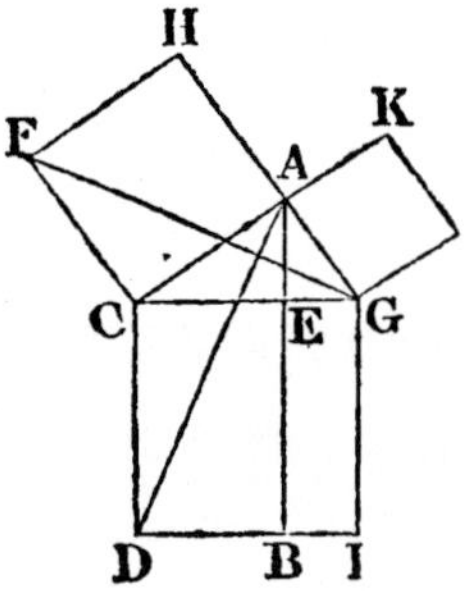

Draw AB parallel to CD, the side of the square on the hypotenuse, then the parallelogram CB is double any triangle on the same base and between the same parallels (114, 117), as ACD; but ACD=FCG, their angles at C being each equal to ACG increased by a right angle, FC to AC, and GC to DC. Again, GAH is a right line(95), parallel to CF, therefore the triangle FCG is half of the square CH on the same base, and CH=CB, since they are the doubles of equal triangles. In the same manner it may be shown that GK=GB; therefore the whole CDIG is equal to the sum of CH and GK.

119. PROBLEM. To find a common measure of any two quantities.

Subtract the less continually from the greater, the remainder from the less, the next remainder from the preceding one, as often as possible, and proceed till there be no further remainder; then the last remainder will be the common measure required. For since it measures the preceding remainder, it will measure the preceding quantities in which that remainder was contained, together with itself, and which, increased at each step by these remainders, makes up the original quantities.

For example, if the numbers 54 and 21 be proposed, $54-21-21=12$, $21-12=9$, $12-9=3$, $9-3-3-3=0$, therefore 3 is the common measure, for it measures 9, and $9+3$ or 12, and $12+9$ or 21, and $2\times21+12$ or 54.

SCHOLIUM. Hence it is obvious, that there can be no greater common measure of the two quantities than the quantity thus found; for it should measure the difference of the two quantities, and all the successive remainders down to the last, therefore it cannot be greater than this last. It must also be remarked, that in some cases no accurate common measure can be found, but the error, or the last remainder, in this process, may always be less than any quantity that can be assigned, since the process may be continued without limit. That there are incommensurable quantities, may be thus shown: every number is either a prime number, that is, a number not capable of being composed by multiplication of other numbers, or it is composed by the multiplication of factors, which are primes. Let the number a be composed of the prime numbers bcd, or $a=bcd$, then $aa=bcd.bcd=bb.cc.dd$ and each prime factor of aa occurs twice; so that every square number must be composed of factors in pairs; and a square number multiplied by a number which is not composed of factors in pairs cannot be a square number: for instance, $2aa$ or $3aa$ cannot be a square number, since the factors of 2 are only 1.2, and of 3, 1.3, and not in pairs: therefore the square root of 2 or 3 cannot be expressed by any fraction, for the square of its numerator would be twice or thrice the square of its denominator. But the ratio of the hypotenuse of a triangle to its side may be that of $\surd 2$ or $\surd 3$ to 1; so that quantities numerically incommensurable may be geometrically determined.

120. Theorem. Triangles and parallelograms of the same height are proportional to their bases.

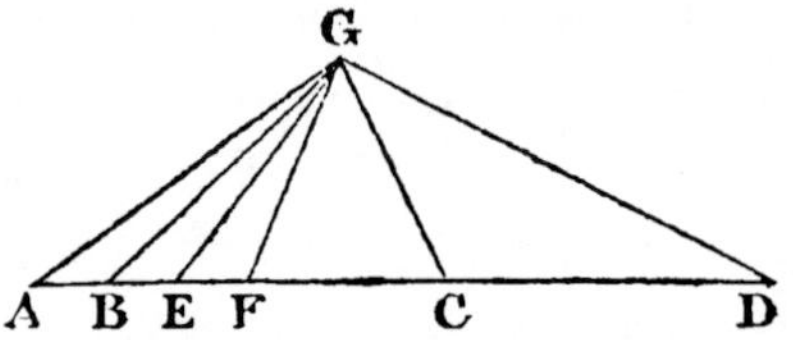

Let AB be a common measure of AC and AD, and let AB=BE =EF; join GB, GE, GF, then the triangles AGB, BGE, EGF, are equal, and the triangle AGD is the same multiple of AGB that AD is of AB; and AGC is the same multiple of AGB that AC is of AB, or AGD : AGB=AD : AB, and AGC : AGB=AC : AB: hence, dividing the first equation by the equal members of the second (18), AGD : AGC=AD : AC, and 2AGD : 2AGC=AD : AC, therefore the parallelograms, which are double the triangles, are also proportional.

Scholium. The demonstration may easily be extended to incommensurable quantities. For if it be denied that AC : AD= AGC : AGD, let AC : AD be the greater, and let the difference be $\frac{1}{n}$, then $\frac{AC}{AD}-\frac{1}{n}=\frac{AGC}{AGD}=\frac{n.AC}{n.AD}-\frac{AD}{n.AD}=\frac{n.AC-AD}{n.AD}$. Let $m.AD$ be that multiple of AD which is less than $n.AC$, but greater than $n.AC-AD$, then a triangle on the base $m.AD$ will be equal to $m.AGD$, which will be less than $n.AGC$, the triangle on $n.AC$; now multiplying the former equation by $\frac{n}{m}$, $\frac{n.AGC}{m.AGD}=\frac{n.AC-AD}{m.AD}$, and $n.AGC.m.AD=m.AGD.(n.AC-AD)$; but the first factors have been shown to be respectively greater than the second, therefore their products cannot be equal, and the supposition is impossible.

121. Theorem. The homologous sides of equiangular triangles are proportional.

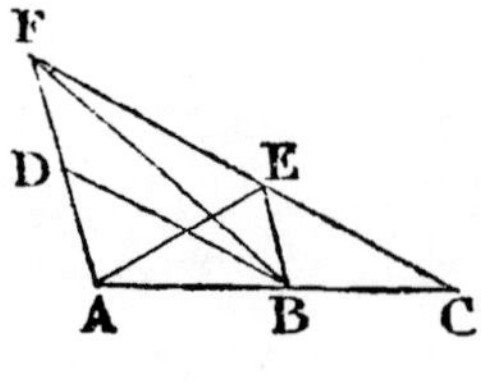

Let the homologous sides AB, BC, of the equiangular triangles ABD, BCE, be placed contiguous to each other in the same line, then AD || BE, and BD || CE; produce AD, CE, till they meet in F, and join AE and BF. Then the triangles FAE, EAC, are proportional to their bases FE, EC, and the triangles AFB, BFC, to AB,

BC (120). But FAE=AFB (117), and EAC=EBC+EAB=EBC +EFB=BFC, therefore FAE : EAC=AFB : BFC, and FE : EC =AB : BC; but FE=DB (114). In the same manner it may be shown that the other homologous sides are proportional.

SCHOLIUM. Hence equiangular triangles are also called similar.

122. THEOREM. Equal and equiangular parallelograms have their sides reciprocally proportional.

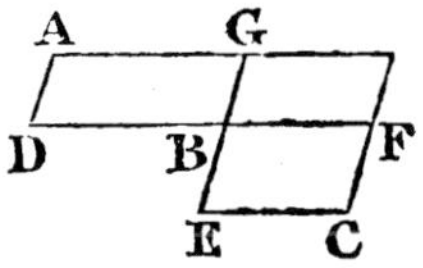

If AB=BC then DB : BE=BF : BG. For DB : BF=AB : GF(120)=BC : GF=BE : BG (120); or DB : BE=BF : BG.

123. THEOREM. Equiangular parallelograms, having their sides reciprocally proportional, are equal.

For they may be placed as in the last proposition, and the demonstration will be exactly similar.

SCHOLIUM. Hence is derived the common method of finding the contents of rectangles; let a and b be the sides of a rectangle, then $1 : a :: b : ab$, and the rectangle is equal to that of which the sides are 1 and ab, or to ab square units. The rectangle contained by two lines is therefore equivalent to the product of their numeral representatives.

124. THEOREM. Equiangular parallelograms are to each other in the ratio compounded of the ratios of their sides.

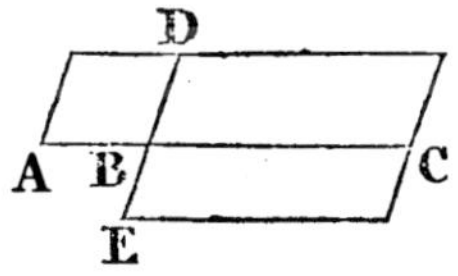

Or in the ratio of the rectangles or numeral products of their sides. For since AB : BC= AD : DC (120), and DC : CE=DB : BE, multiplying the former equation by the members of the latter, AB.DB : BC.BE=AD.CE.

125. THEOREM. Similar triangles, and figures composed of similar triangles, are in the ratio of the squares of their homologous sides.

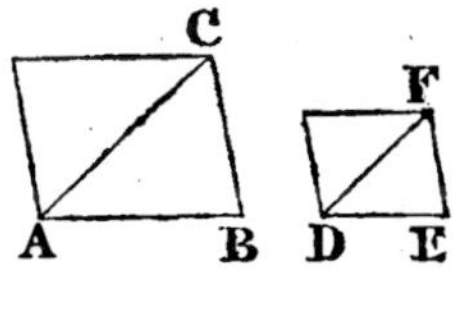

Since similar triangles are the halves of equiangular parallelograms, which are in the ratio compounded of the ratios of their sides (124), the triangles are in the same ratio, or

ABC : DEF=AB.BC : DE.EF ; but AB : DE=BC : EF (121), therefore ABC : DEF=AB.AB : DE.DE, or ABq : DEq. And the same may be proved of similar polygons, by composition (32).

126. DEFINITION. An indefinite right line, meeting a circle and not cutting it, is called a tangent.

127. THEOREM. A right line, passing through any point of a circle, and perpendicular to the radius at that point, touches the circle.

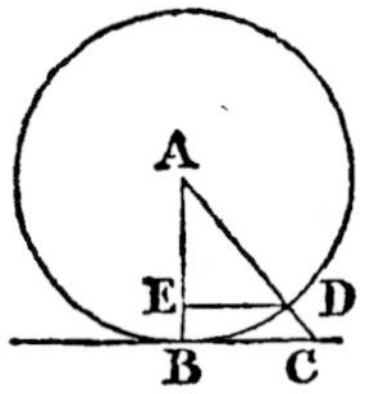

Since the perpendicular AB is shorter than any other line AC, that can be drawn from A to BC (103), it is evident that AC is greater than the radius AD, and that C, as well as every other point of BC, besides B, is without the circle; therefore BC does not cut the circle, but touches it.

128. DEFINITION. BC is called the tangent of the arc BD, or the angle BAD.

129. DEFINITION. AC is the secant of BD, or BAD.

130. DEFINITION. DE perpendicular to AB, is the sine of BD or BAD.

131. DEFINITION. AE is the cosine of BD or BAD.

132. DEFINITION. EB is the verse sine of BD or BAD.

SCHOLIUM. The circle is practically supposed to be divided into 360 equal parts, called degrees; each of these into 60 minutes; a minute into 60 seconds; and the division may be continued without limit; thus $60''=1'$, $60'=1^{\circ}$, and 90° make a right angle. Some modern calculators divide the quadrant into 100 equal parts, and subdivide these decimally, or rather centesimally.

133. THEOREM. The angle subtended at the centre of a circle, by a given arc, is double the angle subtended at the circumference.

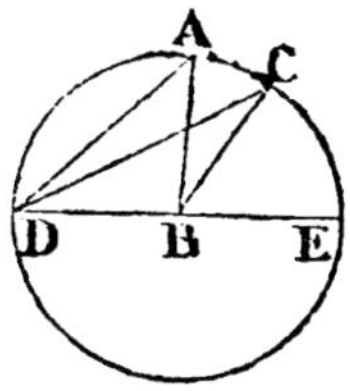

Let ABC and ADC be subtended by AC. Draw the diameter DBE, then ∠ABE=ADB+BAD (108)=2ADB (87). Also ∠CBE=2CDB; therefore ABE—CBE=2ADB—2CDB, or ABC=2ADC. In a similar manner it may be proved in other positions.

134. Theorem. The angle contained by the tangent and any chord, at the point of contact, is equal to the angle contained in the segment on the opposite side of the chord.

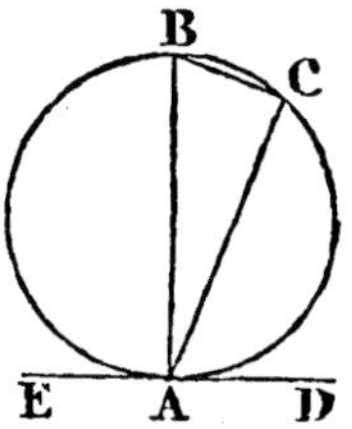

Draw the diameter AB, and join BC; then ∠BCA is equal to half the angle subtended at the centre by the semicircle AB, or to a right angle, and ABC and BAC make together another right angle (93), therefore deducting BAC, ABC=CAD. And it appears also from the last proposition, that the angle, contained in the lesser segment CA, is equal to the complement of ABC to two right angles, or to CAE.

135. Problem. To draw a tangent to a circle from a given point without it.

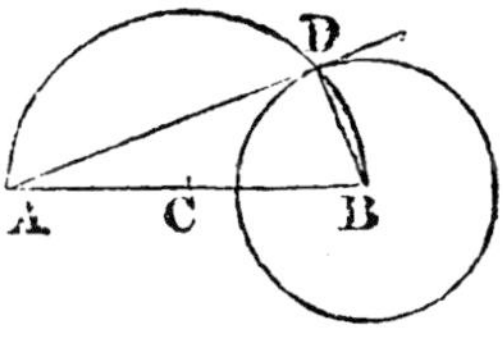

Join AB, bisect it in C, and on C draw a circle, with the radius CB, intersecting the former circle in D, then AD shall touch the circle. For the angle ADB, in a semicircle, is a right angle (134, 127), and BD is the radius of the given circle.

136. Theorem. In equal circles, equal angles stand on equal arcs.

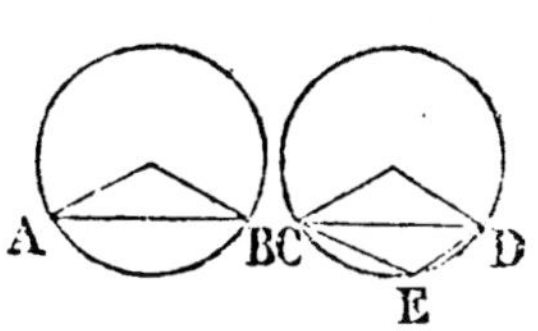

For the chords of equal angles are èqual (86), and the segments cut off by them contain equal angles (133); and if a segment equal to AB be supposed to be described on the chord CD, and on the same side with CED, it must coincide with CED, for since, at each point of each arc, CD subtends the same angle, the points of one arc can never be within those of the other (90); the arcs are therefore equal.

Scholium. Hence it may easily be shown, that multiple and proportionate angles are subtended by multiple and proportionate arcs.

137. Theorem. If two chords of a given circle intersect each other, the rectangles contained by the segments of each are equal.

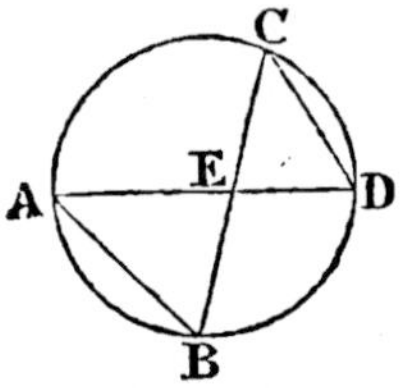

Join AB and CD. Then $\angle AEB = DEC$ (96), and $\angle BAE = DCE$ (133), both standing on BD, therefore the triangles AEB, CED, are similar, and AE : CE :: EB : ED (121), therefore AE.ED = CE.EB (123).

138. Theorem. The rectangle, contained by the segments of a right line, intercepted by a circle and a given point without it, is equal to the square of the tangent drawn from that point.

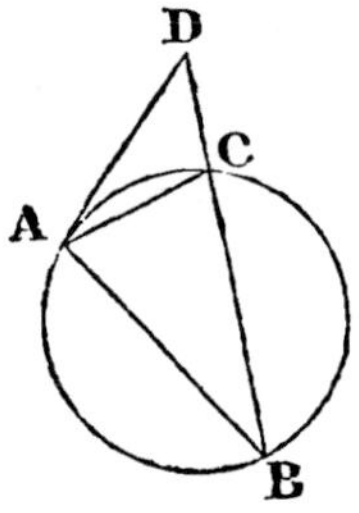

Join AB, AC; then $\angle ABC = CAD$ (134), and the angle at D is common, therefore the triangles ABD, CAD, are similar, and BD : AD :: AD : CD (121), whence BD.DC = ADq (123).

139. Theorem. In every triangle, the sides are as the sines of their opposite angles, the radius being given.

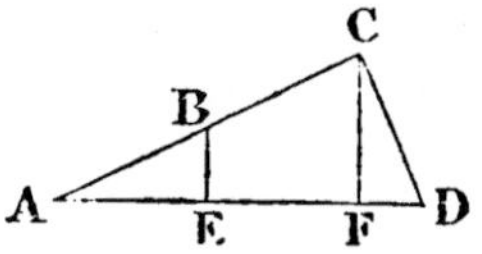

Take AB = CD, and draw BE and CF perpendicular to AD, then they are the sines of the angles A and D, to the radius AB or CD (130), and by similar triangles, AC : CF :: AB : BE (121), or CD : BE. And the same may be shown of the other sides and angles.

140. Theorem. The sine of the sum, or difference, of any two arcs, is equal to the sum or difference of the sines of the separate arcs, each being reduced in the ratio of the radius to the cosine of the other arc.

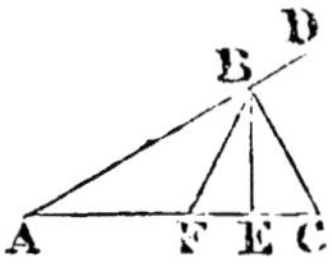

Let AB and BC be the sines of any two angles, ACB, BAC, then AC will be the sine of their sum CBD, or of ABC. Now making BE perpendicular to AC, AC=AE+EC, and rad. : cos. BAC::AB : AE, and rad. : cos. ACB::BC : CE (139). Again, make EF=EC ; then it is plain that AF will represent the sine of ABF, the difference of ACB or CFB and BAC (108).

141. Theorem. The ratio of the evanescent tangent, arc, chord, and sine, is that of equality.

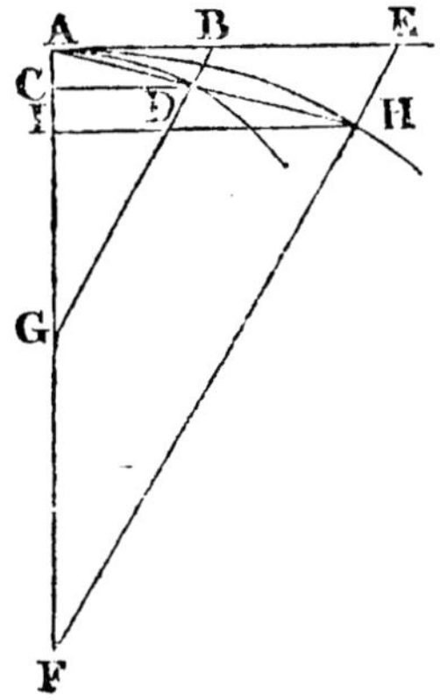

Let AB be the tangent, and CD the sine of the arc AD. Let AE be taken at pleasure in the tangent, and EF be always parallel to DG, the radius of AD, and on the centre F, draw the circle AH; join AH, then since ∠EAD=$\frac{1}{2}$AGD=$\frac{1}{2}$AFH, the chord AH will coincide with the chord AD (133, 134). And when DA vanishes, DG coinciding with AG, EF will be parallel to AF, and the angle EAH will vanish, therefore AH will coincide with AE, and with IH parallel to the sine CD; and by similar triangles the ratio of AB, AD, and CD, is the same as that of AE, AH, and IH, and is ultimately that of equality. But the arc AD is nearer to the chord AD than the figure ABD, and it has no contrary flexure, therefore it is longer than the line AD (79), and shorter than ABD (80), until their difference vanishes, and it coincides with both.

Scholium. The same is obviously true of any curve coinciding at a given point with any circle; and all the elements agree as well in position as in length.

141, B. Theorem. The fluxion of the area of any figure is equal to the parallelogram contained by the ordinate and the fluxion of the absciss. See 190.

142. Theorem. The fluxion of the arc being constant, the fluxion of the sine varies as the cosine.

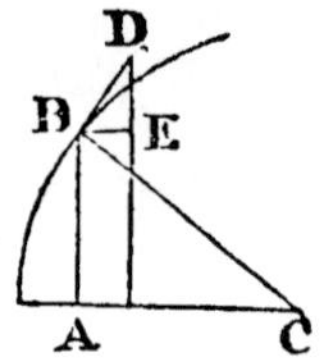

The fluxion of the arc is equal to that of the tangent, since their evanescent increments coincide (141). Let AB be the sine, AC the cosine, BD the increment of the tangent, DE that of the sine: then ∠ABC=EBD (16), and the triangles ABC, EBD, are similar, and BD is to DE as BC to AC; but the ultimate ratio of the increments is that of the fluxions, therefore the fluxion of the tangent, or of the arc, is to that of the sine as the radius to the cosine. The same may easily be inferred from the theorem for finding the sine of the sum of two arcs (140).

143. THEOREM. The area of a circle is equal to half the rectangle contained by the radius and a line equal to the circumference.

Suppose the circle to be described by the revolution of the radius: the elementary triangle, to which the fluxion of the circle is proportional (141), is equal to the contemporaneous increment of the rectangle, of which the base is equal to the circumference, and the height to half the radius: consequently the whole areas are equal (47).

144. THEOREM. The circumferences of circles are in the ratio of their diameters.

Supposing the circles to be concentric, and to be described by the revolution of different points of the same right line, the ratio of the fluxions, and consequently that of the whole circumferences, will be the ratio of the radii, or of the diameters (47).

SCHOLIUM. The diameter of a circle is to its circumference nearly as 7 to 22, and more nearly as 113 : 355, or 1 : 3.14159265359; hence the radius is equal to $57.29578^{\circ}=3437.7467'=206264.8''$; and, the radius being unity, $1^{\circ}=.017453293$, $1'=.000290888$, and $1''=$ .000004848.

145. DEFINITION. A straight line is perpendicular to a plane, when it is perpendicular to every straight line meeting it in that plane.

146. DEFINITION. A plane is perpendicular to a plane, when all the straight lines, drawn in one of the planes,

perpendicular to the common section, are perpendicular to the other plane.

147. DEFINITION. The inclination of a straight line to a plane is the angle, contained by that line, and another straight line drawn from its intersection with the plane to the intersection of a perpendicular let fall from any point of the line upon the plane.

148. DEFINITION. The inclination of two planes is the inclination of two lines, one in each plane, perpendicular to the common section.

149. DEFINITION. Parallel planes are such as never meet, although indefinitely produced.

150. DEFINITION. A solid angle is made by the meeting of two or more plane angles, in different planes.

151. DEFINITION. Similar solid figures are such as have all parts of their surfaces similar and similarly placed; and all their sections, in similar directions, respectively similar.

152. DEFINITION. A pyramid is a solid contained by a plane basis and other planes meeting in a point.

153. DEFINITION. A prism is a solid contained by planes of which two that are opposite, are equal, similar, and parallel, and all the rest parallelograms.

154. DEFINITION. A cube is a solid contained by six equal squares.

155. DEFINITION. A solid of revolution is that which is described by the revolution of any figure round a fixed axis.

156. DEFINITION. A sphere is described by the revolution of a semicircle on its diameter as an axis.

157. DEFINITION. A cone is a solid described by the revolution of an indefinite right line passing through a vertex, and moving round a circular basis.

158. DEFINITION. A cylinder is a solid, described by the revolution of a right angled parallelogram about one side.

159. THEOREM. Two straight lines cutting each other are in one plane.

For a plane passing through one of them may be supposed to revolve on it as an axis until it meet some point of the other; and then the second line will be wholly in the plane (62).

160. THEOREM. If two planes cut each other, their section is a straight line.

For the straight line joining any two points of the section must be in each plane (62), and must, therefore, be the common section of the planes.

161. THEOREM. A straight line, making right angles with two other lines at the point of their intersection, is at right angles to the plane passing through those lines.

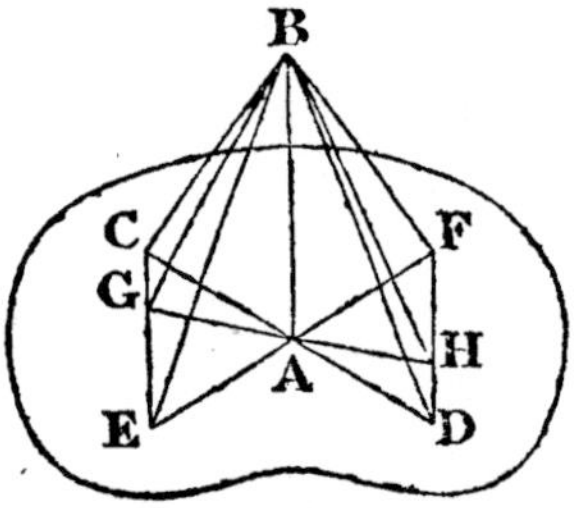

Let AB be perpendicular to CD and EF intersecting each other in A: take AC at pleasure, and make AC=AD=AE=AF; draw through A any line GH, and join CE, DF; then the triangles ADH, ACG are equal and equiangular, AH=AG and DH=CG; but since the triangles CBE, DBF, are equal, and equiangular, the angles BCG and BDH are equal, and the triangle BCG=BDH, BG=BH, and the triangles ABG, ABH, are equal and equiangular: consequently the angle BAG=BAH, and both are right angles: and the same may be proved of any other line passing through A; therefore AB is perpendicular to the plane passing through CD and EF (145).

162. THEOREM. Three straight lines, which meet in one point, and are perpendicular to one line, are in one plane.

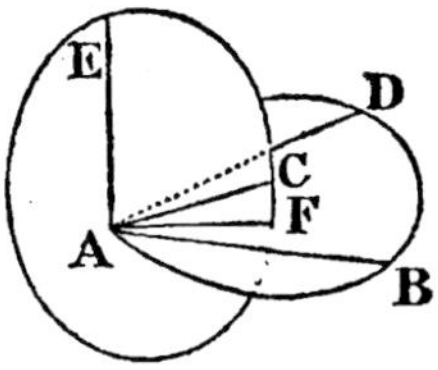

Let AB, AC, and AD meet in A, and be perpendicular to AE, then they are all in one plane. For if either of them AC is out of the plane which passes through the other two, let a plane pass through AE and AC, and let it cut the plane of AB and AD in AF, then the angle EAF is a right angle (161), and EAF=EAC, the greater to the less: which is impossible.

163. THEOREM. Two straight lines, which are perpendicular to the same plane, are parallel to each other; and two parallel lines are always perpendicular to the same planes.

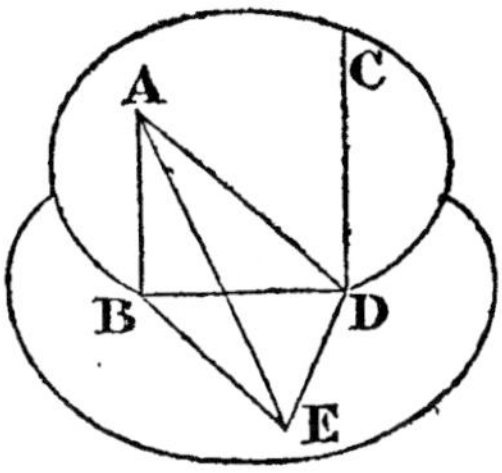

Let AB, CD, be perpendicular to the plane BED: draw DE at right angles to BD, and equal to AB, then the hypotenuses AD, BE, will be equal, and the triangles ABE, EDA, having all their sides equal, will be equiangular, and the angle ADE will be a right angle: consequently DE is perpendicular to the plane BC (161), and to DC (162), and AB is in the same plane with DC: and ABD and BDC being right angles, AB || CD.

Again, if AB || CD, and AB is perpendicular to the plane BED, the triangles ABE and EDA being equiangular, ADE is a right angle: therefore CDE is a right angle (161); but CDB is a right angle (105), therefore CD is perpendicular to BED.

164. THEOREM. Straight lines, which are parallel to the same straight line, not in the same plane, are parallel to each other.

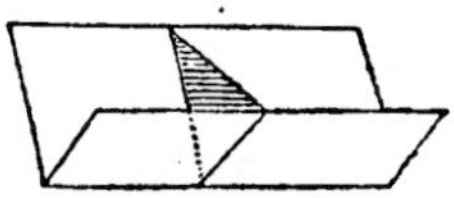

From any point in the third line, draw perpendiculars to the two first, and let a plane pass through these perpendiculars: then the third line is perpendicular to this plane (161); consequently the first and second are perpendicular to it, and therefore parallel to each other (163).

165. THEOREM. If the legs of two angles, not in the same plane, are parallel, the angles are equal.

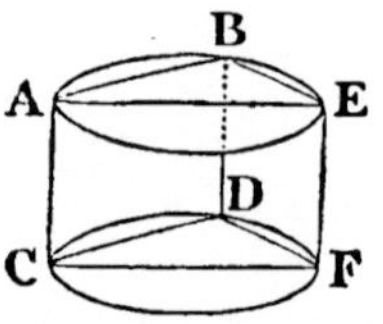

Let AB||CD, and BE||DF, then ∠ABE= CDF. Take AB=BE=CD=DF: then AC||= BD||=EF (109), and AE=CF (109); therefore ABE and CDF are equal and equiangular.

166. PROBLEM. To draw a line, perpendicular to a plane, from a given point above it.

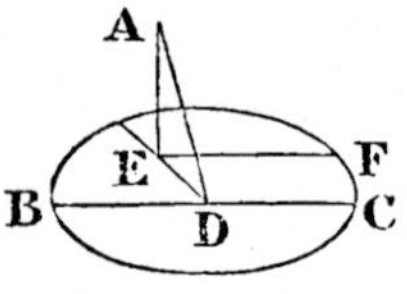

From the point A let fall on any line BC in the given plane a perpendicular AD; draw DE perpendicular to BC in the same plane, and from A draw AE perpendicular to DE: then AE will be perpendicular to the plane BEC; for if EF be parallel to BC, it will be perpendicular to the plane ADE (163), and consequently to AE; therefore AE, being perpendicular to DE and EF, will be perpendicular to the plane passing through them.

167. PROBLEM. From a given point in a plane, to erect a perpendicular to the plane.

From any point above the plane let fall a perpendicular on it, and draw a line parallel to this from the given point: this line will be the perpendicular required.

168. THEOREM. If two parallel planes are cut by any third plane, their sections are parallel lines.

For if the lines are not parallel, they must meet; and, if they meet, the planes in which they are situated must meet, contrarily to the definition of parallel planes.

169. DEFINITION. A parallelepiped is a solid contained by six planes, three of which are parallel to the other three.

170. THEOREM. The opposite planes of every parallelepiped are equal and equiangular parallelograms.

The opposite sides of all the figures are parallel, because they are the sections of one plane with two parallel planes (168): the corresponding sides of two opposite planes being, for the same reason, parallel to each other, contain equal angles (165), and they are also equal, as being the opposite sides of parallelograms; consequently the opposite figures are the doubles of equal triangles, and are, therefore, equal parallelograms.

171. Theorem. If a prism be divided by a plane, parallel to its two opposite surfaces, its segments will be to each other as the segments of any of the divided surfaces or lines.

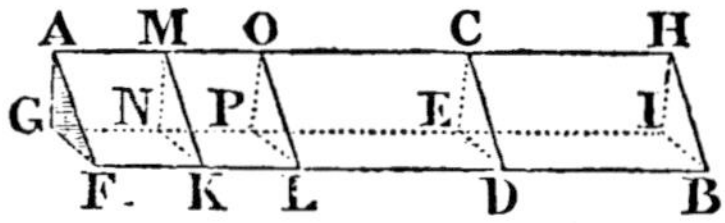

Let the prism AB be divided by the plane CDE parallel to AFG and BHI. Find FK a common measure of FD and DB (119), make KL =FK, and let the planes KMN, LOP be parallel to AFG; then the prisms AK, ML may be shown to be contained by similar and equal figures similarly situated, in the same manner as it is shown of parallelepipeds, and there is no imaginable difference between these prisms: they are therefore equal; and the prism AD is the same multiple of AK that FD is of FK, and AB the same multiple of AK that FB is of FK, or AD : AK=FD : FK, and AB : AK=FB : FK, whence AD : AB=FD : FB, and the prisms are in the same ratio as the segments of the line FB, or of the parallelogram GB (27).

If the segments are incommensurable, they are still in the same ratio, for it may be shown that the ratio of the prisms is neither greater nor less than that of the lines.

172. Theorem. Parallelepipeds on the same base, and contained between the same planes, are equal.

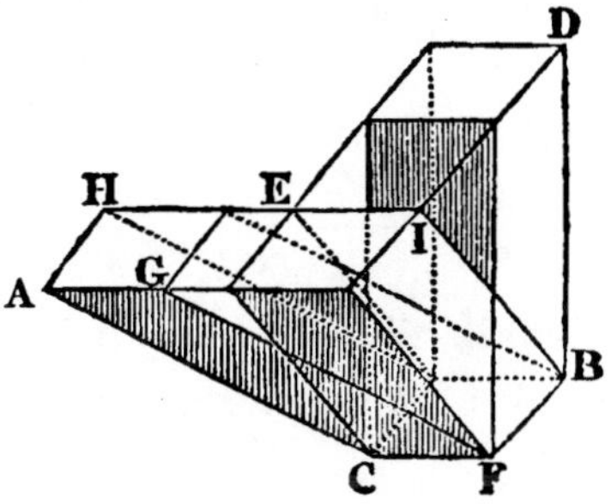

The parallelepiped AB is equal to CD standing on the same base BC, and terminated by the plane AED. For each is equal to the parallelepiped EF; since the triangular prism GB is similar and equal to the triangular prism HC, and deducting these from the solid HCI, the remainders AB and EF are equal. And in the same manner it may be shown that CD=EF; therefore AB=CD.

173. Theorem. Parallelepipeds on equal bases, and of the same height, are equal.

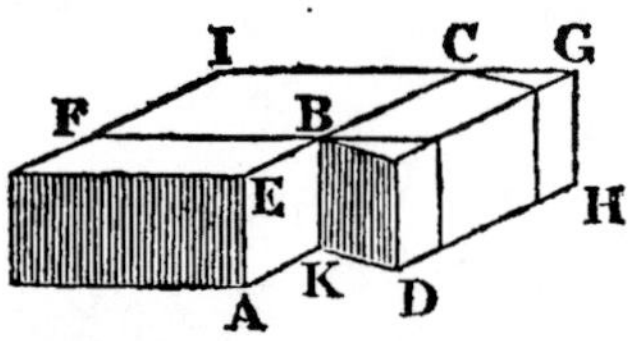

Each parallelepiped is equal to the erect parallelepiped on the same base. Let one of these be so placed, that the plane of one of the sides AB may coincide with the plane BC of the other parallelepiped CD, and that EBC may be a straight line. Then producing FB, and making CG parallel to it, the parallelepiped BH will be equal to CD (172). Now, completing the parallelepiped IK, as the parallelogram CF is to EF, so is KI to AF (171); and as CF to BG, so is KI to BH, but EF is equal to the base of AF, and BG to the base of CD, they are therefore equal, and the parallelepipeds AF and BH are equal, and AF=CD.

174. Theorem. Paralellepipeds, of the same height, are to each other as their bases.

For one of them is equal to a parallelepiped of the same height on an equal base which forms a single parallelogram with the base of the other; and this is to the other in the ratio of the bases (171); consequently the first two are in the same ratio.

175. Theorem. Parallelepipeds are to each other in the joint ratio of their bases and their heights.

For one of them is to a third parallelepiped of the same height with itself, but on the basis of the second, in the ratio of the bases, and the third is to the second in the ratio of the heights, consequently the first is to the second in the joint ratio of the bases and the heights. Thus, a and b being the bases, c and d the heights, e, f, and g the three parallelepipeds, $a : b :: e : g$, and $c : d :: g : f$; $ac : bd = e : f$.

Scholium. Hence is derived the common mode of finding the content of a solid, by multiplying the numerical representatives of its length, breadth, and height, and thus comparing it with the cubic unit of the measure.

176. THEOREM. Similar parallelepipeds are in the triplicate ratio of their homologous sides.

For the joint ratio of the bases and heights is the same as the triplicate ratio of the sides.

177. THEOREM. A plane, passing through the diagonals of two opposite sides of a parallelepiped, divides it into two equal prisms.

The diagonals are parallel, because the lines in which they terminate are parallel and equal, and every line and angle of the one prism is equal to the corresponding line and angle of the other prism; consequently the prisms are equal. Thus AB=CD, AE=CF, DE=BF, the angle EAB=DCF, EAH=GCF, and BAH=DCG.

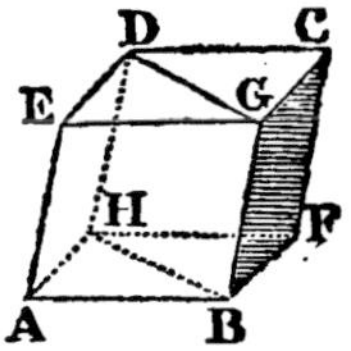

178. THEOREM. Prisms are to each other in the joint ratio of their bases and their heights.

Triangular prisms are in the same ratio as the parallelepipeds on bases twice as great, of which they are the halves; and all prisms may be divided into triangular prisms, by planes passing through lines similarly drawn on their ends, and they will be equal together to the half of a parallelepiped on a basis twice as great; consequently two such prisms are in the same ratio as the parallelepipeds.

179. THEOREM. All solids, of which the opposite surfaces are planes, and the sides such that a straight line may be drawn in them, from any point of the circumference of the ends, parallel to a given line, are to each other in the joint ratio of their bases and their heights.

For if they are terminated by rectilinear figures, the solids are prisms; and if they are terminated by curvilinear figures, they will always be greater than prismatic figures, of which the bases are inscribed polygons, and less than figures of which the bases are circumscribed polygons; and if the proposition be denied, it will always be possible to inscribe a prism in one of the solids, which shall be greater than any solid, bearing to the other solid a ratio assignably less than the ratio determined by the proposition, and to circumscribe a prism less than any solid bearing a ratio assignably greater. Such solids may not improperly be called cylindroids.

180. Theorem. The fluxion of any solid, described by the revolution of an indefinite line, passing through a vertex, and moving round any figure in a plane, is equal to the prismatic or cylindroidal solid, of which the base is the section parallel to the given plane, and the height the fluxion of the height.

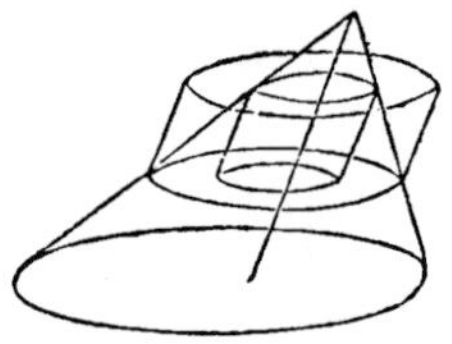

In any increment of the solid, which is cut off by planes determining the increment of the height, suppose a prismatic or cylindroidal solid to be inscribed, of which the base is equal to the upper surface of the segment, and the sides such that a line may always be drawn in them parallel to a given line passing through the vertex and the basis of the solid: and let another solid be similarly described on the lower surface of the segment as a basis: then it is obvious that the increment is always greater than the inscribed solid, and less than the circumscribed; and that when the increment is diminished without limit, its two surfaces are ultimately in the ratio of equality, and the increment coincides with the cylindroid described on its basis. Such solids may be termed in general pyramidoidal.

181. Theorem. All pyramidoidal solids are equal to one third of the circumscribing prismatic or cylindroidal solids of the same height.

The area of each section of such a figure, parallel to the basis, is proportional to the square of its distance from the plane of the vertex. For each section is either a polygon similar to the basis, or it may have polygons inscribed and circumscribed, which are similar to polygons inscribed and circumscribed in and round the basis, and which may differ less from each other in magnitude than any assignable quantity, consequently each section is as the square of any homologous line belonging to it, or, by the properties of similar triangles, as the square of the distance from the vertex, or from the plane of the vertex. If, then, the area of the base be a, the whole height b, and the distance of any section from the plane of the vertex

x, the area of the section will be $\frac{xx}{bb}.a$, and the fluxion of the solid $\frac{a}{bb}x^2\dot{x}$, of which the fluent is $\frac{1}{3}\frac{a}{bb}x^3$, and when $x=b$, the content is $\frac{1}{3}ax$, which is one third of the content of the whole prismatic or cylindroidal solid. Hence a pyramid is one third of the circumscribing prism, and a cone one third of the circumscribing cylinder.

182. Theorem. The fluxion of any solid is equal to the parallelepiped, of which the base is equal to the section of the solid, and the height to the fluxion of its height.

For every part of a solid may be considered as touching some pyramidoidal solid, and having the same fluxion: and the fluxion expressed by a cylindroid is equal to a parallelepiped, on the same base, and of the same height.

183. Theorem. The curve surface of a sphere is equal to the rectangle contained by its verse sine and the sphere's circumference.

The fluxion of the surface is obviously equal to the rectangle contained by the fluxion of the circumference and the circumference of the circle of which the radius is the sine; it varies, therefore, as the sine; but the fluxion of the cosine or of the verse sine varies as the sine, consequently the surface varies as the verse sine. Now, where the tangent becomes parallel to the axis, the fluxion of the surface becomes equal to the rectangle contained by the sphere's circumference, and the fluxion of the verse sine: hence the whole surface of any segment is equal to the whole rectangle contained by its verse sine and the sphere's circumference; and the surface of the whole sphere is four times the area of a great circle.

184. Theorem. The content of a sphere is two thirds of that of the circumscribing cylinder.

The fluxion of the sphere is to that of the cylinder as the square of the sine to the square of the radius; or if the fluxion of the cylinder be $aab\dot{x}$, a being the radius, and x the verse sine, that of the sphere will be $(2ax-xx)b\dot{x}$, or $2abx\dot{x}-bxx\dot{x}$, of which the fluent is $abx^2-\frac{1}{3}bx^3$; which, when $x=a$, becomes $\frac{2}{3}a^3b$, while the content of the cylinder is a^3b.

185. THEOREM. When a picture is projected on a plane, by right lines supposed to be drawn from each point to the eye, the whole image of every right line, produced without limit, is a right line drawn from its intersection with the plane of projection, to its vanishing point, or the point where a line drawn from the eye, parallel to the given line, meets the plane of projection; and this image is divided, by the image of any given point, in the ratio of the portion of the line, intercepted by that point and the picture, to the line drawn from the eye to the vanishing point; so that if any two parallel lines be drawn from the ends of the whole image, and the distances of the eye and of the given point be laid off on them respectively, the line, joining the points thus found, will determine the place of the required image of the point.

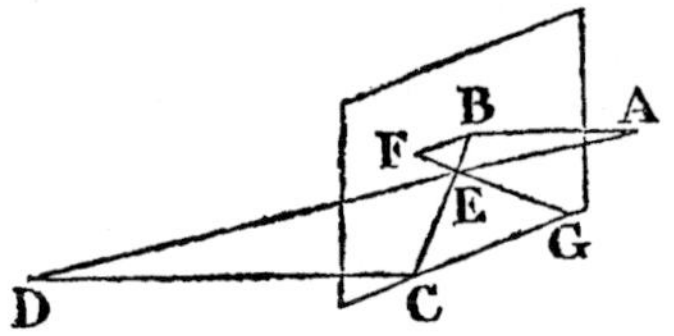

For A being the eye, and B the vanishing point of the line CD; AB and CD, being parallel, are in the same plane, and AD is also in that plane (62); and BC is the intersection of this plane with that of the picture; therefore E, the image of the point D, is always in the line BC; and AB : CD :: BE : EC; and taking the parallel lines BF, CG, in the same ratio, FG will also cut BC in E. When AB is perpendicular to the plane, B is called the point of sight, and is the vanishing point of all lines perpendicular to the plane of the picture: and the vanishing point of any other line may be found by setting off from B a line equal to the tangent of its inclination to the perpendicular line, the radius being AB.

SCHOLIUM. When a line becomes parallel to the plane of the picture, the distance of its vanishing point becomes infinite, and the image is, therefore, parallel to the original. In this case, the magnitude of the image may be determined by means of lines drawn in any other direction through the extremities of the original line. In the orthographical projection, the images of all parallel lines whatever become parallel, the distance of the eye, and consequently that cf the vanishing point, becoming infinite.

186. Definition. The subcontrary section of a scalene cone is that which is perpendicular to the triangular section of the cone, passing through the axis, and perpendicular to the base, and which cuts off from it a triangle similar to the whole, but in a contrary position.

187. Theorem. The subcontrary section of a scalene cone is a circle.

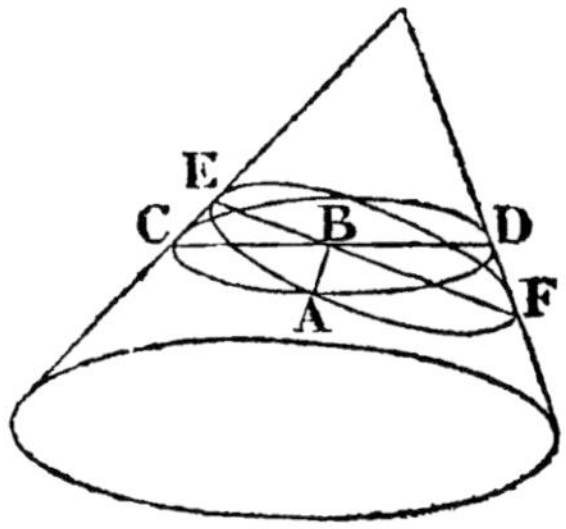

Through any point A of the section, let a plane be drawn parallel to the base; then its section will be a circle, as is easily shown by the properties of similar triangles; and the common section of the planes will be perpendicular to the triangular section of the cone to which they are both perpendicular; consequently, ABq=CB.BD; but since the triangles CBE, FBD are equiangular and similar, CB : BE :: BF : BD, and CB.BD=BE.BF=ABq; therefore EAF is also a circle.

188. Theorem. The stereographic projection of any circle of a sphere, seen from a point in its surface, on a plane perpendicular to the diameter passing through that point, is a circle.

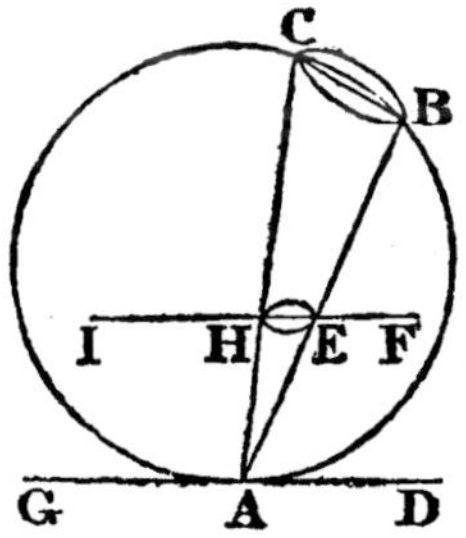

Let ABC be a great circle of the sphere passing through the point A and the centre of the circle to be projected, then the angle ACB=BAD=BEF, and ABC=CAG=CHI, and the triangle AHE is similar to ABC, and the plane ABC is perpendicular to the plane BC and the plane HE, therefore HE is a subcontrary section of the cone ABC, and is consequently a circle.

189. DEFINITION. Any parallel right lines, intercepted between a curve and a given right line, are called ordinates; and each part of that line, intercepted between an ordinate and a given point, is the absciss corresponding to that ordinate.

" 190." [141,B.] THEOREM. The fluxion of the area of any figure is equal to the parallelogram contained by the ordinate and the fluxion of the absciss.

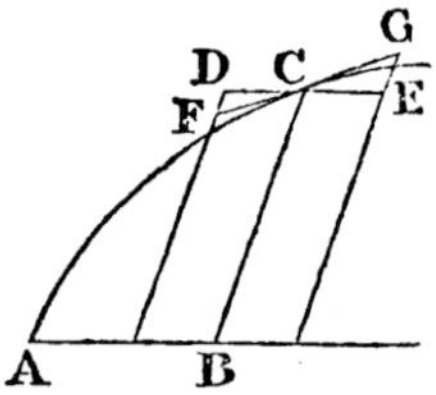

Let AB be the absciss, and BC the ordinate, through C draw DCE || AB, and take DC = DE = half the increment of AB, then the simultaneous increment of the figure ABC will ultimately coincide with the figure FCGEB, since the curve ultimately coincides with its tangent (141), but the triangles CDF, CEG, are equal, therefore the parallelogram DBE is ultimately equal to the increment of ABC. And if any other line than DE represent the fluxion of AB, as DE is to this line, so is the parallelogram DBE to the parallelogram contained by BC and this line: therefore that parallelogram is the fluxion of ABC (46).

SCHOLIUM. Those, who prefer the geometrical mode of representation, may deduce from this proposition a demonstration of the theorem for determining the fluxion of the product of two quantities (48); for every rectangle may be diagonally divided into two such figures as are here considered, and the sum of their fluxions, according to this proposition, will be the same with the fluxion of the rectangle determined by that theorem. It is obvious that this theorem ought not to have followed article 180.

191. DEFINITION. A flexible line being supposed to be applied to any curve, and to be gradually unbent, the curve, described by its extremity, is called the involute of the first curve, and that curve the evolute of the second.

192. DEFINITION. The radius of curvature of the in-

volute is that portion of the flexible line which is unbent, when any part of it is described.

193. THEOREM. The radius of curvature always touches the evolute, and is perpendicular to the involute.

If the radius of curvature did not touch the evolute, it would make an angle with it, and would, therefore, not be unbent; and if the evolute were a polygon composed of right lines, each part of the involute would be a portion of a circle, and its tangent, therefore, perpendicular to the radius: but the number of sides is of no consequence, and if it became infinite, the curvature would be continued, and the curve would still at each point be perpendicular to the radius of curvature.

194. THEOREM. The chord, cut off in the ordinate by the circle of curvature, is directly as the square of the fluxion of the curve, and inversely as the second fluxion of the ordinate, that is, as the fluxion of its fluxion.

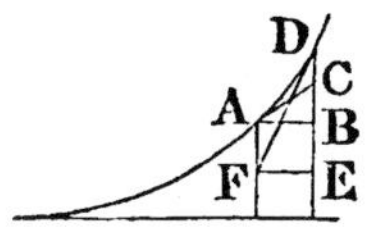

The constant fluxion of the absciss being equal to AB, the fluxion of the ordinate, at A, is BC, at D, DE, consequently its increment is CD+BE, or CD+AF, twice the sagitta of the arc AD: and the chord is equal to the square of AC divided by CD, and it is, therefore, always in the direct ratio of the square of the fluxion of the curve, and the inverse ratio of the second fluxion of the ordinate. See 268.

195. THEOREM. When the curve approaches infinitely near to the absciss, the curvature is simply as the second fluxion of the ordinate.

For the fluxion of the curve becomes equal to that of the absciss, and the perpendicular chord to the diameter.

196. DEFINITION. If the sum of two right lines, drawn from each point of a curve to two given points, is constant, the curve is an ellipsis, and the two points are its foci.

197. **Definition.** The right line passing through the foci, and terminated by the curve, is the greater axis, and the line bisecting it at right angles, the lesser axis.

198. **Theorem.** A right line passing through any point of an ellipsis, and making equal angles with the right lines drawn to the foci, is a tangent to the ellipsis.

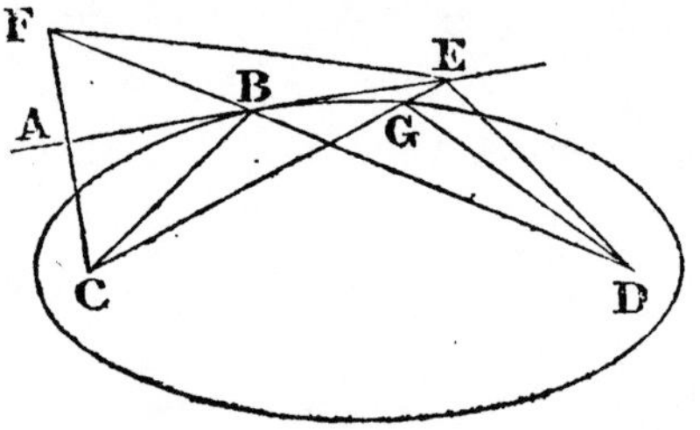

Let AB make equal angles with BC and BD, then it will touch the ellipsis in B. Let E be any other point in AB. Produce DB, take BF=BC, and join CF, then AB bisects the angle CBF, and CAB is a right angle. Join EC, ED, EF, GD, then EC=EF, and EC+ED=EF+ED, and is greater than DF (79), or BC+BD, or GC+GD, therefore E is without the ellipsis, and AB touches it.

199. **Theorem.** The right lines, drawn from any point of the ellipsis to the foci, are to each other as the square of half the lesser axis to the square of the perpendicular from either focus, on the tangent at that point.

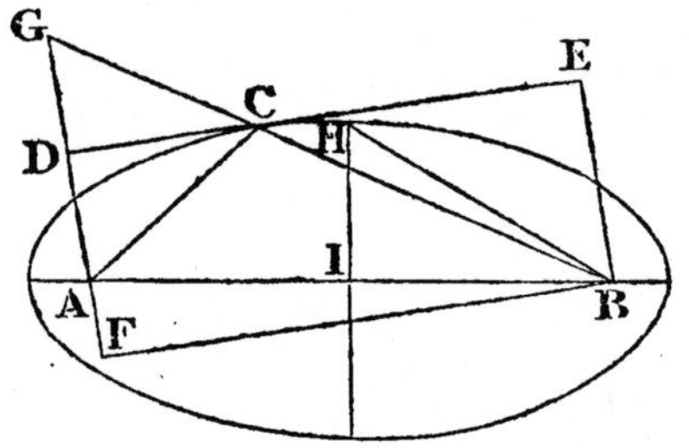

Let A and B be the foci, C the point of contact, and AD the perpendicular to the tangent CD, draw BE and BF parallel to AD and CD, produce AD each way, and let it meet BF and BC in F and G. Then since ∠ACD=BCE=DCG, CG=AC; and BG =AC+BC. And BFq=BGq—FGq=BAq—FAq (118), therefore BGq—BAq=FGq—FAq; but (FG+FA). (FG—FA)=FGq—FAq; and FG+FA=2FD=2BE, and FG—FA=AG=2AD; also BG=2BH, and BA=2BI, whence BGq—BAq=4HIq, therefore BE.AD=HIq, and $BE=\frac{HIq}{AD}$, but BE : BC : : AD : AC, and BE=

$$AD.\frac{BC}{AC}=\frac{HIq}{AD}, \text{ or } \frac{BC}{AC}=\frac{HIq}{ADq}.$$

200. Theorem. The chord of the circle of equal curvature with an ellipsis at any point, passing through the focus, is equal to twice the harmonic mean of the distances of the foci from the given point, or to the product of the distances divided by one fourth of the greater axis.

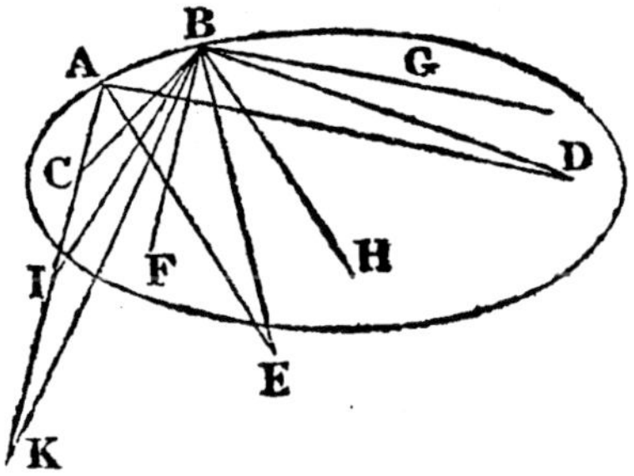

Let AB be an evanescent arc of the ellipsis, coinciding with the tangent, then the radius of curvature bisecting always the angle CAD or CBD, the point B, in which the radii AE and BE meet, will ultimately be the centre of the circle of equal curvature. Let BF, BG, be parallel to AC, AD; then BH, bisecting FBG, will be parallel to AE: but EBH=CBF+FBH−CBE=CBF+$\frac{1}{2}$FBG−$\frac{1}{2}$CBD=CBF−$\frac{1}{2}$CBF+$\frac{1}{2}$DBG= $\frac{1}{2}$ (CBF+DBG)=$\frac{1}{2}$(ACB+ADB). Now, in the triangles ABC, ABD, as AC is to the sine of ABC, so is AB to the sine of ACB, and as AD is to the sine of ABD, so is AB to the sine of BDA; but the sines of ABC and ABD are ultimately equal; consequently ACB and ADB are inversely as AC and AD, or as their reciprocals, and EBH or AEB, which is the half sum of ACB and ADB, is as the mean of those reciprocals: let BI be the reciprocal of that mean, or the harmonic mean of AC and AD, then the angle AIB=AEB; for the evanescent angles ACB, AIB, or their sines, are reciprocally as AC, AI, since these angles have the same side AB opposite to them in the triangles ABC, ABI, and their equals BC, BI are opposite to the same angle BAC; for the same reason, taking BK=2BI, AKB is half of AEB; consequently K is in the circle of curvature, and BK is its chord.

201. Theorem. The square of the perpendicular, falling on the tangent of an ellipsis from its focus, is to the square of the distance of the point of contact from the focus, as a third proportional to the axes is to the focal chord of curvature.

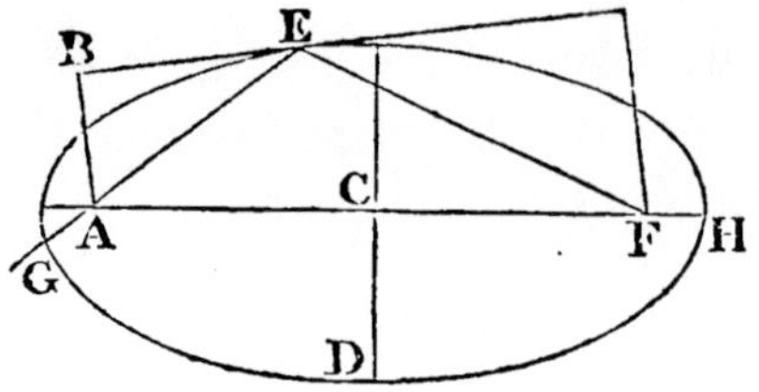

It has been shown that ABq : CDq : : AE : EF (199), therefore ABq : AEq : : CDq : AE.EF; but the chord of the circle of equal curvature, EG, is $\frac{2AE.EF}{CH}$, and AE.EF $=\frac{1}{2}$EG.CH, therefore ABq : AEq : : CDq : $\frac{1}{2}$ EG.CH : : $2\frac{CDq}{CH}$. EG.

SCHOLIUM. It may easily be demonstrated that a perpendicular to the normal of the curve, or to the line perpendicular to its tangent, passing through the point where it meets the axis, bisects the focal chord of curvature, and that a perpendicular, falling from the same point on the chord, cuts off a constant portion from it, equal to the third proportional to the semiaxes.

202. THEOREM. The square of any ordinate of an ellipsis, parallel to the lesser axis, is to the rectangle contained by the segments of the greater axis, as the square of the lesser axis to the square of the greater.

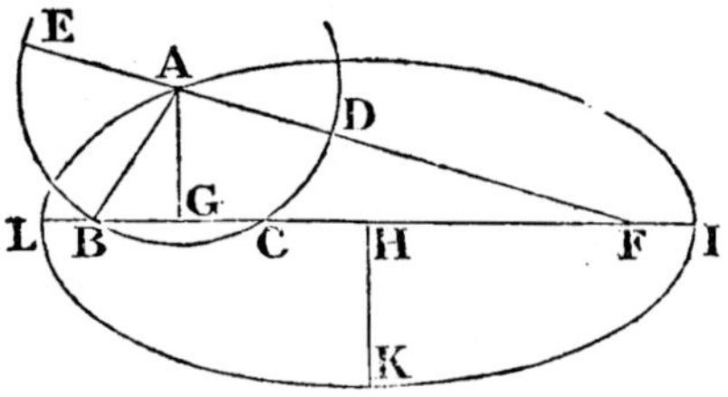

On the centre A describe the circle BCDE, passing through the focus B; then EF : BF : : CF : DF (138). Call HI,a, HB,b, AB,x, GH,z, then EF$=2a$, BF$=2b$, CF$=$2BH$-$2BG$=$ 2GH$=2z$, DF$=$EF$-$ED$=2a-2x$, and $2a : 2b : : 2z : 2a-2x$, $a : b : : z : a-x$, $a : a+b : : z : z+a-x$ $: : a+z : 2a-x+b+z$ (32); also $a : a-b : : z : z-(a-x) : : a-z : 2a-x-(b+z)$, and by multiplying the terms, $aa : aa-bb : : (a+z).(a-z) : (2a-x)^2-(b+z)^2$, or HIq.HKq : : IG.GL : AFq$-$GFq, or AGq.

203. THEOREM. The area of an ellipsis is to that of its circumscribing circle, as the lesser axis to the greater.

For since the square of the ordinate is to the rectangle contained by the segments of the axis, or to the square of the corresponding ordinate of the circle (137), as the square of the lesser axis to that of the greater, the ordinate itself is to that of the circle in the constant ratio of the lesser axis to the greater. For if four quantities are proportional, their squares are proportional, and the reverse. But the fluxions of the areas are equal to the rectangles contained by these ordinates and the same fluxion of the absciss (190), they are, therefore, in the constant ratio of the ordinates, and the corresponding areas are also in the same ratio (47).

204. DEFINITION. If the square of the absciss is equal to the rectangle contained by the ordinate and a given quantity, the curve is a parabola, and the given quantity its parameter.

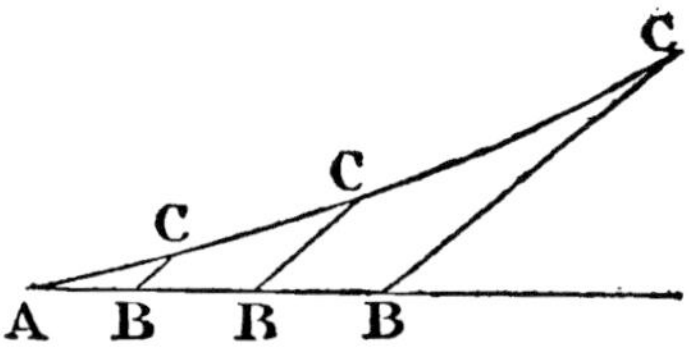

SCHOLIUM. Thus ABq=P.BC. If the axes of an ellipsis are supposed infinite, it becomes a parabola, for since $\frac{b^2}{a^2}=\frac{y^2}{ax-xx}$, if a becomes infinite, xx vanishes in comparison with ax, and $\frac{b^2}{a^2}=\frac{y^2}{ax}$, $\frac{b^2}{a}x=y^2$, and $\frac{b^2}{a}$ is the parameter of the parabola; and the distance from the focus is in a constant ratio to the square of the perpendicular falling on the tangent.

205. DEFINITION. When the ordinate is as any other power of the absciss than the second, the curve is still a parabola of a different order.

Thus when the ordinate is as the third power of the absciss, the curve is a cubic parabola.

206. THEOREM. If any figure be supposed to roll on another, and any point in its plane to describe a curve, that curve will always be perpendicular to the right line joining the describing point and the point of contact.

Suppose the figures rectilinear polygons; then the point of contact will always be the centre of motion, and the figure described will consist of portions of circles meeting each other in finite angles, so that each portion will be always perpendicular to the radius, though no two radii meet in the point of contact. And if the number of sides be increased without limit, the polygons will approach infinitely near to curves, and each portion of the curve described will still be perpendicular to the line passing through the point of contact.

207. **Definition.** A circle being supposed to roll on a straight line, the curve described by a point in the circumference is called a cycloid.

208. **Theorem.** The evolute of a cycloid is an equal cycloid, and the length of its arc is double that of the portion of the tangent cut off by the vertical tangent.

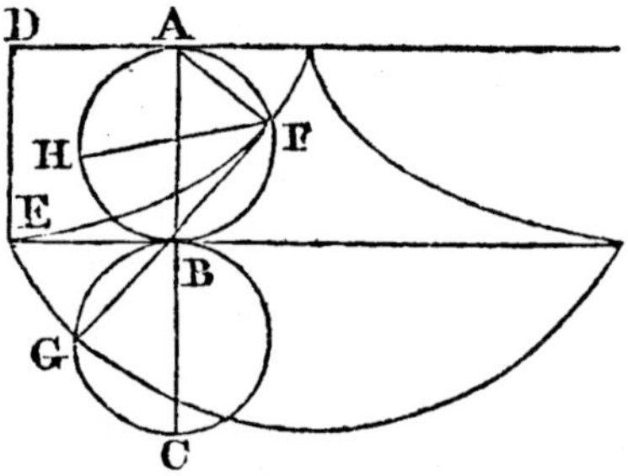

Let two equal circles AB, BC, rolling on the parallel bases DA and EB, at the distance of a diameter of the circles, describe with the points F and G the equal cycloids EF and EG. Draw the diameter FH; then H will be the point that coincided with D, and HA=DA=EB= arc BG, and the remainders AF and GC are equal, therefore ∠ABF=CBG (133), and FBG is a right line (96). But FG is perpendicular to AF (134), therefore it touches EF (206), and it is always perpendicular to EG (206); therefore EG will coincide with the involute of EF, for they set out together from E, and are always perpendicular to the same line FG (193), which they could not be if they ever separated. Consequently the curve EF is always equal to FG (192), or 2FB, twice the portion of the tangent cut off by EB.

209. **Theorem.** The fluxion of the cycloidal arc is to that of the basis, as the evolved radius to the diameter of the generating circle

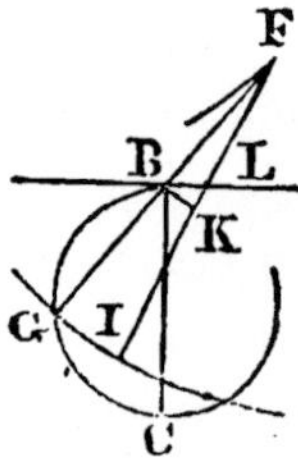

For the increment GI=2BK, and BK : BL : : BG BC, and 2BK : BL : : FG : BC, which is therefore the ratio of the fluxions.

Scholium. If the fluxion of the base be constant, that of the curve will vary as the distance of the describing point from the point of contact.

210. **Definition.** If the absciss be equal to the arc of a given circle, and the perpendicular ordinate to the corresponding sine, the curve will be a figure of sines.

211. **Definition.** If a second figure of sines be added, by taking ordinates equal to the cosines, the pair may be called conjugate figures of sines.

212. **Theorem.** The radius of curvature of the figure of sines at the vertex is equal to the ordinate.

For the fluxion of the base becoming ultimately equal to that of the absciss in the corresponding circle, while the ordinates are also equal, the curve ultimately coincides with a portion of that circle.

213. **Theorem.** The area of each half of the figure of sines is equal to the square of the vertical ordinate.

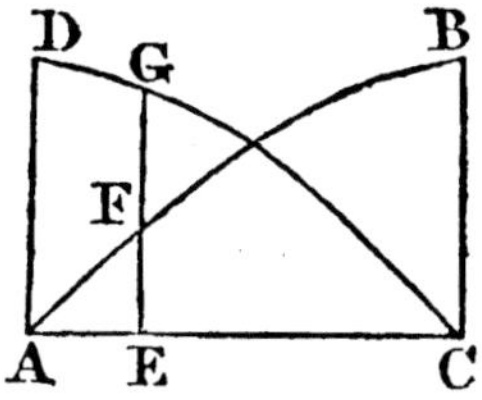

For the fluxion of the absciss being constant, that of the sine varies as the cosine (142), therefore the fluxion of the ordinate of the figure of sines may always be represented by the corresponding ordinate of the conjugate figure. Let AB, CD, be the conjugate figures, then EF will represent the fluxion of EG, and, since the arc and sine are ultimately equal, the fluxion of EG at C will be equal to that of the absciss, therefore BC will always represent the constant fluxion of the absciss. But the fluxion of the area AEF is the rectangle, under the fluxion of the absciss AE and the ordinate EF; that is, the rectangle under BC and the fluxion of EG, and the fluent BC.(AD—EG) is, therefore, equal to the area, which at C becomes BCq.

214. **Definition.** Each ordinate of the figure of sines being diminished in a given ratio, the curve becomes the harmonic curve.

Scholium. The ordinates being diminished in a constant proportion, their increments and fluxions are diminished in the same proportion, the fluxion of the base remaining constant.

215. THEOREM. The radius of curvature at the vertex of the harmonic curve is to that of the figure of sines, on the same base, as the greatest ordinate of the figure of sines to that of the harmonic curve.

For, taking any equal evanescent portions of the vertical tangents, the radii will be inversely as the sagittae, which are similar portions of the corresponding ordinates, and are therefore to each other in the ratio of those ordinates.

216. THEOREM. The figure, of which the ordinates are the sums of the corresponding ordinates of any two harmonic curves, on equal bases, but crossing the absciss at different points, is also a harmonic curve.

The absciss of the one curve being x, that of the other will be $a+x$, and the ordinates will be $b.(\sin. x)$ and $c. (\sin. a+x)$; now $\sin. a+x=(\cos. x).(\sin. a)+(\cos. a).(\sin. x)$ and the joint ordinate will be $(b+c.(\cos. a)).(\sin. x).+c.(\sin. a).(\cos. x)$; if, therefore, d be the angle of which the tangent is $\frac{c.(\sin. a)}{b+c.(\cos. a)}$ its sine and cosine will be in the ratio of $c.(\sin. a.)$ to $b+c(\cos. a)$, and $(\cos. d).(\sin. x)+(\sin. d).(\cos. x)$, will be to the ordinate in the constant ratio of $\sin. d$ to $c.(\sin. a)$; but $(\cos. d).(\sin. x)+(\sin. d).(\cos. x)$ is the sine of $d+x$; consequently the newly formed figure is a harmonic curve.

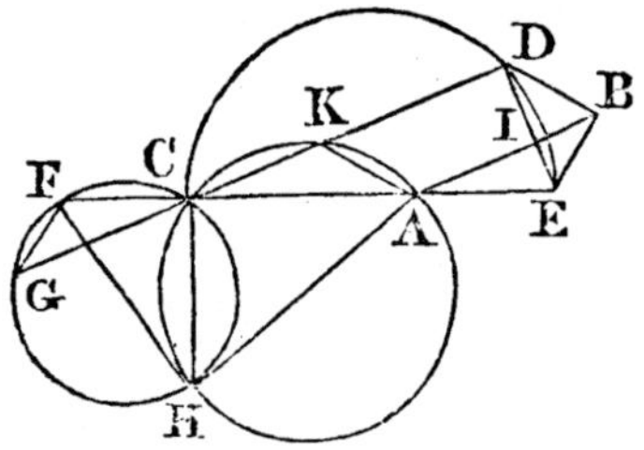

The same may be shown geometrically, by placing two circles, having their diameters equal to the greatest ordinates of the separate curves, so as to intersect each other in an angle equal to the angular distance of the origin of the curves: then a right line revolving round their intersection, with an equable velocity, will have segments cut off by each circle equal to the corresponding ordinate, and the sum or difference of the segments will be the joint ordinate: and if a circle be described through the point of intersection, touching the common chord of the two circles, and having its radius equal to the distance of their centres, this circle will always cut off in the revolving line a portion equal to the ordinate. For if AB be made parallel to CD, and EB to FG, ∠ABE=CGF=CHF: but EIB is a right angle, as well as HCF, and EI : IB::FC : CH::AE : CH, since AF is equal to twice the distance of the centres, which bisect AH and FH, and therefore to CE, and FC=AE, or EI : AE::IB : CH; but EI : AE::ID : AC, therefore IB : CH::ID : AC, and the triangles ACH, DIB, are similar, and ∠DBI=CHA=DKA, and AD is a parallelogram, consequently KD=AB=CG.

If the circle CG be supposed to revolve round C, the intersection H will always show the angular distance of the point in which the curve crosses the axis; and the distance of the centres will be equal to the greatest ordinate. If, therefore, the circles are equal, the greatest ordinate will also vary as the chord of an arc increasing equably, or as the ordinate of the harmonic curve.

ELEMENTARY ILLUSTRATIONS

OF

THE CELESTIAL MECHANICS.

Book the First.

OF THE GENERAL LAWS OF EQUILIBRIUM AND MOTION.

[*Divisions of the Original.*]

" CHAPTER I.

OF THE EQUILIBRIUM AND COMPOSITION OF FORCES ACTING ON A MATERIAL POINT.

§ 1, 2. *Of motion, force, and the composition and decomposition of forces.* M. C. P. 3. (249.)

§ 3. *Equation of the equilibrium of a point subjected to various forces. Method of determining pressure. Theory of momenta, or rotatory pressures.* P. 9. (250, 256.)

CHAPTER II.

OF THE MOTION OF A MATERIAL POINT.

§ 4. *Of the law of inertia, uniform motion, and velocity.* P. 14. (221.)

§ 5, 6. *Investigation of the relation between force and velocity, which in nature are proportional to each other. Results of this law.* P. 15.

§ 7. *Equation of the motion of a body actuated by any number of forces.* P. 19. (264.)

§ 8. *General expression of the square of the velocity.* (264, Cor.) *The body describes a curve in which the sum of the products of the velocity into the elements of space for the whole curve is a minimum.* P. 21. (266.)

§ 9. *Determination of the pressure of a moving point on a surface or a curve. Centrifugal force.* P. 23. (272.)

§ 10. *Motion of a gravitating point in a resisting medium. Law of resistance required for the description of a particular curve.* (273.) *Case where there is no resistance.* P. 25. (274 . . .)

§ 11. *Motion of a body in a spherical surface. Time of the oscillation. Very small oscillations isochronous.* P. 28. (280.)

§ 12. *Investigation of the curve in which the isochronism is perfect, in a resisting medium, with resistances proportional to the first two powers of the velocity.* P. 31. (282)"

The order of the subsequent sections is preserved unaltered.

CHAPTER I.

[OF MOTION, FORCE, AND PRESSURE.

SECTION I. *Of undisturbed motion.*

217. AXIOM. Like causes produce like effects, or, in similar circumstances, similar consequences ensue

SCHOLIUM 1. This axiom has always been essentiallv concerned in every improvement of natural philosophy, but it has been more and more employed, ever since the revival of letters, under the name INDUCTION. It is the most general and the most important law of nature; it is the foundation of all analogical reasoning, and it is collected from constant experience, by an indispensable and unavoidable propensity of the human mind.

SCHOLIUM 2. It does not appear that we can have any other accurate conception of causation, or of the connexion of a cause with its effect, than a strong impression of the observation, from uniform experience, that the one has constantly followed the other. We do not know the intimate nature of the connexion by which gravity causes a stone to fall, or how the string of a bow urges the arrow forwards; nor is there any original absurdity in supposing it possible, that the stone might have remained suspended in the air, or that the bowstring might have passed through the arrow as light passes through glass. But it is obvious that we cannot help concluding the stone's weight to be the immediate and necessary cause of its fall, and that every heavy body will fall unless supported; and the pressure of the string to be the necessary cause of the arrow's motion, and that if we shoot, the arrow will fly; and if we hesitated to make these conclusions, we should often pay dear for our scepticism. This explanation is sufficient to show the identity of the two expressions, that "like causes produce like effects," and that "in similar circumstances, similar consequences ensue." And such is the ground of argument from experience, the simplest principle of reasoning, after pure mathematical truths; which appear to be so far prior to experience, as their contradiction always implies an absurdity repugnant to the imagination.

SCHOLIUM 3. In the application of induction, the greatest caution and circumspection are necessary; for it is obvious that, before we can infer with certainty the complete similarity of two contingent events, we must be perfectly well assured that we are acquainted with every circumstance which can have any relation to their causes. The error of some of the ancient schools consisted principally in the want of sufficient precaution in this respect; for although Bacon is, with great justice, considered as the author of the most correct method of induction, yet, according to his own statement, it was chiefly the guarded and gradual application of the mode of argument, that he laboured to introduce. He remarks, that the Aristotelians, from a hasty observation of a few concurring facts, proceeded immediately to deduce universal principles of science, and fundamental laws of nature, and then derived from these, by their syllogisms, all the particular cases, which ought to have been made intermediate steps in the inquiry. Of such an error we may easily find a familiar instance. We observe, that, in general, heavy bodies fall to the ground unless they are supported; it was therefore concluded that all heavy bodies tend downwards: and since flame was most frequently seen to rise upwards, it was inferred that flame was naturally and absolutely light. Had sufficient precaution been employed in observing the effects of fluids on falling and on floating bodies, in examining the relations of flame to the circumambient atmosphere, and in ascertaining the specific gravity of the air at different temperatures, it would readily have been discovered, that the greater weight of the colder air was the cause of the ascent of the flame; flame being less heavy than common air, but yet having no spontaneous tendency to ascend. And accordingly the Epicureans, whose arguments, as far as they related to matter and motion, were often more accurate than those of their contemporaries, had corrected this error; for we find in the second book of Lucretius a very just explanation of this phenomenon.

> "See with what force yon river's crystal stream
> Resists the weight of many a massy beam;
> To sink the wood the more we vainly toil,
> The higher it rebounds with swift recoil.
> Yet that the beam would of itself ascend
> Will no man rashly venture to contend.
> Thus too the flame possesses weight, though rare,
> Nor mounts but when compelled by heavier air."

218. DEFINITION. Motion is the change of rectilinear distance between two points.

SCHOLIUM 1. Allowing the accuracy of this definition, it appears that two points at least are necessary to constitute motion; that in all cases, when we are inquiring whether or no any body or point is in motion, we must recur to some other point with which we can compare it, and that if a single atom existed alone in the universe, it could neither be said to be in motion nor at rest. This may seem in some measure paradoxical, but it is the necessary consequence of admitting the definition, and the paradox is only owing to the difficulty of imagining the existence of a single atom, unsurrounded by innumerable points of space which we represent to ourselves as immoveable.

SCHOLIUM 2. It has been for want of a precise definition of the term motion, that many authors have fallen into confusion with respect to absolute and relative motion. For the definition of motion, as the change of rectilinear distance between two points, appears to be the definition of what is commonly called relative motion; but, on a strict examination, we shall find, that what we usually call absolute motion is merely relative to some space, which we imagine to be without motion, but which may very often be so in imagination only. The space, which we call quiescent, is in general that which is in the neighbourhood of the earth's surface: yet we well know, from astronomical considerations, that every point of the earth's surface is perpetually in motion, and that the direction of its motion is even continually varying: nor are there any material objects accessible to our senses, which we can consider as absolutely motionless, or even as completely motionless with regard to each other, since the continual variation of temperature, to which all bodies are liable, and the minute agitations, arising from the motions of other bodies with which they are connected, will always tend to produce some imperceptible change of their distances.

SCHOLIUM 3. These minute changes are neglected in the elementary operations of practical geometry: it must not, however, be forgotten that they exist, and it is right to make it one of the postulates, which are the basis of mathematical demonstration, "that the length of a straight line be capable of being identified, whether by the effect of any object on the senses, or merely in imagination, so that it may remain invariable" (76): although this postulate has more generally been tacitly understood than expressed.

SCHOLIUM 4. When, therefore, we assert that a body is absolutely at rest, we only mean to express its relation to some comparatively large space in which it is contained: for that there exists a body, or even a point, absolutely at rest, in as strict a sense as an absolutely straight line may be conceived to exist, we cannot positively affirm; and if such a quiescent body or point did exist, we have no criterion by which it could be distinguished. Supposing a ship to move at the rate of three miles in an hour, and a person on board to walk or to be drawn towards the stern at the same rate, he would be relatively in motion, with respect to the ship, yet we might very properly consider him as absolutely at rest: but he would, on a more extended view, be at rest only in relation to the earth's surface; for he would still be revolving round the axis of the earth with that surface, and with the whole earth round the sun: and with the sun and the whole solar system he would perhaps be slowly moving among the starry worlds which surround us. Now with respect to any effects within the ship, all the subsequent relations to exterior objects are of no consequence whatever, and the change of his rectilinear distance, from the various parts of the ship, is all that needs to be considered in determining those effects. In the same manner, if the ship appear, by comparison with the water only, to be moving through it with the velocity of three miles an hour, and the water be moving at the same time in a contrary direction at the same rate, in consequence of a tide or current, the ship will be at rest with respect to the shore, but the mutual actions and relations of the ship and the water will be the same, as if the water were actually at rest, and the ship in motion. Laplace (§ 1. P. 3.) views this subject in the more popular light, and employs much mathematical reasoning, to deduce from it the principles, here laid down as fundamental. (§ 4. P. 14. §. 5. P. 15.)

219. Definition. A space or surface, of which all the points remain spontaneously at equal distances from each other, is said to be quiescent, or at rest within itself.

Scholium. The term "spontaneously" is introduced, in order to exclude, from the definition of a quiescent space, any surface, of which the points are only retained at rest by means of a centripetal force, while they revolve round a common centre; for, with respect to such a revolving space or surface, the motions of any body will deviate from the laws which govern them in other cases.

220. Definition. When a point is considered as in motion with respect to a quiescent space, the right line, joining any two of its proximate places, is called its direction, and such a point is often simply denominated a moving point.

Scholium. Supposing the point to remain continually in one right line drawn in the quiescent space, that line is always the line of its direction; if it describes several right lines, each line is the line of its direction as long as it continues in it; but if its path becomes curved, we can no longer consider it as perfectly coinciding at any time with a right line, and we must recur to the letter of the definition, by supposing a right line to be drawn through two successive points in which it is found; and then if these points be conceived to approach each other without limit, we shall have the line of its direction. Now, such a line is called in geometry a tangent, for it meets the curve, but does not cut it, provided that the curvature be continued without contrary flexure (126).

221. THEOREM. A moving point never quits the line of its direction without a new disturbing cause.

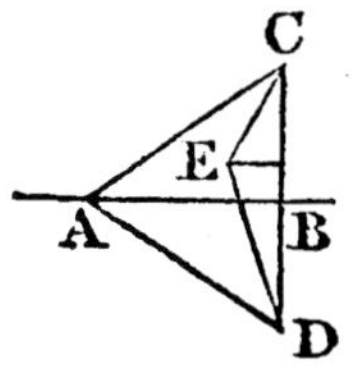

A right line being the same with respect to all sides, since it must remain wholly at rest if it be supposed to turn round any two of its points (60), there can be no imaginable reason why the point should incline to one side more than to another. Let AB be the direction of the motion of A in the plane ABC, and let CB and DB be equal and perpendicular to AB, then the triangles ABC and ABD are equal (86), and A is similarly related to C and D. But if A depart from AB, and be found in any point out of it, as E, ED will be greater than EC (103), and A will be no longer similarly related to C and D, contrarily to the general law of induction (217).

SCHOLIUM. This argument appears to be sufficiently satisfactory to give us ground for asserting, that the law of motion, here laid down, may be considered as independent of experimental proof. It was once proposed as a prize question by the Academy of Sciences at Berlin, to determine whether the laws of motion were necessary or accidental; that is, whether they were to be considered as mathematical or as physical truths. Maupertuis, then president of the academy, endeavoured to deduce them from a metaphysical principle of the minimum of action, which is of a very complicated and almost fanciful nature; and the intricacy of his theory tends only to envelope the subject in unnecessary obscurity; while the fundamental laws of motion appear to be easily demonstrable from the simplest mathematical truths, granting only the homogeneity or similarity of matter with respect to motion, and allowing the general axiom, that like causes produce like effects. If, however, any person thinks differently, he is at liberty to call these laws experimental axioms, collected from a comparison of various phenomena: for we cannot easily reduce them to direct experiments, since we can never remove from our apparatus the action of all disturbing causes; for either gravitation, or the contact of surrounding bodies, will interfere with all the motions which we can examine.

222. Definition. The times, in which a point, moving without disturbance, describes equal parts of the line of its direction, are called equal times.

223. Theorem. The equality of times being estimated by any one undisturbed motion, all other points, moving without disturbance, will describe equal portions of their lines of direction in equal times.

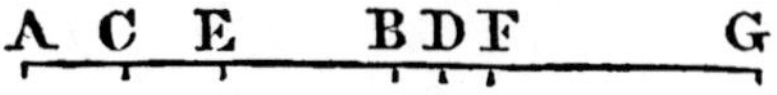

Let A and B be moving in the same line, and while A describes AC, let B describe BD; then while A describes CE=AC, B will describe DF=BD. For suppose AC=2BD, and let AG=2AB, then AB and BG have been equally decreased in one instance, and the relations remaining the same, they will still be equally decreased (217): the relative motion of A and *B* being equal to that of B and G, and any absolute motion being no way determinable, there can be no reason why the one should be otherwise affected than the other; therefore CE will be twice DF: and a similar mode of reasoning may be extended to all other cases, where the proportion of the motions is less simple.

SCHOLIUM 1. Having established the permanency of the rectilinear direction of undisturbed motion, we come to consider its uniformity. Here the idea of time enters into our subject; and we must have some measure of equal times, which cannot be merely intellectual, and must therefore be estimated by some changes in external objects. Of these changes, the simplest aud most convenient is the apparent motion of the sun, or rather of the stars, derived from the actual rotation of the earth on its axis, which is not, indeed, an undisturbed rectilinear motion, but which is equally applicable to every practical purpose: and hence we obtain, by astronomical observations, the well known measures of the duration of time, implied by the terms day, hour, minute, and second.

SCHOLIUM 2. Now, the equality of times being thus estimated from any one motion, all other bodies, moving without disturbance, will describe equal successive parts of their lines of direction in equal times. And this is the second law of motion, which, with the former law, constitutes Newton's first axiom or law of motion; "that every body perseveres in its state of rest or uniform rectilinear motion, except so far as it is compelled by some force to change it." This second law appears to be strictly deducible from the axioms and definitions which have been premised, and principally from the consideration of the relative nature of motion, and the total deficiency of any criterion of absolute motion: it is also confirmed by its perfect agreement with all experimental observations, although it is too simple to admit of an immediate proof. For we can never place any body in such circumstances, as to be totally exempt from the operation of all accelerating or retarding causes; and the deductions from such experiments, as we can make, would require, in general, for the accurate determination of the necessary corrections, a previous assumption of the law which we wish to demonstrate.

SCHOLIUM 3. When, indeed, we consider the motion of a projectile, we have only to allow for the disturbing force of gravitation, which so modifies the effect, that the body deviates from a right line, but remains in the same vertical plane; whence we may infer, that, in the absence of the force of gravitation, the body would continue to move in every other plane in which its motion began, as well as in the vertical plane, since in that case all planes would be indifferent to it; it would, therefore, necessarily remain in their common intersection, which could only be a straight line: so that, by thus combining argument with observation, we may obtain a confirmation of the law of the rectilinear direction of undisturbed motion, founded in great measure on direct experiment. The uniformity of undisturbed motion, is, however, still less subjected to immediate examination; yet, from a consideration of the nature of friction and resistance, combined with the laws of gravitation, we may ultimately show the perfect coincidence of the theory with experiment.

SCHOLIUM 4. The tendency of matter to persevere in the state of rest, or of uniform rectilinear motion, is called its inertia, or sometimes, very improperly, its *vis inertiae*. But the properties of matter, as such, belong to physical rather than to mathematical science: and we are, at present, considering the motions of a supposed inert point only.

224. THEOREM. If any number of points move in parallel lines, describing equal spaces in equal times, they are quiescent with respect to each other; and if all the points of a plane move in this manner on another plane, either plane will be in rectilinear motion with respect to the other.

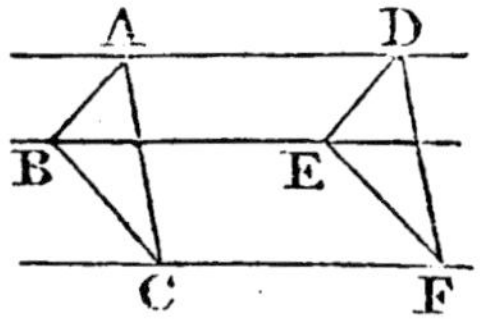

Let A, B, and C describe in a given time the equal parallel lines AD, BE, CF, then AB=DE, EF=BC, and DF=AC (109), and the points are mutually quiescent (218, 219). It is also obvious, that if two points have equal and parallel motions, the whole of the plane will also have a similar motion.

225. DEFINITION. If a plane be in rectilinear motion with respect to another plane, in contact with it, and if, besides this general motion of the plane, any point be supposed to have a particular motion in it, this point will have two motions with respect to the other plane, one in common with its plane, and the other peculiar to itself; and the joint effect of these motions, with respect to the

other plane, is called the result of the two motions.

226. THEOREM. The result of two motions, with respect to a quiescent space, is the diagonal of the parallelogram of which the two sides would be described by the separate motions; and any motion may be considered as the result of any other motions thus composing it.

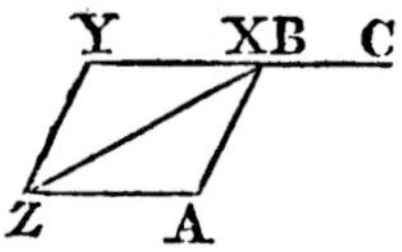

Let A, B, and C be three quiescent points, and let Z, Y, and X be three points in another plane which moves in the direction AZ, or BY; then the point A has a rectilinear motion ZA with respect to the plane ZYX. Now, while AZ is described by Z, let A have a motion in its own plane equal to AB; then it will have two motions with respect to ZYX, by the joint effect of which it will arrive at X in that plane; and if the motions are both equable, it may be shown, by the properties of similar triangles, that it describes the diagonal ZX. But it is of no consequence to the relative motion of A and ZXY which, or whether either, be imagined to be absolutely at rest: therefore, in general, the result of two motions, in a quiescent space, is the diagonal of the parallelogram of which the sides would be described by the separate motions: and the motion, thus produced, is precisely the same as if it were derived from a simpler cause.

SCHOLIUM 1. The existence of two or more motions at the same time, in the same body, is not at first comprehended without some difficulty. But it is, in fact, only a combination or separation of relations that is considered; in the same manner as by combining the relation of son to father, and brother to brother, we obtain the relation of nephew to uncle, so by combining the motion of a man walking in a ship, with the motion of the ship, we determine the relative velocity of the man with respect to the earth's surface.

SCHOLIUM 2. When an arm is made to slide upon a bar, and a thread, fixed to the bar, is made to pass, over a pulley at the end of the arm next the bar, to a slider which is moveable along the arm, the slider moves on the arm with the same velocity as the arm on the bar; but if the thread, instead of being fixed to the slider, be passed again over a pulley attached to it, and then brought back to be fixed to the arm, the motion of the slider will be only half that of the arm; and this will be true in whatever position the arm be fixed. Here we have two motions in the slider, one in common with the arm, and the other peculiar to itself, which may be either equal or unequal to the first; and by tracing a line on a fixed plane, with a point attached to the slider, we may easily examine the joint result of both the motions.

SCHOLIUM 3. The line described by the tracing point of this apparatus will be precisely the same, whether it is simply drawn along by the hand in the given direction, or made to move on the arm with a velocity equal to that of the arm, or when the arm is in a different position, with only half that velocity. The line AB, for example, may be either simply drawn in the direction AB, or it may be traced by the equal motions AC and AD of the arm and its slider, or by the unequal motions AE and AF.

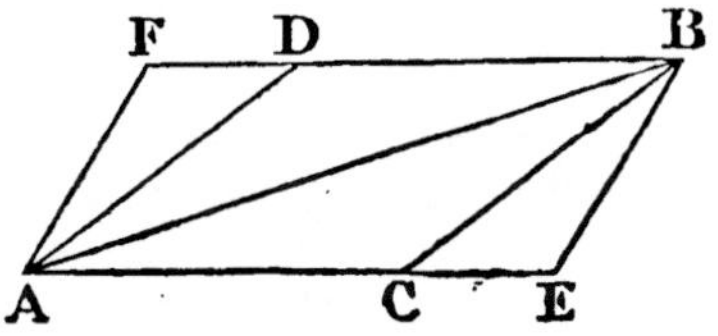

SCHOLIUM 4. There is some difficulty in imagining a slower motion to contain, as it were, within itself, two more rapid motions opposing each other: but, in fact, we have only to suppose ourselves adding or subtracting mathematical quantities, and we must relinquish the prejudice, derived from our own feelings, which associates the idea of effort with that of motion. When we conceive a state of rest as the result of equal and contrary motions, we use the same mode of representation, as when we say, that a cipher is the sum of two equal quantities with opposite signs; for instance, plus ten and minus ten make nothing.

SCHOLIUM 5. The law of motion, here established, differs but little, in its enunciation, from the original words of Aristotle, as they stand in his Mechanical Problems. He says, that "if a moving body has two motions, bearing a constant proportion to each other, it must necessarily describe the diameter of a parallelogram, of which the sides are in the ratio of the two motions." It is obvious, that this proposition includes the consideration not only of uniform motions, but also of motions which are similarly accelerated or retarded: and we should scarcely have expected, that, from the time at which the subject began to be so clearly understood, an interval of two thousand years would have elapsed, before the law began to be applied to the determination of the velocity of bodies actuated by deflecting forces, which Newton has so simply and elegantly deduced from it.

SCHOLIUM 6. In the laws of motion, which are the chief foundation of the Principia, their great author introduces at once the consideration of forces; and the first corollary stands thus: "a body describes the diagonal of a parallelogram by two forces acting conjointly, in the same time, in which it would describe its sides, by the same forces acting separately." It appears, however, to be more natural and perspicuous to defer the consideration of force until the simpler doctrine of motion has been separately examined.

227. THEOREM. Any equable motions, represented by the sides of a triangle or polygon, supposed to take place in the same moveable point, in directions parallel to those sides, and in the order of going round the figure, destroy each other, and the point remains at rest.

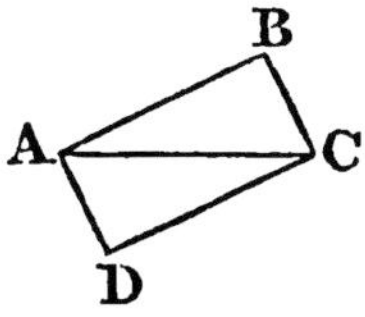

For two sides of the triangle, AB, BC, are sides of the parallelogram ABCD, therefore by the motions AB, BC, or AB, AD, A would arrive at C, while by the motion CA it would be brought back to A in the same time; and all the motions being equable, it will always remain in A: and, in the same manner, the proof may be extended to a figure with any greater number of sides. The truth of the proposition will also appear by considering several successive planes as moving on each other, and the point A as moving in the last: or we may imagine each motion to take place in succession for an equal small interval of time; then the point would describe a small polygon similar to the original one, and would be found, at the end of the whole of the small intervals, in its original situation.

SCHOLIUM. When the motions to be combined are numerous and diversified, it is often convenient to resolve each motion into three parts, reduced to the directions of three given lines perpendicular to each other: and, in this mnnner, the general result of any number of motions may be obtained, by addition and subtraction only. Thus, if a bird ascended in an oblique direction, we might describe its flight by estimating its progress northwards or southwards, eastwards or westwards, and at the same time upwards, as accurately as if we ascertained the immediate bearing and angular elevation of its path, and its velocity in the direction of its motion.

SECTION II. OF SIMPLE ACCELERATING FORCES.

228. DEFINITION. Any immediate cause of a change of motion is called a force.

SCHOLIUM 1. The word force ought to be very strictly confined to a cause which produces motion in a body at rest, or which increases, diminishes, or modifies it in a body which was before in motion. Thus, the power of gravitation, which causes a stone to fall to the ground, is called a force; but when the stone, after descending down a hill, rolls along a horizontal plane, it is no longer impelled by any force, and its relative motion continues unaltered, until it is gradually destroyed by the retarding force of friction. It was truly asserted by Descartes, that the state of motion is equally natural with that of rest, and that when a body is once in motion, it requires no foreign power to sustain its velocity. Since, however, the inertia of one body may easily become the cause of motion in another which is impelled by it, the term force is not uncommonly employed as almost synonymous with motion, and hence has arisen the incorrect notion of the *vis inertiae*, and of the force possessed by a moving body: but we must be careful to recollect that this sense of the term force is only so far correct, as it is applied to the power of causing motion in another body, and not to the motion of any one body considered separately.

SCHOLIUM 2. It is a necessary condition in the definition of force, that it be the cause of a change of motion with respect to a quiescent space. For if the change were only in the relative motion of two points, it might happen without the operation of any force: thus, if a body be moving without disturbance, its motion with respect to another body, not in the line of its direction, will be perpetually changed; and this change, considered alone, would indicate the existence of a repulsive force: and, on the other hand, two bodies may be subjected to the action of an attractive force, while their distance remains unaltered, in consequence of the centrifugal effect of a rotatory motion: the inertia here becoming a relative force, which tends to increase the distance of the body from a point out of the line of its direction, with an accelerated motion, unless counteracted by an attractive force.

SCHOLIUM 3. The muscular exertion of an animal, the unbending of a bow, and the impulsion produced by the apparent contact of a moving body, are familiar instances of the actions of forces. We must not imagine that the idea of force is naturally connected with that of labour or difficulty; this association is only derived from habit, since our voluntary actions are in general attended with a certain effort, leaving an impression almost inseparable from that of the force which it calls into action.

SCHOLIUM 4. It is natural to inquire, in what immediate manner any force acts, so as to produce motion; for instance, by what means the earth causes a stone to gravitate towards it. In some cases, indeed, we are disposed to imagine that we understand better the nature of the action of a force, as, when a body in motion strikes another, we conceive that the impenetrability of matter is a sufficient cause for the communication of motion, since the first body cannot continue its course without displacing the second; and it has been supposed, that if we could discover any similar impulse, which might be the cause of gravitation, we should have a perfect idea of its operation. But the fact is, that even in cases of apparent impulse, the bodies impelling each other are not actually in contact; and if any analogy between gravitation and impulse be ever established, it will not be by referring them both to the impenetrability of matter, but to the intervention of some common agent, which must probably be imponderable. It was observed by Newton, that a considerable force was necessary to bring two pieces of glass into a degree of contact, which still was not quite perfect; and Robison has estimated this force at a thousand pounds for every square inch. These extremely minute intervals have been ascertained by observations on the colours of the thin plate of air included between the glasses; and when an image of these colours is exhibited by means of the solar microscope, it is very easily shown that the glasses are separated from each other, by the operation of this repulsive force, as soon as the pressure of the screws which confine them is diminished; the rings of colours, dependent on their distance, contracting their dimensions accordingly. Hence it is obvious, that whenever two pieces of glass strike each other, without exerting a pressure equivalent to a thousand pounds for each square inch, they may affect each other's motion without actually coming into contact. It might perhaps be imagined, that this repulsion depended on some particles of air adhering to the glass; but the experiment has been found to succeed equally well in the vacuum of an air pump. We must, therefore, be contented to acknowledge our total ignorance of the intimate nature of forces of every kind; and we have, at present, only to examine the effect of forces, considered with regard to their magnitude and direction, without inquiring into their origin.

229. DEFINITION. When the increase or diminution of the velocity of a moving body is uniform, its cause is called a uniform force; the increments of space, which would be described in any given time with the initial velocities, being always equally increased or diminished.

SCHOLIUM 1. The word velocity appears to be sufficiently understood from common usage, although it is not easy to give a correct definition of it. The velocity of a body may be said to be the quantity or degree of its motion, independently of any consideration of its mass or magnitude; and it might always be measured by the space described in a certain portion of time, for instance, a second, if there were no other motions than undisturbed or uniform motions: but the velocity may vary very considerably within the second, and we must, therefore, have some other measure of it than the space actually described in any finite interval of time. If, however, the times be supposed infinitely short, the elements of space described may be considered as the true measures of velocity. These elements, though conceived to be smaller than any assignable quantity, may yet be accurately compared with each other; and the reason that they afford a true criterion of the velocity is this, that the change produced in the velocity, during an evanescent interval of time, must be absolutely inconsiderable in comparison with the whole velocity; so that the element of space becomes a true measure of the temporary velocity, in the same manner as any larger portion of space may be the measure of a uniform velocity.

SCHOLIUM 2. In this country it has been usual, at least till very lately, to preserve the geometrical accuracy introduced by the great inventor of the method of fluxions, and to call "any finite quantities, in the ratio of the velocities of increase and decrease of two or more magnitudes," the fluxions of these magnitudes (46). Thus, if we call the increments of x and y, $\acute{x}$ and $\acute{y}$, we have, for the fluxions, any magnitudes $\dot{x}$ and $\dot{y}$, so assumed, that $\dot{x} : \dot{y}$ shall be equal to $\acute{x} : \acute{y}$ when these increments become evanescent. On the continent, it has been more common to write dx and dy for $\acute{x}$ aud $\acute{y}$, considered as actually evanescent. It has been observed by Euler, at the beginning of his Integral Calculus, that the language of the English is the more correct, but that the continental notation is the more convenient. His words are these: "Quas enim nos quantitates variabiles vocamus, eas Angli, nomine magis idoneo, quantitates fluentes vocant, et earum incrementa infinite parva seu evanescentia fluxiones nominant, ita ut fluxiones ipsis idem sint, quod nobis differentialia. Haec diversitas loquendi ita jam usu invaluit, ut conciliatio vix unquam sit expectanda: equidem Anglos in formulis loquendi lubenter imitarer, sed signa, quibus nos utimur, illorum signis longe anteferenda videntur." Art. 6. In fact, however, the English do *not* call the *evanescent* increments fluxions, any more than a mile is an evanescent quantity, when we speak of a velocity of a mile an hour. There are certainly some cases in which the fluxional notation is inconvenient; thus, when we have occasion to write $\mathrm{d}\delta x = \delta \mathrm{d}x$, it would be impossible to express this equation without deviating from that method; we might, indeed, write $(\delta x)^{\cdot} = \delta \dot{x}$, but we still introduce a heterogeneous character. It is, however, a great inclegance, to say the least, not to distinguish a characteristic from a multiplying quantity by a difference of type; for dx means, according to all analogy, the product of d and x: and it is much more intelligible to write $\mathrm{d}x$, as Lacroix and many others have done, instead of dx, as it is generally printed in the works of Laplace. It must always be understood, then, that $\mathrm{d}x$, as well as $\dot{x}$, denotes a finite quantity proportional to an evanescent element: but when we use other characteristics of variation, such as δ or Δ, it is not always necessary to limit their signification so precisely: and it will sometimes be convenient to employ the mark D for an element of matter, considered as evanescent, and Δx for an evanescent increment of x, corresponding to the fluxion $\mathrm{d}x$.

SCHOLIUM 3. Now, a uniform force is a force that uniformly increases the velocity of a moving body. For example, if the velocities, at the beginning of any two separate seconds, be such that the body would describe one foot and ten feet in the respective seconds, and the spaces actually described become two feet and eleven feet, each being increased one foot, the accelerating force must be denominated uniform: it must also be uniform, in the still stricter sense of the definition, if the velocities, at the end of the second, have been so increased, that the body would describe two and eleven feet respectively in another second, if they continued their motion unaltered.

SCHOLIUM 4. The power of gravitation, acting at or near the earth's surface, may, without sensible error, be considered as a uniform force. Thus, if a body begins to fall from a state of rest, it acquires in a single second a velocity of $32\frac{2}{11}$ feet in a second; and in two seconds a velocity of $64\frac{4}{11}$ feet: having described in the first second $16\frac{1}{11}$ feet or 16·09, and in the second $32\frac{2}{11}+16\frac{1}{11}=48\frac{3}{11}$. The decrease of the force of gravitation, in proportion to the square of the distances from the earth's centre, is barely perceptible, at any heights within our reach, by the nicest tests that we can employ. See 288.

230. THEOREM. The velocity, produced by any uniformly accelerating force, is proportional to the magnitude of the force, and the time of its operation, conjointly.

For, the time and the velocity both flowing equably, their finite increments will be in a constant ratio (229, 47), and the velocity being the measure of the force, the velocity generated in a given time must also be proportional to the force. It may also readily be shown, by the composition of motion, that a double action must produce a double velocity: for when the equal sides of a parallelogram, representing two separate motions, approach to each other, and at last coincide in direction, the diagonal of the parallelogram, representing their joint effect, becomes equal to the sum of the sides: and the action of two independent forces must be truly represented by the two sides of the parallelogram, which represent them separately, otherwise they would not be independent, nor could their combination be called a double force. If we call the accelerating force *a*, the time *t*, and the velocity produced *v*, we shall have *v* proportional to *at*, and $\frac{at}{v}$ a constant quantity; or, if this quantity be called unity, $at=v$.

SCHOLIUM 1. The v of Laplace is sometimes employed as denoting the number of metres described in a decimal second, or $\cdot864^s$, which is also the number of myriometers described in a decimal hour, or the tenth of a day (§ 4. P. 15.): but it is often more convenient for computation to make v the number of English feet described in an ordinary second.

SCHOLIUM 2. The machine, invented by Mr. Atwood, furnishes us with a very convenient mode of making experiments on accelerating forces. The velocity, produced by the undiminished force of gravity, is much too great to be conveniently submitted to experimental examination; but by means of this apparatus, we can diminish it in any degree that is required. Two boxes, which are attached to a thread passing over a pulley, may be filled with different weights, which counterbalance each other, and constitute, together with the pulley, an inert mass, which is put into motion by a small weight added to one of them. The time of descent is measured by a second or half second pendulum, the space described being ascertained by the place of a moveable stage, against which the bottom of the descending box strikes: and when we wish to determine immediately the velocity acquired at any point, by measuring the space uniformly described in a given time, the accelerating force is removed, by means of a ring, which intercepts the preponderating weight, and the box proceeds with a uniform velocity, except so far as the friction of the machine retards it. By changing the proportion of the preponderating weight to the whole weight of the boxes, it is obvious that we may change the velocity of the descent, and thus exhibit the effects of forces of different magnitudes. Now, that the velocity generated is proportional to the time of the action of the force, or that the force of gravitation, at least when thus modified, is properly called a uniform accelerating force, may be shown by placing the moveable ring so as to intercept the same bar successively at two different points; thus the space uniformly described in a second, by the box alone, is twice as great, when the force is withdrawn after a descent of ten half seconds, as it is after a descent of five. And if we chose to vary the weight of the bar, we might show, in a similar manner, that the velocity generated in a given time is proportional to the force employed.

231. THEOREM. The increment of space described is as the increment of the time, and as the velocity, conjointly.

This is evident from the definition of velocity (45); and calling the space described x, and its increment x', we have $x'=vt'$, or $\Delta x=v\Delta t$; if we make the unities of time and space equivalent. This proposition is true of all increments, when the motion is uniform, but when variable, of evanescent increments only.

232. THEOREM. The space described, by means of a uniformly accelerating force, is as the square of the time of its action; it is also equal to half the space which would be described in the same time with the final velocity; and if the forces vary, the spaces are as the forces, and the squares of the times, conjointly: or $x=\frac{1}{2}at^2$

Since the velocity v is expressed by at, the product of the force and the time (230), and since $x'=vt'$ (231), or substituting fluxions for increments, $\dot{x}=v\dot{t}$, or(231) $dx=vdt$ and $vdt=atdt$, and the fluent x is equal to $\frac{1}{2}at^2$ (49) or $\frac{1}{2}vt$. Consequently x varies as t^2, and v being the velocity acquired at the end of the time t, the space described by it in that time would be vt, instead of $\frac{1}{2}vt$, the space actually described with the accelerated motion.

SCHOLIUM. The law, discovered by Galileo, that the space described is as the square of the time of descent, and that it is also equal to half the space which would be described in the same time with the final velocity, is one of the most useful and interesting propositions in the whole science of mechanics. Its truth is easily shown in a popular manner, by comparing the time with the base, and the velocity with the perpendicular of a right angled triangle gradually increasing in length and height, the area of which will represent the space described. We may also observe, by means of Atwood's machine, that a quadruple space is always described in a double time, by the continued operation of any constant accelerating force.

233, A. THEOREM. The times are as the square roots of the spaces directly, and of the forces inversely; they are also as the spaces directly, and the final velocities inversely.

Since $x=\frac{1}{2}at^2$, $t=\sqrt{\frac{2x}{a}}$; but $v=at$, $x=\frac{1}{2}vt$, and $t=\frac{2x}{v}$.

233, B. THEOREM. The final velocities are also as the spaces directly, and the times inversely

That is, $v=at=\frac{2x}{t}$ (233, A).

234. THEOREM. The forces are as the spaces directly, and the squares of the times inversely, beginning from the state of rest: they are also as the squares of the velocities directly, and as the spaces inversely.

Since $x=\frac{1}{2}at^2$, $a=\frac{2x}{tt}$; and since $v^2=a^2t^2$, $a=\frac{vv}{att}$ $=\frac{vv}{2x}$.

SCHOLIUM. Thus it may be shown by experiment, that if a body falls through one foot in a second by means of a certain force, it will require a quadruple force to make it fall through the same space in half a second; and that, in general, where the spaces are equal, the forces are as the squares of the velocities.

235. THEOREM. The fluxions of the squares of the velocities are as the fluxions of the spaces, and as the forces conjointly, whether the forces be uniform or variable.

In the evanescent time t', the variation of the force vanishes in comparison with the whole, so that it may be considered as a uniform accelerating force, and $v'=at'$ (230); consequently $\mathrm{d}v=a\mathrm{d}t$: but $\mathrm{d}x=v\mathrm{d}t$ (231); therefore $a\mathrm{d}t\mathrm{d}x=v\mathrm{d}t\mathrm{d}v$, and $a\mathrm{d}x=v\mathrm{d}v=\frac{1}{2}\mathrm{d}\,(v^2)$ (49).

SCHOLIUM. This proposition is one of the most important of the discoveries of Newton; and it is of consequence to bear in mind, that wherever the space and the force remain the same, whether the force be uniform or not, the squares of any two velocities, with which a body enters the space, will receive equal additions during the passage through it.

236. THEOREM. In considering the effects of a retarding force, the body may be supposed to be at rest in a moveable plane, and the motion generated by the force may be deducted from that of the plane.

In this case a being negative, we have $v=b-at$, and $\mathrm{d}x=v\mathrm{d}t=b\mathrm{d}t-at\mathrm{d}t$, whence $x=bt-\frac{1}{2}at^2$, bt being the space described by the initial velocity, and $\frac{1}{2}at^2$ being deducted from it by the effect of the retarding force.

SCHOLIUM. The degrees, by which an ascending body loses its motion, are the same as those by which it is again accelerated at the same points, when it has acquired its greatest height and again descends. We may thus calculate to what height a body will rise, when projected upwards with a given velocity, and retarded by the force of gravitation. Since the force of gravitation produces or destroys a velocity of 32 feet in every second, an initial velocity of 320 feet, for instance, will be destroyed in 10 seconds; and in 10 seconds a body would fall through 100 times 16 feet, or 1600 feet, which is therefore the height, to which a velocity of 320 feet in a second will carry a body, moving without resistance in a vertical direction. We may also obtain the same result by squaring one eighth of the velocity; thus one eighth of 320 is 40, of which the square is 1600, the height corresponding to the given velocity; and this velocity is sometimes called the velocity due to the height, being found by multiplying its square root by 8; thus $\sqrt{1600}\times 8=320$.

237. THEOREM. If two forces act in the same right line on a moveable body, varying inversely as the square of its distance from two given points, situated at the distance a from each other, the magnitudes of the forces being expressed by b and c at the distance d, the square of the velocity generated in the passage of the body, between any two points of which the distances from the first centre are successive values of x, is the difference of the corresponding values of $2d^2\left(\frac{b}{x}+\frac{c}{a\pm x}\right)$.

The sum of the forces, acting on the body, is $b\frac{dd}{xx}+c\frac{dd}{(a\pm x)^2}$, and since $v\dot{v}=$"$a\dot{x}$" (235), $v\dot{v}=\frac{bdd}{xx}\dot{x}\pm\frac{cdd}{(a\pm x)^2}\dot{x}$, and $\frac{vv}{2}=-\frac{bdd}{x}-\frac{cdd}{a\pm x}$, consequently $vv=-+\left(\frac{2bdd}{x}+\frac{2cdd}{a\pm x}\right)$: and if $c=0$, $v=d\sqrt{\frac{2b}{x}}$.

SCHOLIUM. This proposition is not altogether entitled to a place among the elementary doctrines of motion, having arisen from an inquiry into the origin of the meteoric stones: but it serves as a very good illustration of the utility of the 235th article. In the case of a body projected from the moon towards the earth, $d=20\,900\,000$ feet, $a=60d$, $b=32.2$ feet, the velocity produced in a second at the earth's surface; and $c=\frac{1}{70}b$, nearly; then taking $x=\frac{219}{220}a$, at the moon's surface, and $\frac{84}{94}a$, at the point where the force becomes neutral, we have $\frac{2bdd}{a}\left(\frac{1}{219}\right.$

$+\frac{1}{70})\times 220$ and $\frac{2bdd}{a}(\frac{1}{84}+\frac{1}{700})\times 94$, of which the difference is $\frac{5\cdot788\, bdd}{a}$, or $.09646\, bd$, and its square root about 8070 feet. Hence, if the velocity of a projectile from the moon exceed 8070 feet, it may pass the neutral point, and descend to the earth, where its velocity will become more than 36000 feet in a second.

SECTION III. OF PRESSURE AND EQUILIBRIUM.

238. "281." DEFINITION. A pressure is a force counteracted by another force, so that no motion is produced.

SCHOLIUM. Thus we continually exert a pressure by means of our weight, upon the ground on which we stand, the seat on which we sit; and the bed on which we sleep; but at the instant when we are falling or leaping, we neither exert nor experience a pressure on any part.

239. "282." DEFINITION Equal and proportionate pressures are such, as are produced by forces, which would generate equal and proportionate motions in equal times.

240. "283" THEOREM. Two contrary pressures will balance each other, when the motions, which the forces would separately produce in contrary directions, are equal; and one pressure will counterbalance two others, when it would produce a motion equal

and contrary to the result of the motions, which would be produced by the other forces.

If we conceive the forces to act alternately, during equal evanescent intervals of time, then the one will at each step destroy the preceding effect of the other, and there will be no motion left: then if we suppose this action to be doubled, the forces will become a continual pressure, and the total effect will still be the state of rest.

241 "284." THEOREM. If a body remain at rest by means of three pressures, they must be related in magnitude as the sides of a triangle parallel to the directions.

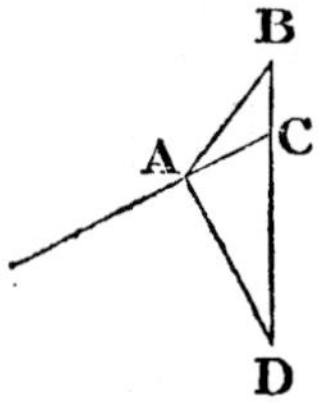

This proposition is the immediate consequence of the law of the composition of motion (226, 240). Suppose the body A, for example, to be suspended by the thread AB, on the inclined plane AC, to which AD is perpendicular, BD being the direction of gravity. Then in order that the force BD may be destroyed, it must be opposed by an equal force DB, and if DB be composed of forces acting in the directions DA, AB, the forces must be as those sides of the triangle, or as the sides of the parallelogram of which DB is the diagonal; and the same is true of any other pressures.

SCHOLIUM 1. This extension of the laws of the composition of motion to that of pressure seems to be free from any material objection. For since we measure forces by the motions which they produce, the composition of forces seems to be obviously included in the doctrine of the composition of motions; and when we combine these forces according to the laws of motion, there can be no question that the resulting motion is truly determined in all cases, whatever may be its magnitude, nor can any reason be given why it should be otherwise, when this motion is evanescent, and the force becomes a pressure.

SCHOLIUM 2. The proposition may be familiarly illustrated by a simple experiment; we attach three weights to as many threads, united in one point, and passing over three pullies; then by drawing any triangle, of which the sides are in the directions of the threads, or in directions parallel to them, we may always express the magnitude of each weight by the length of the side of the triangle corresponding to its thread.

SCHOLIUM 3. The laws of pressure have however been deduced by some of the most celebrated mathematicians, independently of those of motion, from the principle of the equality of the effects of equal causes; and such a demonstration may be found in an improved form, in the article Dynamics of the First Supplement of the Encyclopædia Britannica, contributed to that publication by the late Professor Robison; but its steps are still tedious and intricate. It will however be necessary, in conformity with the plan of this work, to insert here the demonstration of Laplace, which is sufficiently conclusive, though less simple than conld perhaps be desired: and it will be convenient to premise some lemmas, which are but very slightly connected with the immediate subjects of discussion. Every lemma is indeed an interruption of systematic order, and is inadmissible in a completely methodical treatise; but in following the steps of another author, this interruption may sometimes become indispensable.

242. DEFINITION. A series of units, a series of natural numbers, a series of their sums, and a series formed of the sums of all the numbers of any preceding series, are called figurate numbers of the first, second, and other higher orders respectively.

243. LEMMA A. The figurate number, of which the place is M, in the order N, is equal to $\frac{M(M+1)(M+2)\ldots(M+N-2)}{1.2.3\ldots(N-1)}$.

For, the two successive values of this expression, taken for $M-1$ and for M, are $\frac{(M-1)M(M+1)\ldots(M+N-3)}{1.2.3\ldots(N-1)}$ and $\frac{M(M+1)(M+2)\ldots(M+N-2)}{1.2.3\ldots(N-1)}$, and their difference is $(N-1).\frac{M(M+1)\ldots(M+N-3)}{1.2\ldots(N-1)}=\frac{M(M+1)\ldots(M+(N-1)-2)}{1.2\ldots(N-2)}$, which is the Mth figurate number of the order $N-1$, according to the definition: and when $N=2$, we obtain the natural series of integers. Since also $\frac{M(M+1)\ldots(M+N-2)}{1.2.3\ldots N}$ $=\frac{1.2.3\ldots(M+N-2)}{1.2.3\ldots(M-1).1.2\ldots(N-1)}$, it is obvious that M and N are equally concerned in the expression, and the number which occupies the place M in the order N is the same as the number N of the order M.

244. LEMMA B. The binomial or rather dinomial quantity $(1+x)=1+Nx+N\cdot\frac{N-1}{2}x^2+$ $N\cdot\frac{N-1}{2}\cdot\frac{N-2}{3}x^3+\ldots$

By actual multiplication, we find

$$\begin{array}{ll} & 1+x \\ \times x - & \quad x+x^2 \\ \hline & 1+2x+x^2 \\ \times x = & \quad x+2x^2+x^3 \\ \hline & 1+3x+3x^2+x^3, \text{ and the coefficients are} \end{array}$$

(N)

1,1	each being obtained by adding together
1,2, 1	two contiguous coefficients of the pre-
1,3, 3, 1	ceding lines; whence it follows, that each
1,4, 6, 4, 1	of the vertical columns must contain a
1,5,10,10, 5, 1	series of figurate numbers of an order
1,6,15,20,15,6,1	indicated by its distance from the begin-
	ning, the place of the coefficients in the

order being lowered by one at each step, so that for any horizontal line answering to the power N, we have I, N_2, $(N-1)_3$, $(N-2)_4$ ... denoting the place of the figurate number by the letters N, (N—1). . , and the order by the figures below. Now, the third coefficient, $(N-1)_3$, putting 3 for " N", is $\frac{M(M+1)}{1.2.}$, and then substituting N—I for M, $\frac{(N-1)N}{1.2}$: in the same manner the fourth coefficient $(N-2)_4$, or $\frac{M(M+1)(M+2)}{1.2.3}$, becomes $\frac{(N-2)(N-1)N}{1.2.3}$; and the subsequent terms may be shown in a similar manner to follow the same law.

SCHOLIUM. This demonstration is only strictly applicable to integral and positive powers, such as are very properly denoted, in the article Fluents of the Supplement of the Encyclopædia Britannica, by small Roman capitals: it may be extended without much difficulty to other cases: but for the present purpose, that of showing the analogy to the laws of differences, the integral powers are sufficient. See 278.

245. LEMMA C. If $\Delta u, \Delta^2 u \ldots$ be the successive finite differences of the quantities $u, u_1, u_2 \ldots$ we shall have $u_n = u + n\Delta u + n \cdot \frac{n-1}{2}\Delta^2 u + \ldots$, and $\Delta^n u = u_n - nu_{n-1} + n \cdot \frac{n-1}{2} u_{n-2} -$

In the first place

$$\begin{array}{lll} \Delta u = u_1 - u & \Delta^2 u = \Delta u_1 - \Delta u & \Delta^3 u = \Delta^2 u_1 - \Delta^2 u \\ \Delta u_1 = u_2 - u_1 & \Delta^2 u_1 = \Delta u_2 - \Delta u_1 & \Delta^3 u_1 = \Delta^2 u_2 - \Delta^2 u_1 \\ \Delta u_2 = u_3 - u_2 & \Delta^2 u_2 = \Delta u_3 - \Delta u_2 & \\ \Delta u_3 = u_4 - u_3 & & \Delta^4 u = \Delta^3 u_1 - \Delta^3 u \end{array}$$

Hence,

$$u_1 = u + \Delta u$$

$$u_2 = u_1 + \Delta u_1 \quad = u + \Delta u + \Delta(u + \Delta u) \quad = u + \Delta u + \Delta u + \Delta^2 u = u + 2\Delta u + \Delta^2 u$$

$$u_3 = u_2 + \Delta u_2 = u + 2\Delta u + \Delta^2 u \quad (+\Delta u_2) + \Delta(u_2) \quad \Delta u + 2\Delta^2 u + \Delta^3 u =$$

$$u + 3\Delta u + 3\Delta^2 u + \Delta^3 u$$

Now the steps of this operation are just the same as if we multiplied each time by $1 + \Delta$, though the symbol Δ^n is not exactly a power of Δ: but we may always make $\Delta u_1 = \Delta u + \Delta^2 u$ when u^1 is $= u + \Delta u$, which is in itself sufficiently evident, and is also shown by the equation $\Delta^2 u = \Delta u_1 - \Delta u$ whence $\Delta u_1 = \Delta u + \Delta^2 u$. The process is thus obviously similar to that of involution, and the law of the coefficients must be the same (244.) This method of reasoning, applied to the eye only, has been much extended by Lagrange, Arbogast, and others.

Again $\Delta u = u_1 - u -$

$$\Delta^2 u = \Delta(u_1 - u) = \Delta u_1 - \Delta u = u_2 - u_1$$
$$- u_1 + u =$$
$$\overline{u_2 - 2u_1 + u}$$

$$\Delta^3 u\Delta = \Delta(\Delta^2 u) = \Delta u_2 - 2\Delta u_1 + \Delta u$$
$$- u_3 - u_2$$
$$- 2u_2 + 2u_1$$
$$+ u_1 - u =$$
$$\overline{u_3 - 3u_2 + 3u_1 - u}$$

Here the operation of the characteristic Δ at each step doubles the number of terms to be added together, the coefficients being always formed, as in involution, by the addition of two contiguous ones of the former step: consequently the same law prevails as in the dinomial theorem.

246. LEMMA D. If a constant finite difference of x be called h, and any other difference h', the difference of u, corresponding to h, being Δu, that which corresponds to h' will be

$$\Delta' u = \frac{h'}{h}\Delta u + \frac{h'(h'-h)}{1.2\,h^2}\Delta^2 u + \frac{h'(h'-h)(h'-2h)}{1.2.3\,h^3}\Delta^3 u + \ldots$$

Since $u_n = u + n\Delta u + n.\frac{n-1}{2}\Delta^2 u + \ldots$ (245), if we suppose $a + nh = x$, x representing an absciss of which u is an ordinate, and a the initial value; or, in other words, u being a function of x, and the difference Δu corresponding to the difference $h = \Delta x$, substituting for n its value $\frac{x-a}{h}$, we shall have $u_n = u + \frac{x-a}{h}.\Delta u + \frac{x-a}{h}.\left(\frac{x-a}{h} - 1\right)\frac{\Delta^2 u}{.1.2} + .$

and substituting h' for $x-a$, and $\Delta' u$ for $u_n - u$, $\Delta' u = \frac{h'}{h}$.

$$\Delta u + \frac{h'(h'-h)}{1.2\,h^2}.\Delta^2 u + \frac{h'(h'-h)(h'-2h)}{1.2.3\,h^3}.\Delta^3 u + \ldots$$

SCHOLIUM. This proposition, which was invented by Newton, may be applied with great convenience to some cases of interpolations, the constant difference of time being h, and the variations of any other quantities depending on it being Δu and $\Delta' u$. For this purpose, if we make the fraction $\frac{h'}{h} = m$, the theorem will become $\Delta' u = m\Delta u - m.\frac{1-m}{2}\Delta^2 u + m.\frac{1-m}{2}.\frac{2-m}{3}\Delta^3 u - \ldots$; and these three terms will be abundantly sufficient for almost all cases that can occur in practice.

247. LEMMA E. Supposing the quantity x to vary gradually and uniformly, and h to be any finite difference of x, the corresponding finite difference of another quantity u, depending on it, will be $\Delta u' = h.\frac{\mathrm{d}u'}{\mathrm{d}x} + \frac{h^2}{1.2}.\frac{\mathrm{d}^2 u'}{\mathrm{d}x^2} + \frac{h^3}{1.2.3}.\frac{\mathrm{d}^3 u'}{\mathrm{d}x^3} + .$, u' being the initial value of the quantity u.

If we suppose the constant finite difference h of the preceding proposition to become evanescent, we shall have $\frac{\Delta u}{h} = \frac{\mathrm{d}u}{\mathrm{d}x}$ (46), $\frac{\Delta^2 u}{h^2} = \frac{\mathrm{d}^2 u}{h^2}$, and the equation will become $\Delta' u = h' \frac{\mathrm{d}u}{\mathrm{d}x} + \frac{h'h'}{1.2}\Delta^2 u + \ldots$, since $h'-h$, $h'-2h$ may be considered as simply equal to h', when h vanishes: and we may write Δ and h for Δ' and h', in the same sense. The initial value of u has sometimes been distinguished by a capital letter (*Phil.* Trans. 1819); Mr. Wronsky marks it by a point, $\dot{u}$, at least when u is supposed to vanish; but we must not altogether forget that this is the Newtonian character for a fluxion; the point, if it were thought necessary, might be written under the letter, $\underset{.}{u}$, or a prosodial mark might be employed instead of it, as u, or rather $\breve{u}$, which would partly explain itself; as indeed u' may be said to do.

COROLLARY 1. If h be an arc $=s$ and u its sine, making $u'=0$, we have $\frac{du'}{ds} = \cos s = 1$; $\frac{d^2u'}{ds^2} = -\sin s = 0$; $\frac{d^3u'}{ds^3} = -1$ .., whence $\sin s = s - \frac{1}{2.3}s^3 + \frac{1}{2..5}s^5 - ..$

COROLLARY 2. In the same manner $\cos s = 1$ $\frac{s^2}{1.2} + \frac{x^4}{1..4} - ..$

COROLLARY 3. If $u=a^x$, since $\frac{du}{dx} - a^x \text{hl}a$ (51), $\frac{d^2u}{dx^2} = a^x \text{hl}^2 a, \ldots$; putting $x'=0$, and $a^{x'}=1$, we have $a^x = 1 + \text{hl}a.x + \text{hl}^2 a.\frac{x^2}{1.2} + \text{hl}^3 a.\frac{x^3}{1..3} +$

COROLLARY 4. If $u=\text{hl}(a+x)$, $\frac{du'}{dx} = \frac{1}{a}$, $\frac{d^2u'}{dx^2} = -\frac{1}{a^2}$, $\frac{d^3u'}{dx^3} = \frac{2}{a^3}, \ldots$; and $\text{hl}\,(a+x) = \text{hl}a + \frac{x}{a} - \frac{x^2}{2a^2} + \frac{x^3}{3a^3} - \ldots$ Hence $\text{hl}\,(1+x) = x - \frac{x^2}{2} + \frac{x^3}{3} - \ldots$; and $\text{hl}(1-x) = -x$ $\frac{x^2}{2} - \frac{x^3}{3} - \ldots$; consequently $\text{bl}\frac{1+x}{1-x} = 2\ (x + \frac{x^3}{3} + \frac{x^5}{5} \ldots)$;

and if $\frac{1+x}{1-x}=z$, $x=\frac{z-1}{z+1}$; whence the hyperbolical logarithm of any number z may be readily found.

SCHOLIUM 1. Though the Taylorian theorem may be called a universal solvent of all analytical difficulties, yet considerable judgment is required, as with other universal remedies, for its proper application; and accident, perhaps, rather than talent, will often point out a device which will obtain from it unexpected results. There are however two general observations, respecting the employment of this theorem, which it will be proper to bear in mind. The first is, that where several variable quantities are concerned in a problem, it will be right to consider which of them is the most capable of affording a converging expression for the others by a series of its powers; thus, in the case of atmospherical refraction, the change of density of the medium may be easily obtained in terms of the refraction, supposed to be given, while the series for expressing the refraction in terms of the density is of little or no use; although the celebrated author of the theorem imagined, that he had sufficiently solved the problem of refraction, by determining a few of its first coefficients. The second observation is, that the employment of the theorem frequently requires a beginning to be made with a series obtained by the method of indeterminate coefficients; and that it may then be applied with advantage to the completion of the computation, when the series thus found loses its convergency: but in this case, we must not attempt to continue the series from the differences of its terms, since its convergence would be little affected by this operation, but we must revert to the original equation, which furnished the series by a different method. Taking for an example the equation $\int yx\mathrm{d}x = sx = x.\frac{\mathrm{d}x}{\surd(\mathrm{d}x^2+\mathrm{d}y^2)}$, putting $s = ax + bx^3 + cx^5 + \ldots$ and $\frac{\mathrm{d}y}{\mathrm{d}x} = \frac{s}{\surd(1-ss)} = s(1+\frac{1}{2}s^2+\ldots)$, and substituting for the powers of s, we obtain a value of y which affords that of $\int yx\mathrm{d}x$, and by comparing its terms with those of the value of sx, we determine the successive coefficients. (Suppl. Enc. Brit. Art. Cohesion). But the series is often inconvenient for want of convergence: we may therefore supply its defects by means of the Taylorian theorem, taking the successive fluxions of s at the point of the curve where we find it necessary to abandon the series: thus $yx\mathrm{d}x = s\mathrm{d}x + x\mathrm{d}s$, $y = \frac{s}{x} + \frac{\mathrm{d}s}{\mathrm{d}x}$, $\frac{\mathrm{d}s}{\mathrm{d}x} = y - \frac{s}{x}$; $\mathrm{d}\frac{\mathrm{d}s}{\mathrm{d}x} = \mathrm{d}y - \frac{\mathrm{d}s}{x} + \frac{s\mathrm{d}x}{xx}$ or, if $1-s^2=u^2$, $\frac{\mathrm{dd}s}{\mathrm{d}x^2} - \frac{s}{u} - \frac{y}{x} + \frac{2s}{xx}$, and $\mathrm{d}u$ being $= -\frac{s\mathrm{d}s}{u}$, and $\frac{\mathrm{d}u}{\mathrm{d}x} = \frac{ss}{ux} - \frac{sy}{u}$, $\frac{\mathrm{d}^3s}{\mathrm{d}x^3} = \frac{y}{u} - \frac{2s}{ux} - \frac{s^3}{u^3x} + \frac{s^2y}{u^3} + \frac{3y}{xx} - \frac{6s}{x^3}$, or, if $\frac{s}{u} - t$, $\frac{\mathrm{d}^3s}{\mathrm{d}x^3} = \frac{y}{u} - \frac{2t}{x} - \frac{t^3}{x} + \frac{t^2y}{u} + \frac{3y}{xx} - \frac{6s}{x^3}$; that is, since $1+t^2 = \frac{1}{u^2}$, $\frac{y}{u^3} - \frac{2t}{x} - \frac{t^3}{x} + \frac{3y}{xx} - \frac{6s}{x^3}$; and the fourth fluxion may be found in a similar manner, if its value be required: but the first three will be fully sufficient, provided that the curve be divided into small parts, even though they may be much larger than those which Laplace has employed in the Connaissance des Tems for 1810: and this method will probably be found at least as convenient as the much more elaborate process of Mr. Ivory. (Suppl. Enc. Br. IV). We may take, for another example of a difficulty precisely similar, the equation $\frac{dy}{y}=\int yx^2 dx.\frac{dx}{xx}-dz$, (Phil. Tr. 1819) the series, which it affords, losing its convergence when x becomes large: here we find; $\frac{dz}{dx}=w$, putting $\int yx^2 dx = wx^2$; $\frac{ddz}{dx^2}=\frac{dw}{dx}=y-\frac{2\int yx^2 dx}{x^3}=y-\frac{2w}{x}$; $\frac{d^3z}{dx^3}-\frac{dy}{dx}-\frac{2dw}{xdx}+\frac{2w}{xx}=$ $-wy-2\frac{y}{x}+\frac{4w}{xx}+\frac{2w}{xx}=\frac{6w}{xx}-\left(w+\frac{2}{x}\right)y$; and lastly $\frac{d^4z}{dx^4}-$ $\frac{4y}{x^2}-\frac{12w}{x^3}+wy\left(w+\frac{4}{x}\right)-y^2$.

SCHOLIUM 2. An important inversion of the Taylorian theorem will be found at the end of this Book.

248. LEMMA F. Whenever one quantity is dependent on another, their evanescent increments are ultimately in a constant proportion to each other.

It is not sufficient to observe that, if $y=ax+bx^m+cx^n+dx^p+\ldots$ the fluxion dy is $=dx\,(a+mbx^{m-1}+ncx^{n-1}+pdx^{p-1}+\ldots)$ the quantity multiplying dx being constant with regard to any small changes of the value of x and y; but it must also be shown, that the evanescent increment of any quantity being supposed to be increased or diminished in any given ratio, while it still remains evanescent, that of another quantity depending on it will be increased or diminished in the same ratio; and this is not demonstrable from the properties of the fluxions, strictly so called; but it may be understood by observing that, whatever be the form of the curve representing y by its ordinates, while the absciss is x, a very small portion of it may always be considered as approaching infinitely near to a straight line, and the increment of the ordinate will be, for an infinitely small space, proportional to that of the absciss, whether it be doubled or quadrupled, or in any way subdivided. The truth of the proposition is however shown more generally and conclusively by means of the invaluable theorem of Taylor, demonstrated in Lemma E, for the increment $\Delta u'$ of the ordinate, beginning from u', is to the increment h of the absciss in the constant ratio of $\frac{du'}{dx}$ to 1, as long as the increment h remains so small, that its square and its higher powers may be supposed to vanish in comparison with itself.

SCHOLIUM. It is however necessary to except the case in which the first fluxion of one of the quantities compared becomes $=0$. (See 249, Sch. 2).]

249. THEOREM 240, of the Composition of Forces, demonstrated in Laplace's manner.

CASE 1. The forces x and y, acting at right angles to each other, will produce a joint result z, of which the magnitude is expressed by the diagonal of the rectangle xy. For we may obviously suppose x to be composed of two forces, x' and x'', also at right angles to each other, and in the proportion of x to y, since the same law must apply to forces similarly related, whatever their magnitude may be; and the result x must be derived from x'. and x'' in the same manner as z from x and y; consequently we have $x'=\frac{x}{z}x$ and $x''-\frac{y}{z}x$. Now

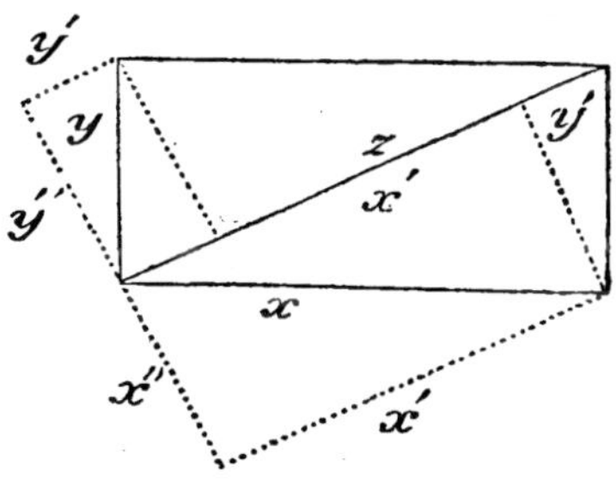

if x' be in the direction of z, x'' must be perpendicular to it; and supposing y to be similarly composed of y' and y'', y' being in the direction of z, and y'' perpendicular to it; x'' must be equal and contrary to y''; and x' and y' together must be equal to z: but $y'=\frac{y}{z}y$, and $y''-\frac{x}{z}y$: so that $x'+y'=\frac{x}{z}x+\frac{y}{z}y=z$, and $x^2+y^2=z^2$; consequently z is equal to the diagonal of the rectangle, the sides of which are x and y.

It must however be shown that z coincides with this diagonal in position as well as in magnitude. For this purpose we must consider one of the forces y as increasing from nothing to its actual magnitude, and we must trace the effects of its combination with x through the intermediate steps. Now if an elementary force δy be combined with a finite force x, the variation of the angular direction of the result, which may be called $\delta\theta$, will be inversely as x and directly as some constant multiple or submultiple of δy, since the evanescent increments of two quantities, related to each other, are initially in a constant ratio, (248), so that the chord of the angle $\delta\theta$ may be called $k\delta y$, and the angle itself $\frac{k\delta y}{x}$: the elementary chord $k\delta y$ obviously depending on x and on the variation of the angle $\delta\theta$, in such a manner, that δy may be expressed by $\frac{x\delta\theta}{k}$, and $\delta\theta$ by $\frac{k\delta y}{x}$. It is indeed sufficiently obvious that the chord can in this case be no other than δy itself, since a force in the direction of the radius could scarcely influence another in the direction of the circumference, but Laplace does not think it right to take this for granted without proof. We have therefore initially $\delta\theta = \frac{k\delta y}{x}$. In any other situation of the result z, we must suppose the element δy to be resolved into two portions, one in the direction of z, which only affects its magnitude, the other perpendicular to it, which determines the increment of the angle $\delta\theta$ from z, in the same manner as δy determined it in the first instance from x. Now the portion of the force δy perpendicular to z is $\frac{x}{z}\delta y$: consequently $\delta\theta = \frac{kx\delta y}{zz}$; or, since x

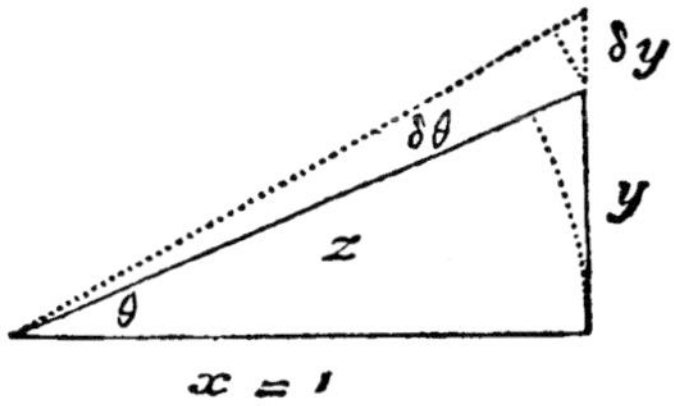

is here considered as invariable, or $=1$, $\delta\theta = \frac{k\delta y}{zz}$. But the fluxion of the angle φ, of which the tangent is y, is $\frac{dy}{\sec^2\varphi}$, as is readily understood from considering the relative situation of the increments; consequently, since z^2 has been shown to be equal to x^2+y^2, $\frac{\delta y}{z^2} = \frac{\delta y}{1+yy} = \delta$ ang tang y, and $\theta = k$ ang tang $y+c$. But since $\theta=0$ when $y=0$, c vanishes, and $\theta = k$ ang tang y: and when y is infinite, $\theta = 90^\circ$, since z coincides with y, consequently k must be $=1$, and $\theta=$ ang tang y. So that z must coincide with the diagonal of the rectangle in position as well as in magnitude.

SCHOLIUM 1. Laplace has supposed both the forces x and y to vary together: but this is evidently an unnecessary complication.

SCHOLIUM 2. The principle of the proportional variation of evanescent increments must not be applied without some caution, for in the present investigation, if it had been required to express the relation of δy to δx while y remained evanescent, the proposition would have failed, since in this case δx is initially $=0$, and varies at first as the square of δy, the first term of the Taylorian theorem, on which the reasoning is founded, here vanishing altogether: but when the application is clearly understood, the argument is readily admitted almost as an axiom.

Case 2. When the forces concerned are not at right angles to each other, they may both be referred to orthogonal coordinates, and if their projections in any three such directions be a, b, and c, these lines will represent the respective portions of the force $\sqrt{(a^2+b^2+c^2)}$: and if the second force be represented by similar ordinates a', b', and c', the forces may be combined by adding together their constituent portions, as reduced to the same directions, giving together $a+a'$, $b+b'$, and $c+c'$, which may again be combined into a single force; and this force z will always be represented by the diagonal of the parallelogram, formed by lines representing the two former, x and y: and in the same manner any greater number of forces may be combined, by reducing them to three orthogonal directions, and by adding together their respective results.

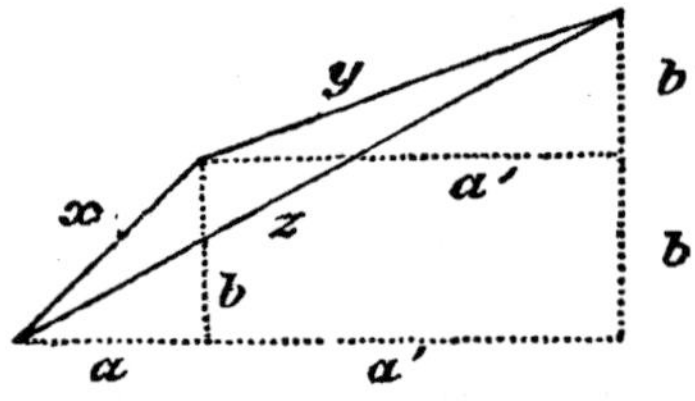

250. Theorem. When several forces act on the same moving point, if we suppose the place of the point to be changed in any manner whatever to a minute distance, the product of the joint force into this distance

will be equal to the sum of the products of each separate force into the respective distances or variations in the lines of their directions.

Or, $V\delta u = \Sigma S\delta s$; (a);

V being the joint force, δu the variation of the distance in its direction, and $\Sigma S\delta s$ the sum of the quantities $S\delta s$ obtained by multiplying each force S by the variation of the line of its direction δs. §. 2. P. 7.

Let s be the distance of the point M from the origin of one of the forces S, and let the position of M be determined by the three coordinates x, y, and z, and that of the origin of the force by a, b, and c; we have then $s = \sqrt{\left\{(x-a)^2 + (y-b)^2 + (z-c)^2\right\}}$, and the portions of S acting in the directions of x, y, and z will be $S.\frac{x-a}{s}$, $S.\frac{y-b}{s}$, and $S.\frac{z-c}{s}$ respectively: and it is obvious that, if s be made to vary by the alteration of x alone, δs will be to δx as $x-a$ to s, and $\frac{\delta s}{\delta x} = \frac{x-a}{s}$. According to the mode of notation commonly adopted, the coefficient of this partial variation is compendiously expressed by the notation $\frac{\delta s}{\delta x}$, though it would be more correct to write it $\frac{\delta_x s}{\delta x}$, or even $\frac{\delta' s}{\delta x}$, in order that the same symbol might not be employed for a partial and a total variation: and it is easily found, by taking the fluxion of s,

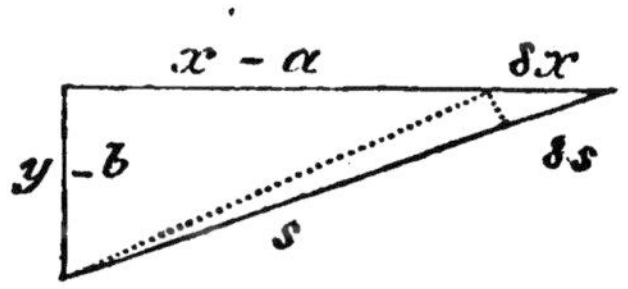

that $\frac{\mathrm{d}'s}{\mathrm{d}x}$ or $\frac{\delta' s}{\delta x} = \frac{x-a}{s}$. If there be a second force S', and s' be the distance of M from any fixed point in its direction, we have in a similar manner $S'.\frac{\delta' s'}{\delta x}$ for the portion of this force acting in the direction of x; and employing the characteristic Σ for the sum of all the forces thus determined, we have $\Sigma S.\frac{\delta' s}{\delta x}$ for the whole force in the direction of x. Now if V be the result of all the forces $S, S', S'', \ldots$ thus combined, and u the distance of any point in its direction from M, we have $V.\frac{\delta' u}{\delta x}$ for the portion of this force, which acts in the direction of x, and which must be equal to $\Sigma S.\frac{\delta' s}{\delta x}$, by the supposition: and by comparing, in the same manner, the forces in the directions y and z, we obtain $V.\frac{\delta' u}{\delta y} = \Sigma S.\frac{\delta' s}{\delta y}$, and $V.\frac{\delta' u}{\delta z} = \Sigma S.\frac{\delta' u}{\delta z}$; and then, adding these partial variations, we obtain $V\delta u = \Sigma S.\delta s$, an equation which may be said to contain the three former, because, since the variations are perfectly arbitrary, we may make any two of them vanish, and the third will remain alone on both sides of the equation.

251. Corollary 1. When the point remains in equilibrium, the sum of the products of each force, multiplied by the elementary variation of its distance, is equal to nothing.

Or $\Sigma S.\delta s = 0$; since $V = 0$. (b)

252. COROLLARY 2. For the equilibrium of a point resting on a given surface, we may either comprehend the reaction of the surface among the forces S, or, with greater convenience, call the direction of the reaction r, and the force R, and we shall have $\Sigma S\delta s + R\delta r = 0$. (c)

253. COROLLARY 3. For a canal, or a curved line, which may be considered as a combination of two curved surfaces, the reaction of the second surface being called R', and the perpendicular to it r', we have $\Sigma S\delta s + R\delta r + R'\delta r' = 0$. (d)

Whatever the direction of the canal may be, its resistance may be conceived to be the result of the reactions R and R' of the two surfaces which determine its form, since the resistance being perpendicular to the curve, it must be in the same plane with the forces, which are perpendicular to the surfaces, of which it is the intersection.

254. SCHOLIUM 1. If we suppose the arbitrary variations δx, δy, δz, to take place in the direction of the surface to which the body is confined, we shall have $\delta r = 0$, and the equation $\Sigma S\delta s = 0$ will still be true: but the variations of s must then be taken so as to be limited to the given surface by means of its equations, and they cannot be all arbitrary. In the same manner we may make δr and $\delta r'$ both vanish when the motion is confined to a canal or a single curve, but in that case any one of the variations of s will determine the other two. It is, however, more convenient to

retain the variations δr and $\delta r'$, and to substitute for them their values derived from the nature of the surface, since we are thus enabled to determine the pressure.

SCHOLIUM 2. Now, if a, b, and c be the coordinates of the origin of the perpendicular r, for the part of the surface in question, without any regard to this origin remaining as a fixed point, we have the equation $r^2=(x-a)^2+(y-b)^2+(z-c)^2$, supposing only that a, b, and c remain constant for an elementary portion of the surface, as they must do in all cases: we have, then, for δx, $2r\delta' r=2(x-a)\delta x$, and $\frac{\delta' r}{\delta x}=\frac{x-a}{r}$, and in the same manner $\frac{\delta' r}{\delta y}=\frac{y-b}{r}$, and $\frac{\delta' r}{\delta z}=\frac{z-c}{r}$; and since $\left(\frac{x-a}{r}\right)^2+\left(\frac{y-b}{r}\right)^2+\left(\frac{z-c}{r}\right)^2=1$, we have consequently $\left(\frac{\delta' r}{\delta x}\right)^2+\left(\frac{\delta' r}{\delta y}\right)^2+\left(\frac{\delta' r}{\delta z}\right)^2=1$.

[SCHOLIUM 3. The substance of these scholia is expressed by the author in a form somewhat different; and in order that no injustice may be done to the symmetry of his system, it will be proper to insert his reasoning in its original form, with some explanatory remarks. "Let $u=0$ be the equation of the surface, then the two equations $\delta r=0$ and $\delta u=0$ will both be true together, which implies that δr may be $=N\delta u$, N being a function of x, y, and z. In order to determine this function, the coordinates of the origin of r may be called a, b, and c, we shall then have $r=\sqrt{\left\{(x-a)^2+(y-b)^2+(z-c)^2\right\}}$, whence we obtain $\left(\frac{\delta r}{\delta x}\right)^2+\left(\frac{\delta r}{\delta y}\right)^2+\left(\frac{\delta r}{\delta z}\right)^2=1$, and consequently $N^2\left\{\left(\frac{du}{dx}\right)^2+\left(\frac{du}{dy}\right)^2+\left(\frac{du}{dz}\right)^2\right\}=1$; so that if we make $\lambda=R:\surd$

$\left\{\left(\frac{du}{dx}\right)^2+\left(\frac{du}{dy}\right)^2+\left(\frac{du}{dz}\right)^2\right\}$, the term $R\delta r$ of the equation (c) (252) will become $\lambda.\delta u$, and the equation will become $0=\Sigma S\delta s+\lambda.\delta u$, in which the coefficients of the variations δx, δy, δz must be made to vanish separately; so that it affords three separate equations, which, however, are only equivalent to two, since they contain the indeterminate quantity λ." Now, supposing the equation of the surface $u=0$ to be $r^2-x^2-y^2-z^2=0$, as in the sphere, the natural sense of the symbol δu is $2r\delta r-2x\delta x-2y\delta y-2z\delta z$, which must be $=0$: but it must here be understood as relating only to the variations of x, y, and z, exclusively of r, that is, $\delta u=\frac{\delta' u}{\delta x}\delta x+\frac{\delta' u}{\delta y}\delta y+\frac{\delta' u}{\delta z}\delta z$. The subject may be further illustrated by an extract from the Mécanique Analytique of Lagrange, Sect. ii. n. 7, 8.

"Supposing, as is always allowable, the force P to tend to a fixed centre at the distance p, we have $p=\sqrt{\left\{(x-a)^2+(y-b)^2+(z-c)^2\right\}}$, and $pdp=(x-a)dx+(y-b)dy+(z-c)dz$. Now, if p be perpendicular to a given surface, its variation with respect to that surface will vanish, and we have $dp=0$: the surface being spherical if a, b, and c, are constant, but of any other form when they are considered as variable. If now the force P be in general perpendicular to a surface represented by the equation $\alpha dx+\beta dy+\gamma dz=0$, in order to make it coincide with the equation $(x-a)dx+(y-b)dy+(z-c)dz=0$, which results from the supposition $dp=0$, we must make $\frac{\alpha}{\gamma}=\frac{x-a}{z-c}$, and $\frac{\beta}{\gamma}=\frac{y-b}{z-c}$, whence $x-a=\frac{\alpha}{\gamma}(z-c)$, and $y-b=\frac{\beta}{\gamma}(z-c)$, and substituting these values in the value of dp, it becomes $dp=$

$\frac{\alpha \mathrm{d}x + \beta \mathrm{d}y + \gamma \mathrm{d}z}{\surd(\alpha^2 + \beta^2 + \gamma^2)}$," [since $(x-a)^2 + (y-b)^2 + (z-c)^2 = \left\{ \left(\frac{\alpha}{\gamma}\right)^2 + \left(\frac{\beta}{\gamma}\right)^2 + 1 \right\} (z-c)^2$, and $\frac{x-a}{p} = \frac{\alpha}{\gamma} \cdot \frac{z-c}{p} = \frac{\alpha}{\surd(\alpha^2 + \beta^2 + \gamma^2)}$; and $y-b$, and $z-c$ afford, in a similar manner, the terms $\beta \mathrm{d}y$ and $\gamma \mathrm{d}z$.] "We obtain, therefore, the value of $\mathrm{d}p$, whatever may be the form of the equation of the surface, from the equation $\mathrm{d}p = \mathrm{d}u : \surd \left\{ \left(\frac{\mathrm{d}u}{\mathrm{d}x}\right)^2 + \left(\frac{\mathrm{d}u}{\mathrm{d}y}\right)^2 + \left(\frac{\mathrm{d}u}{\mathrm{d}z}\right)^2 \right\}$, calling $\alpha \mathrm{d}x + \beta \mathrm{d}y + y \mathrm{d}z = \mathrm{d}u$, which must be admissible, since the differential equation of a surface must be a complete fluxion: and employing the usual mode of notation for the partial fluxions, in which $\frac{\mathrm{d}u}{\mathrm{d}x} = \alpha$, $\frac{\mathrm{d}u}{\mathrm{d}y} = \beta$, and $\frac{\mathrm{d}u}{\mathrm{d}z} - \gamma$: and $P\mathrm{d}p$, the efficacy of the force P, will be expressed by $P \mathrm{d}u : \surd \left\{ \left(\frac{\mathrm{d}u}{\mathrm{d}x}\right)^2 + \left(\frac{\mathrm{d}u}{\mathrm{d}y}\right)^2 + \left(\frac{\mathrm{d}u}{\mathrm{d}z}\right)^2 \right\}$," the $\lambda . \delta u$ of the Mécanique Céleste.]

[255. DEFINITION. The rotatory pressure of a given force, with respect to any axis, is the product of its magnitude into the distance of its line of direction from that axis.]

256. THEOREM. The rotatory pressure of the result of any number of forces is equal to the sum of the rotatory pressures of the same forces taken separately, with respect to the same axis.

The variations δx, δy, δz being considered as arbitrary and independent, we may substitute, in the equation $V\delta u = \Sigma S\delta s$, for the coordinates x, y, and z, three other quantities depending on them: and then make the coefficients of the variations of these quantities equal to nothing. Thus if we take ϱ, the radius drawn from the origin of the coordinates to the projection of the point M on the plane of x and y, and let ϖ be the angle formed by ϱ with the direction of x, we shall have

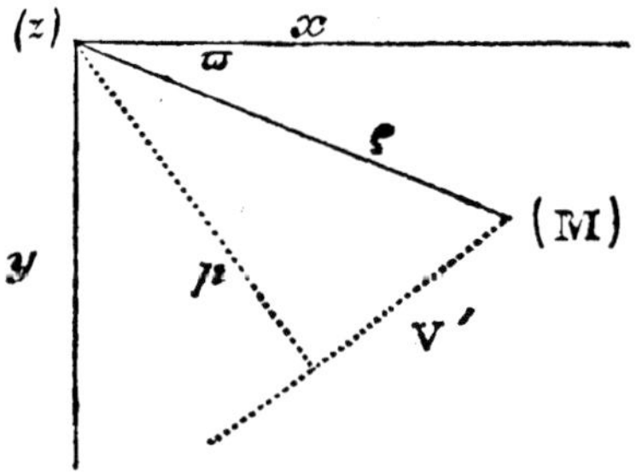

$x = \varrho \cos \varpi$, and $y = \varrho \sin \varpi$: and we may proceed to consider u, and the values of s, as depending on ϱ, ϖ, and z, and take the variation $V\frac{\delta' u}{\delta\varpi} - \Sigma S\frac{\delta' s}{\delta\varpi}$. (*e*) [This supposition is equivalent to taking the variation of the place of M by making it move in a plane parallel to that of x and y, while it remains at an equal distance from the origin of the coordinates, the element of its motion, or its variation, being $\varrho\delta\varpi$,] and the force V, so reduced to this direction, becomes obviously $V\frac{\delta' u}{\varrho\delta\varpi}$ (250). Again, if V' be the portion of V, which acts in the plane of x and y, and p be a perpendicular falling on its direction from the axis perpendicular to x and y, passing through the origin of the coordinates, the portion of V', which acts in the direction of the element $\varrho\delta\varpi$, will be $\frac{p}{\varrho} V$, consequently $\frac{p}{\varrho} V' = V\frac{\delta' u}{\varrho\delta\varpi}$, and $p\, V' = V\frac{\delta' u}{\delta\varpi}$. It follows, therefore, from the definition of rotatory pressure (254), that $V\frac{\delta' u}{\delta\varpi}$ is the rotatory pressure of the result of all the forces combined, which is equal to $\Sigma S \frac{\delta s}{\delta \varpi}$, the sum of the rotatory pressures of the separate forces, with respect to an axis parallel to z, and perpendicular to the plane of x and y; and this axis may be situated in any imaginable direction with respect to the forces concerned, the proposition holding good in all cases.

CHAPTER II.

OF DEFLECTIVE FORCES.

257. DEFINITION. "238." Any force, tending to alter the direction of the motion of a moving body or point, is called a deflective force.

SCHOLIUM. This definition includes not only accelerating forces, which, when directed to a point out of the line of the body's motion, are called central forces, but also the reaction of surfaces or threads, which limit the motion to particular surfaces, and are subject to the same laws, though they are only accelerative in a negative sense, as retarding rather than producing motion: but this distinction is of no consequence, nor could it in all cases be correctly established. It will serve as a useful introduction to the more general and analytical discussion of this subject in Laplace's manner, to premise a simple geometrical illustration of some of the properties of central forces, though they might be deduced as corollaries from the formulas of the Mécanique Céleste.

258. THEOREM "239." The force, by which a body is deflected into any curve, is directly as the square of the velocity, and inversely as that chord of the circle of equal

curvature, which is in the direction of the force; and the velocity in the curve is equal to that which would be generated by the same force, during the description of one fourth of the chord by its uniform action.

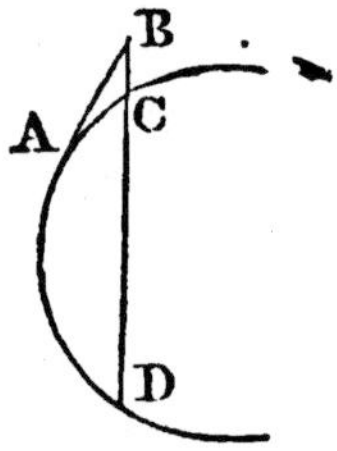

For the force is as the space described by its action, beginning from a state of rest, or as the evanescent sagitta through which the body is drawn from the tangent of the curve in a given instant of time: but the portion AB of the tangent spontaneously described in a given instant is as the velocity, and BC the sagitta is $= \frac{ABq}{BD}$, or ultimately $\frac{ABq}{CD}$, that is, as the square of the velocity directly, and inversely as the chord of curvature of the arc AC.

Now the velocity generated during the description of BC is expressed by 2BC, since the force may be considered for an instant as constant, and the final velocity is measured by twice the space actually described (232): the velocity generated is therefore to the orbital velocity as 2BC to AB, or as 2AB to BD, or as AB to half BD: and if the time were increased in the ratio of AB to half BD, the velocity generated by the force would be equal to the orbital velocity, but in this time half BD would be described by the velocity in the orbit, and half as much, or one fourth of BD, by a uniformly accelerated velocity (232).

259. Corollary 1. "240." When a body describes a circle by means of a force di-

rected to its centre, the velocity is every where equal to that which it would acquire in falling, by the action of the same force, supposed to be uniform, through the length of half the radius: and the force is as the square of the velocity directly, and as the radius inversely.

SCHOLIUM. By means of this proposition we may easily calculate the velocity, with which a sling of a given length must revolve, in order to retain a stone in its place in all positions; supposing the motion to be in a vertical plane, it is obvious that the stone will have a tendency to fall when it is at the uppermost point of the orbit, unless the centrifugal force be at least equal to the force of gravity. Thus if the length of the sling be two feet, we must find the velocity acquired by a heavy body in falling through a height of one foot, which will be eight feet in a second, since $8\sqrt{1}=8$; and this, at least, must be its velocity at the highest point, in order that the string may remain stretched throughout its revolution. With this velocity it would perform each revolution in about a second and a half; but its motion will be greatly accelerated during its descent by the gravitation of the stone.

260. COROLLARY 2. "241." In equal circles the forces are as the squares of the times inversely.

For the velocities are inversely as the times.

SCHOLIUM. It may easily be shown, by the apparatus called a whirling table, that when two sliding stages are equally loaded, one of them, which is made to revolve with twice the velocity of the other, will raise four equal weights at the same instant that the other raises a single one, the velocities being gradually and slowly increased, by turning the handle more and more rapidly, till the stages fly off.

261. COROLLARY 3. "242." If the times are equal, the velocities being as the radii, the forces are also as the radii; and in general, the forces are as the distances directly, and as the squares of the times inversely: and the squares of the times are directly as the distances, and inversely as the forces.

The forces are as the distances directly, and as the squares of the times inversely, because the velocities are as the distances directly, and as the times inversely, and their squares are as the squares of the distances divided by those of the times, and dividing these quantities by the distances (259) we have the distances divided by the squares of the times, whence the other part of the proposition follows.

SCHOLIUM. Thus if one of the stages of the whirling table be placed at twice the distance of the other, it will raise twice as great a weight, when the revolutions are performed in the same time: and again, the same weight revolving in a double time, at the same distance, will have its effect reduced to one fourth, but at a double distance the effect will again be increased to half of its original magnitude, while the time remains doubled.

262. COROLLARY 4. "243." If the forces are inversely as the squares of the distances, the squares of the times are as the cubes of the distances.

For the squares of the times are as the distances directly, and as the forces inversely (261): that is, in this case, as the distances, and as the squares of the distances, or as the cubes of the distances.

263. THEOREM. "244." The right line, joining a revolving body and its centre of attraction, always describes equal areas in equal times, and the velocity of the body is inversely as the perpendicular drawn from the centre to the tangent.

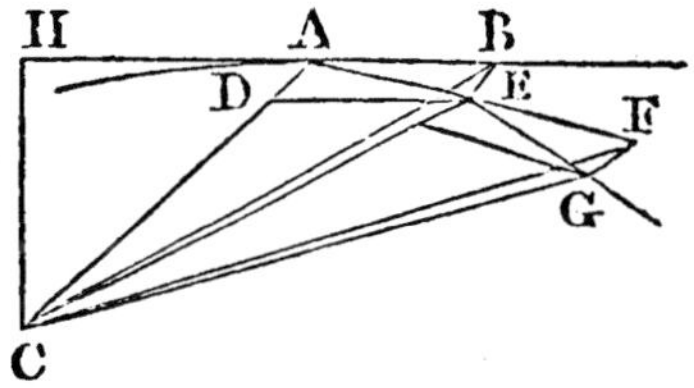

Let AB be a tangent to any curve, in which a body is retained by an attractive force directed to C, and let AB represent its velocity at A, or the space which would be described in an instant of time without disturbance, and AD the space which would be described by the action of C in the same time; then completing the parallelogram, AE will be the joint result (226); again, take EF=AE, and EF will now represent its spontaneous motion in another equal instant of time, and by the action of C it will again describe the diagonal of a parallelogram EG; but the triangles ABC, AEC; AEC, ECF; ECF, ECG, being between the same parallels, are equal (117); and if these triangles be infinitely diminished, and the action of C become continual, they will be the evanescent increments of the area described by the revolving radius, while the body moves in the curvilinear orbit; and the whole areas described in equal times will therefore be equal. And since the constant area $ABC=AB.\frac{1}{2}CH$ (117, 114), $AB=2ABC.\frac{1}{CH}$ and AB, representing the velocity, is always inversely as CH, or $v=\frac{1}{u}$.

SCHOLIUM. Laplace demonstrates this proposition by means of the law which makes the sum of a number of rotatory pressures, which he calls moments, with respect to a given axis, equal to the pressure of the result: observing that whatever is demonstrated of forces and their composition may be applied with equal truth to combinations of motions or velocities. It is true that the same symbols and the same reasoning may generally be applied to forces and to motions; but it appears to be an inversion of the natural order of demonstration to deduce the laws of motion from those of pressure, especially in a case where the real process of nature is so easily traced in the geometrical representation. Laplace observes, however, with respect to the laws of rotatory pressure, (256) that if we project each force and the result of all the forces on a fixed plane, the sum of the rotatory pressures of the constituent forces, with respect to any fixed point in the plane, is equal to the rotatory pressure of the result of all the forces: and drawing to this point a line, which is commonly called the radius vector, but more properly in English the revolving radius, this radius would describe an area in the fixed plane, in virtue of each force acting separately, equal to the product of the line described by the moving body into $\frac{1}{2}$ the perpendicular falling from this fixed point on its direction, and consequently, for any one force or motion, proportional to the time, since the force is conceived to have acted instantaneously, and to have produced a uniform velocity: this area is also expressed by the same product which has been denominated the rotatory pressure, that is the product of the perpendicular into the projection of the force, or into the motion in the given plane: consequently the area described, in virtue of all the motions, is proportional to the projection of the whole force, and the sum of the separate areas is equal to the area described by the radius in virtue of the result of all the motions. Now the addition of any force or forces, directed to or from the given point, can make no difference in the magnitude of the area described round it: because no motion directed to the point would separately cause any area at all to be described. §. 6. P. 18.

COROLLARY. Hence, reciprocally, if a body describes equal areas round a given point, the force by which it is actuated must be directed to that point.]

264. THEOREM. When a moveable point is actuated by a combination of forces, their results being reduced to three orthogonal directions; the time being supposed to flow uniformly, the forces, diminished by quantities proportional to the second fluxions of the spaces described in each direction, and multiplied by the respective variations of the directions, will balance each other.

Since, in the case of equilibrium, $0=\Sigma S\delta s$, when there is no motion, and since any uncompensated force must be employed in producing an increase or diminution of the elocity proportional to its magnitude (230); it follows that so much of the force, in the direction x, as is otherwise uncompensated, must be employed in producing a change of the velocity v expressed by dv, in the elementary portion of time expressed by dt: and if the force be called P, or more properly Pdt, because its effect depends on the elementary portion of time in which it is supposed to act, the unemployed portion may be called $P\mathrm{d}t-\mathrm{d}v=P\mathrm{d}t-\mathrm{d}\frac{\mathrm{d}x}{\mathrm{d}t}$; and the same law will hold good, with respeet to the portions of any number of forces thus remaining unemployed, as if the moving point remained at rest. Consequently the equation $0=\Sigma S\delta s$ (251) will afford us, changing the signs, $0=\delta x\left(\mathrm{d}\frac{\mathrm{d}x}{\mathrm{d}t}-P\mathrm{d}t\right)+\delta y\left(\mathrm{d}\frac{\mathrm{d}y}{\mathrm{d}t}-Q\mathrm{d}t\right)+\delta z\left(\mathrm{d}\frac{\mathrm{d}z}{\mathrm{d}t}-R\mathrm{d}t\right)$; (f). We have also, when the body is at liberty, $P=\frac{\mathrm{dd}x}{\mathrm{d}t^2}$, $Q=\frac{\mathrm{dd}y}{\mathrm{d}t^2}$ and $R=\frac{\mathrm{dd}z}{\mathrm{d}t^2}$.

SCHOLIUM. We must here carefully distinguish the arbitrary variations δx, δy, δz, from the fluxions dx, dy, dz, the former being subject to no conditions whatever, provided that all the forces concerned be comprehended in the equation, while the latter are confined to the expression of the actual motion of the body M.

COROLLARY 1. We are, however, at liberty to assume the variations as equivalent to the fluxions, and to substitute dx, dy, and dz, for δx, δy, and δz, and in this case the equation will become $0=\frac{\mathrm{d}x\mathrm{dd}x}{\mathrm{d}t}-P\mathrm{d}t\mathrm{d}x+$

$\frac{dyddy}{dt}-Qdtdx+\frac{dzddz}{dt}-Rdtdz$, whence, taking the fluent, and dividing by dt, we have $\frac{1}{2}\frac{dx^2+dy^2+dz^2}{dt^2}-\int(Pdx+Qdy+Rdz)$, and $\frac{dx^2+dy^2+dz^2}{dt^2}=2\int Pdx+Qdy+Rdz)$. Now the first member of this equation is the square of the velocity $\frac{ds}{dt}$ or v, and the second may be called $c+2\varphi$, supposing the expression to be a possible fluxion, or capable of integration, which it must be when the forces are in any way dependent on the distance of *M* from their origins, as they generally are in nature: we have then $v^2=c+2\varphi$ (235). *(g)*.

Corollary 2. This supposition of the equality of the variations to the actual evanescent increments of the body's path, is equally applicable to the motion of a body in a given surface: and it follows from the preceding corollary, that the velocity remains unaltered, if no other force be acting on it but the pressure of the surface: as indeed it is obvious that the portion destroyed by the curvature at each step must be infinitely less than that which remains, the hypotenuse of a triangle exceeding its base by a quantity which is only the square of the perpendicular, which may be compared with the evanescent sagitta of the curvature. See 286.

Corollary 3. It is also obvious that the velocity must be the same, by whatever path, or upon whatever surface, the moveable body passes from one given point to another.

265. Lemma. The variation of the difference or of the fluxion is equal to the dif-

ference or the fluxion of the variation, that is $\delta\Delta x = \Delta\delta x$, and $\delta \mathrm{d}x = \mathrm{d}\delta x$.

Supposing two successive values of the ordinate x, corresponding to the abscisses y and $y+h$, to be x and $x + \Delta x$, and the curve or the equation to be so altered, that the ordinates receive the addition characterized by δ; the values corresponding to y and $y+h$ will then be $x+\delta x$, and $x+\Delta x+\delta(x+\Delta x)$ or $x+\Delta x+\delta x+\delta\Delta x$. If we now suppose the latter variation to take place first, and then the former, we shall have x and $x+\delta x$, both corresponding to y; and $x+\Delta x$ and $x+\delta x+\Delta(x+\delta x) = x+\delta x+\Delta x + \Delta\delta x$, corresponding to $y+h$. Now the final effects being the same, by the supposition, which ever change we consider first, it follows that $x+\Delta x+\delta x+\delta\Delta x = x+\delta x+\Delta x+\Delta\delta x$, and consequently $\delta\Delta x = \Delta\delta x$, which is true whatever be the magnitude of the changes, and consequently when they become evanescent, so that we may substitute d for Δ, and $\delta \mathrm{d}x$ will be still equal to $\mathrm{d}\delta x$.

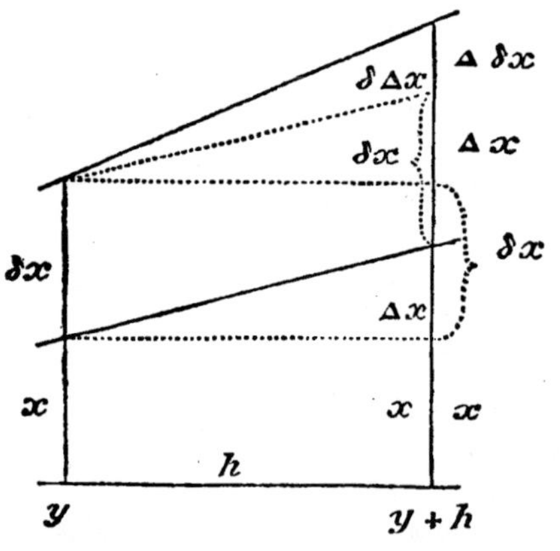

The truth of this propositiou may also be easily understood from a geometrical illustration, the variation or difference, expressed by Δ, arising from the comparison of the ordinates at y and at $y+h$, and the variatiou δ relating to the change produced in the same ordinates by any arbitrary variation of the curve: the same portion of the second ordinate representing $\delta\Delta x$ and $\Delta\delta x$.

Scholium. The foundation of the method of independent variations was laid by Leibnitz, under the name of

" differentiation from curve to curve." For a further illustration of this method, see article 289.]

266. THEOREM. The path of a moveable body is always such, that the fluent $\int v\mathrm{d}s$, taken between its extreme points, is less than in any other curve subjected to the same conditions.

Since $v^2 = c + 2\int(P\mathrm{d}x + Q\mathrm{d}y + R\mathrm{d}z)$ (264, Cor. 1) and since the coefficients of the variation δ and of the fluxion d are the same, it follows that $v\delta v = P\delta x + Q\delta y + R\delta z$; consequently in the equation (*f*, 264) $0 = \delta x\,(\mathrm{d}\frac{\mathrm{d}x}{\mathrm{d}t} - P\mathrm{d}t) + \delta y\,(\mathrm{d}\frac{\mathrm{d}y}{\mathrm{d}t} - Q\mathrm{d}t) + \delta z\,(\mathrm{d}\frac{\mathrm{d}z}{\mathrm{d}t} - R\mathrm{d}t)$, we may substi-stitute $v\delta v\mathrm{d}t$ for $P\delta x\mathrm{d}t + Q\delta y\mathrm{d}t + R\delta z\mathrm{d}t$, and it will become $0 = \delta x\mathrm{d}\frac{\mathrm{d}x}{\mathrm{d}t} + \delta y\mathrm{d}\frac{\mathrm{d}y}{\mathrm{d}t} + \delta z\mathrm{d}\frac{\mathrm{d}z}{\mathrm{d}t} - v\delta v\mathrm{d}t$, (*h*); now if the fluxion of the curve be called d*s*, we shall have $\mathrm{d}s = v\mathrm{d}t$, and $\mathrm{d}s^2 = \mathrm{d}x^2 + \mathrm{d}y^2 + \mathrm{d}z^2$, consequently $v\delta v\mathrm{d}t = \mathrm{d}s\delta v$; and, taking the variation of $\mathrm{d}s^2$, $\mathrm{d}s\delta\mathrm{d}s = \mathrm{d}x\delta\mathrm{d}x + \mathrm{d}y\delta\mathrm{d}y + \mathrm{d}z\delta\mathrm{d}z = v\mathrm{d}t\delta\mathrm{d}s$. Now $\mathrm{d}\,(\mathrm{d}x\delta x) = \mathrm{d}x\mathrm{d}\delta x + \delta x\mathrm{d}^2x$; but since $\mathrm{d}\delta x = \delta\mathrm{d}x$ (265), $\mathrm{d}x\delta\mathrm{d}x = \mathrm{d}x\mathrm{d}\delta x = \mathrm{d}\,(\mathrm{d}x\delta x) - \delta x\mathrm{d}^2x$, and the same substitutions being applied to the coordinates *y* and *z*, the equation for $\mathrm{d}s\delta\mathrm{d}s$ is transformed into $v\delta\mathrm{d}s = \frac{\mathrm{d}(\mathrm{d}x\delta x + \mathrm{d}y\delta y + \mathrm{d}z\delta z)}{\mathrm{d}t} - \delta x\mathrm{d}\frac{\mathrm{d}x}{\mathrm{d}t} - \delta y\mathrm{d}\frac{\mathrm{d}y}{\mathrm{d}t} - \delta z\mathrm{d}\frac{\mathrm{d}z}{\mathrm{d}t}$; which, by means of the equation (*h*) becomes $v\delta\mathrm{d}s = \mathrm{d}\frac{\mathrm{d}x\delta x + \mathrm{d}y\delta y + \mathrm{d}z\delta z}{\mathrm{d}t} - v\delta v\mathrm{d}t$: but $v\delta v\mathrm{d}t = \mathrm{d}s\delta v$; and $\delta(v\mathrm{d}s) = v\delta\mathrm{d}s + \mathrm{d}s\delta v = \mathrm{d}\frac{\mathrm{d}x\delta x + \mathrm{d}y\delta y + \mathrm{d}z\delta z}{\mathrm{d}t}$: and taking the fluent on both sides with respect to d, $\delta\int v\mathrm{d}s = c +$

$\frac{dx\delta x + dy\delta y + dz\delta z}{dt}$. Now at the beginning and end of the body's motion, the variations δx, δy, δz, must necessarily vanish, because it is supposed to move from one fixed point to another, the intermediate points only being supposed to be subjected to an elementary variation: consequently the value of $\delta\int v ds$ between these two points is equal to nothing, and the quantity $\int v ds$ is a minimum, since its variation vanishes.

[SCHOLIUM. The nature of this fluent may be understood by supposing the path to be changed towards the middle by a slight variation δx of x only; the variation $\delta\int v ds$ with respect to the whole portion of the path would then become $\frac{dx\delta x}{dt}$, which is equal to the product of the variation of x into the velocity of the body in its direction at the given point.]

COROLLARY 1. If the body is moving freely on a given surface with a uniform velocity v, we have $\int v ds = v\int ds = vs$, consequently s is also a minimum, and the curve is the shortest that can be described between the two points.

[SCHOLIUM. The curve in this case may always be traced by bending a flattened wire over the surface, which must necessarily determine the minimum, since the two edges of the wire are of the same length, and the variation of the length of the path contained within its limits vanishes: the same curve must also be that which a body would describe spontaneously on the surface, because it lies always in a plane perpendicular to the surface, and the pressure of the surface could cause the motion to be deflected in no other direction. That the curve must be in a plane perpendicular to the surface, is evident from the motion of all bodies rolling on each other, which come into contact in directions at right angles to the surface: and the wire must bend continually, from the position of the tangent into that of the curve, in such a manner that all its points descend perpendicularly upon the surface.

267. LEMMA A. The sum of the squares of the projections of any right line on three orthogonal planes is equal to twice the square of the line.

If the orthogonal ordinates of the line s be a, b, and c, we have $s^2 = a^2 + b^2 + c^2$; now the projection on the plane of a and b is $\surd(a^2 + b^2)$, on that of a and c, $\surd(a^2 + c^2)$; and on that of b and c, $\surd(b^2 + c^2)$; consequently the sum of the squares of the three is $a^2 + b^2 + a^2 + c^2 + b^2 + c^2 = 2s^2$.

268. LEMMA B. The fluxion of an arc is as the radius of curvature and the fluxion of the angular extent conjointly, or $ds = r d\theta$, and the radius is $r = \frac{ds}{d\theta}$.

The angle being measured in equal circles by the arc subtending it, and in different circles being inversely as the radius, when the arc is the same, it becomes evident that the elementary arc of any curve must be as the angle subtended by it, or as its angular extent and the radius of the circle of curvature conjointly, the element of the curve coinciding in length and in curvature with that of the circle of equal curvature at the given point.

269. LEMMA C. The radius of curvature is in general $r = \mathrm{d}x : \mathrm{d}\frac{\mathrm{d}y}{\mathrm{d}s} - - \mathrm{d}y : \mathrm{d}\frac{\mathrm{d}x}{\mathrm{d}s}$.

Since $\mathrm{d}\sin\theta = \cos\theta \mathrm{d}\theta$ (142), we have $\mathrm{d}\theta = \frac{\mathrm{d}\sin\theta}{\cos\theta}$; and since $\mathrm{d}\cos\theta = -\sin\theta\mathrm{d}\theta$, $\mathrm{d}\theta = -\frac{\mathrm{d}\cos\theta}{\sin\theta}$. Now $\sin\theta - \frac{\mathrm{d}y}{\mathrm{d}s}$, and $\cos\theta = \frac{\mathrm{d}x}{\mathrm{d}s}$; whence $\mathrm{d}\theta = \mathrm{d}\frac{\mathrm{d}y}{\mathrm{d}s}.\frac{\mathrm{d}s}{\mathrm{d}x} - \mathrm{d}\frac{\mathrm{d}x}{\mathrm{d}s}.\frac{\mathrm{d}s}{\mathrm{d}y}$, and $r = \frac{\mathrm{d}s}{\mathrm{d}\theta} = \mathrm{d}x : \mathrm{d}\frac{\mathrm{d}y}{\mathrm{d}s} = -\mathrm{d}y : \mathrm{d}\frac{\mathrm{d}x}{\mathrm{d}s}$.

270. LEMMA D. When the fluxion of the curve is constant, the radius of curvature is $r = \frac{\mathrm{d}s\mathrm{d}x}{\mathrm{dd}y} = -\frac{\mathrm{d}s\mathrm{d}y}{\mathrm{dd}x} = \frac{\mathrm{d}s^2}{\sqrt{(\mathrm{d}^2x^2 + \mathrm{d}^2y^2)}}$.

When $\mathrm{d}s$ is constant, we have $\mathrm{d}\frac{\mathrm{d}y}{\mathrm{d}s} = \frac{\mathrm{dd}y}{\mathrm{d}s}$, and $\mathrm{d}\frac{\mathrm{d}x}{\mathrm{d}s} - \frac{\mathrm{dd}x}{\mathrm{d}s}$, and $r = \mathrm{d}z : \mathrm{d}\frac{\mathrm{d}y}{\mathrm{d}s} = \frac{\mathrm{d}s\mathrm{d}x}{\mathrm{dd}y} = -\frac{\mathrm{d}s\mathrm{d}y}{\mathrm{dd}x}$. But since $\mathrm{d}x^2 + \mathrm{d}y^2 = \mathrm{d}s^2$, $\mathrm{d}x\mathrm{d}^2x + \mathrm{d}y\mathrm{d}^2y = 0$, and $\mathrm{d}x^2\mathrm{d}^2x^2 = \mathrm{d}y^2\mathrm{d}^2y^2$, consequently $\frac{\mathrm{d}^2y^2}{\mathrm{d}^2x^2} = \frac{\mathrm{d}x^2}{\mathrm{d}y^2}$, and $\frac{\mathrm{d}^2y^2}{\mathrm{d}^2x^2 + \mathrm{d}^2y^2} = \frac{\mathrm{d}x^2}{\mathrm{d}x^2 + \mathrm{d}y^2}$ (34) $- \frac{\mathrm{d}x^2}{\mathrm{d}s^2}$, and $\frac{\mathrm{dd}y}{\sqrt{(\mathrm{d}^2x^2 + \mathrm{d}^2y^2)}} = \frac{\mathrm{d}x}{\mathrm{d}s}$; whence $\frac{\mathrm{d}s\mathrm{d}x}{\mathrm{dd}y} = \frac{\mathrm{d}s^2}{\sqrt{(\mathrm{d}^2x^2 + \mathrm{d}^2y^2)}}$.

271. LEMMA E. When the curve is referred to three orthogonal ordinates, its fluxion

ds being constant, the radius of curvature is $r = \frac{ds^2}{\sqrt{(d^2x^2 + d^2y^2 + d^2z^2)}}$.

We may first suppose a plane to pass through the tangent of the curve and the ordinate z, which we may call vertical, while x and y are horizontal, and a second plane to pass through the tangent, and to be perpendicular to the first or supposed vertical plane: then if the curve be projected vertically on this plane, the horizontal ordinates z and y will be the same for the projection as for the curve, and the tangent will be the same, and the curves will only differ in length as the tangent differs from the curve, the elements of all three being ultimately in the ratio of equality: the sagitta of curvature, $\frac{\Delta s^2}{2r}$, of the projection, will be horizontal, being perpendicular to the first plane, which is vertical; it is, therefore, the projection of the primitive sagitta on a plane parallel to that of x and y: and the same may be shown of two other projections on the other two orthogonal planes, of x and z, and of y and z, substituting y and x successively in the place of z.

Now, the sagittae of the three projections are $\frac{1}{2}\sqrt{(\Delta^2x^2 + \Delta^2y^2)}$, $\frac{1}{2}\sqrt{(\Delta^2x^2 + \Delta^2z^2)}$, and $\frac{1}{2}\sqrt{(\Delta^2y^2 + \Delta^2z^2)}$, and the sum of their squares is $\frac{1}{2}\sqrt{(\Delta^2x^2 + \Delta^2y^2 + \Delta^2z^2)}$, consequently the primitive sagitta of which they are the projections, is $\frac{1}{2}(\Delta^2x^2 + \Delta^2y^2 + \Delta^2z^2)$, and the true radius of the curve $\frac{\Delta s^2}{\sqrt{(\Delta^2x^2 + \Delta^2y^2 + \Delta^2z^2)}} = \frac{ds^2}{\sqrt{(d^2x^2 + d^2y^2 + d^2z^2)}}$.

Scholium. The characteristic Δ is here substituted for d in speaking of the sagitta, Δ being intended to represent an actual evanescent variation, while the fluxion d is a finite magnitude proportional to it (229). The student will readily understand that dx^2 is generally used for $(dx)^2$ and d^2x^2, Δ^2x^2, for $(d^2x)^2$, $(\Delta^2x)^2$; and not for $d(x^2)$, $d^2(x)^2$, as might have been done without impropriety, if it had been equally convenient.

272. THEOREM. The pressure of a moving body on any curve, derived from its centrifugal force, is expressed by the square of the velocity, divided by the radius of curvature: and the pressure on any surface is expressed by the square of the velocity divided by the radius of curvature of its path, and multiplied by the sine of the inclination of the plane of the curvature to the plane of the surface. (§. 9. P. 23.)

The equation $V\delta u = \Sigma S\delta s$ (250) affords us here the conditions of equilibrium between the forces depending on the curvature, and the pressure; but those forces are $\frac{ddx}{dt^2}$, $\frac{ddy}{dt^2}$, and $\frac{ddz}{dt^2}$, (233, 264), or, since $ds = vdt$, $dt = \frac{ds}{v}$, and $\frac{1}{dt^2} = \frac{v^2}{ds^2}$, $\frac{v^2}{ds^2}.d^2x$, $\frac{v^2}{ds^2}.d^2y$, and $\frac{v^2}{ds}.d^2z$, which must be respectively equal to $V\frac{\delta' u}{\delta x}$, $V\frac{\delta' u}{\delta y}$, and $V\frac{\delta' u}{\delta z}$, or, in this case, putting Λ for the pressure, and r for the perpendicular to the surface, to $\Lambda\frac{\delta' r}{\delta x}$, $\Lambda\frac{\delta' r}{\delta y}$, and $\Lambda\frac{\delta' r}{\delta z}$, since the forces in each direction must balance each other. We have consequently, adding together the squares of each equation, $\Lambda^2\left\{\left(\frac{\delta' r}{\delta x}\right)^2+\left(\frac{\delta' r}{\delta y}\right)^2+\left(\frac{\delta' r}{\delta z}\right)^2\right\}=\left(\frac{v^2}{ds^2}\right)^2(d^2x^2+d^2y^2$

$+d^2z^2)$, and $A\sqrt{\left\{\left(\frac{\delta r}{\delta x}\right)^2+\left(\frac{\delta r}{\delta y}\right)^2+\left(\frac{\delta r}{\delta z}\right)^2\right\}}-\frac{v^2}{ds^2}\sqrt{}$ $(d^2x^2+d^2y^2+d^2z^2)$. But ds being constant, we have $\frac{1}{ds^2}\sqrt{(d^2x^2+d^2y^2+d^2z^2)}-\frac{1}{r}$ (271); and $\left(\frac{\delta r}{\delta x}\right)^2+\left(\frac{\delta r}{\delta y}\right)$ $+\left(\frac{\delta r}{\delta z}\right)^2=1$ (254, Sch. 2): consequently $A=\frac{v^2}{r}$, as has already been inferred, with respect to the central force in a circle, from a simpler mode of reasoning (258); but the coincidence is of use in strengthening the basis of the analytical investigation.

Now, if the surface be spherical, the curve described will obviously be a great circle of the sphere, and its radius of curvature that of the sphere, since the deflection can only be in the direction of the radius, and in the plane in which the body moves. And if a thread be substituted for a surface, the tension of the thread will be equivalent and equal to the pressure on the surface.

The whole pressure on the surface will be obtained, by adding to the centrifugal force any extraneous forces which may be acting on the body. And since the force always acts in the direction of the plane of the body's motion, when that plane is not perpendicular to the surface, the pressure on the surface will obviously be reduced in the proportion of the radius to the sine of the inclination of the plane to the tangent plane; the remaining portion acting in the direction of the surface, and requiring to be counteracted by some other force. But in the absence of such forces, it has been shown that the centrifugal force is simply equal to the pressure on the surface; the plane of the motion is, therefore, in that case, always perpendicular to the surface.

The curve thus described, on a spheroid, has been called the perpendicular to the meridian: and it traces, as has already been observed, (266) the shortest distance between any two places in its direction. It does not, however, remain actually perpendicular to the meridians which it crosses, but is conceived to be traced by levelling, in the same way as a flattened wire would trace it when bent on the spheroid.

[SCHOLIUM. It follows from considering the proportion of the sagitta of curvature in a perpendicular and in an oblique plane, that the radius of curvature must always vary in the direct ratio of the sine of the inclination of the planes, so that the pressure on the plane is the same whether the body move in a great or a lesser circle, the immediate centrifugal force being increased, by the increase of curvature, in the same ratio that its action with regard to the surface is diminished, provided that the velocity be the same in both cases.]

273. THEOREM. If a body move in a resisting medium, and be subject to a uniform gravitation in a vertical direction, its motion will be defined by the equation $\frac{\varsigma}{g}=\frac{\mathrm{d}s\mathrm{d}^3z}{2\mathrm{d}^2z^2}$; s being the space described in the direction of the motion, z the vertical ordinate, x a horrizontal one, ς the resistance, and g the force of gravity, $\mathrm{d}x$ being constant: and if the resistance is as the square of the velocity, and $h=\frac{\varsigma}{vv}$, $\frac{\mathrm{dd}z}{\mathrm{d}x^2}=ae^{2hs}$; a being a constant quantity, and $\mathrm{hl}e=1$.

Resuming the equation (*f*, 264), $0=\delta x\,(\mathrm{d}\,\frac{\mathrm{d}x}{\mathrm{d}t}-P\mathrm{d}t)+\delta y\,(\mathrm{d}\frac{\mathrm{d}y}{\mathrm{d}t}-Q\mathrm{d}t)+\delta z\,(\mathrm{d}\,\frac{\mathrm{d}z}{\mathrm{d}t}-R\mathrm{d}t)$, and supposing z to begin at the highest point of the curve, we may resolve the force of resistance ς into three directions, and it will afford us $-\varsigma\,\frac{\mathrm{d}x}{\mathrm{d}s}$, $-\varsigma\,\frac{\mathrm{d}y}{\mathrm{d}s}$, and $-\varsigma\,\frac{\mathrm{d}z}{\mathrm{d}s}$; consequently $P=-\varsigma\,\frac{\mathrm{d}x}{\mathrm{d}s}$, $Q=-\varsigma\,\frac{\mathrm{d}y}{\mathrm{d}s}$, and $R=-\varsigma\,\frac{\mathrm{d}z}{\mathrm{d}s}+g$. Hence we have $0=\delta x(\mathrm{d}\frac{\mathrm{d}x}{\mathrm{d}t}+\varsigma\frac{\mathrm{d}x}{\mathrm{d}s}\mathrm{d}t)+\delta y(\mathrm{d}\frac{\mathrm{d}y}{\mathrm{d}t}+\varsigma\,\frac{\mathrm{d}y}{\mathrm{d}s}\mathrm{d}t)+\delta z(\mathrm{d}\frac{\mathrm{d}z}{\mathrm{d}t}+\varsigma\,\frac{\mathrm{d}z}{\mathrm{d}s}\mathrm{d}t-g\mathrm{d}t)$: and if the motion is subjected to no further limitation, we have the three equations $0=\mathrm{d}\frac{\mathrm{d}x}{\mathrm{d}t}+\varsigma\,\frac{\mathrm{d}x}{\mathrm{d}s}\mathrm{d}t$; $0=\mathrm{d}\frac{\mathrm{d}y}{\mathrm{d}t}+\varsigma\,\frac{\mathrm{d}y}{\mathrm{d}s}\mathrm{d}t$; and $0=\mathrm{d}\,\frac{\mathrm{d}z}{\mathrm{d}t}+\varsigma\,\frac{\mathrm{d}z}{\mathrm{d}s}\mathrm{d}t-g\mathrm{d}t$. From the two first, we obtain, by multiplication and subtraction, $\frac{\mathrm{d}y}{\mathrm{d}t}.\,\mathrm{d}\,\frac{\mathrm{d}x}{\mathrm{d}t}=\frac{\mathrm{d}x}{\mathrm{d}t}.\,\mathrm{d}\,\frac{\mathrm{d}y}{\mathrm{d}t}$; and, d$t$ being constant, dividing both sides by $\frac{x\mathrm{d}y}{\mathrm{d}t^2}$, we have $\frac{\mathrm{dd}x}{\mathrm{d}x}=\frac{\mathrm{dd}y}{\mathrm{d}y}$, and $\mathrm{hld}x=\mathrm{hld}y+c=\mathrm{hl}\,f\mathrm{d}y$, and $\mathrm{d}x=f\mathrm{d}y$, f being a constant quantity. But since $\mathrm{d}x=f\mathrm{d}y$, the horizontal motion must be rectilinear, and the body must move in a vertical plane, which is indeed sufficiently obvious from the absence of any lateral force. We may, therefore, consider x as situated in this plane, y being always $=0$; and from the two equations $0=\mathrm{d}\frac{\mathrm{d}x}{\mathrm{d}t}+\varsigma\,\frac{\mathrm{d}x}{\mathrm{d}s}\mathrm{d}t$, and $0=\mathrm{d}\frac{\mathrm{d}z}{\mathrm{d}t}+\varsigma\,\frac{\mathrm{d}z}{\mathrm{d}s}\mathrm{d}t-g\mathrm{d}t$, we obtain, making

dx constant, and dt of course variable, $dx\frac{ddt}{dt^2}=\mathfrak{C}\frac{dx}{ds}dt$, and $\mathfrak{C}=\frac{dsddt}{dt^3}$; also $0=\frac{ddz}{dt}-\frac{dzddt}{dt^2}+\mathfrak{C}\frac{dz}{ds}dt-gdt$; consequently $gdt^2=ddz-\frac{dzddt}{dt}+\mathfrak{C}\frac{dz}{ds}dt^2$: but $\frac{ddt}{dt}=\frac{\mathfrak{C}dt^2}{ds}$, so that $gdt^2=d^2z$, and, taking the fluxion, $2gdtd^2t=d^3z$; now since $d^2t=\frac{\mathfrak{C}dt^3}{ds}$, and $dt^2=\frac{ddz}{g}$, we have $d^3z=2gdtd^2t=2gdt\frac{\mathfrak{C}dt^3}{ds}=\frac{2g\mathfrak{C}}{ds}dt^4=\frac{2g\mathfrak{C}}{ds}\left(\frac{ddz}{g}\right)^2=\frac{2\mathfrak{C}d^2z^2}{gds}$, and $\frac{\mathfrak{C}}{g}=\frac{dsd^3z}{2d^2z^2}$, which determines the law of the resistance $\mathfrak{C}$, required for the description of a particular curve.

Now supposing the resistance proportional to the square of the velocity, which is nearly true in a medium of uniform density, $\mathfrak{C}$ being expressed by $h.\frac{ds^2}{dt^2}$, we have $\frac{\mathfrak{C}}{g}=\frac{hds^2}{gdt^2}=\frac{hds^2}{d^2z}$, and $hds=\frac{d^3z}{2ddz}$, since $\frac{hds^2}{ddz}=\frac{\mathfrak{C}}{g}=\frac{dsd^3z}{2d^2z^2}$; hence, taking the fluent, we have $hs=\frac{1}{2}\,\mathrm{hl}d^2z+c$, or $2hs=\mathrm{hl}d^2z+c$, in which, since dx is constant, we must take $\mathrm{hl}d^2z+c=\mathrm{hl}\frac{ddz}{adx^2}$, and since $\mathrm{hl}\,(ae^{2hs})=2hs+\mathrm{hl}a$, we have $\frac{ddz}{dx^2}=ae^{2hs}$

Corollary. If we make $h=0$, and suppose the resistance to vanish, we have $d^2z=adx^2$; the fluent of which is $dz=\int axdx+bdx$, whence $z=\frac{1}{2}ax^2+bx+c$, which is the equation of a parabola (204) b and c being determined by the conditions of projection; and since $d^2z=$

$a dx^2 = g dt^2$, we have $dt^2 = \frac{a}{g} dx^2$, and $t = x\sqrt{\frac{a}{g}} + f'$; but if x, z, and t begin together, $c=0$, and $f' = 0$, consequently $t = x\sqrt{\frac{a}{g}}$, $x = \sqrt{\frac{g}{a}}t$, and $z = \frac{1}{2} ax^2 + bx$, whence $z = \frac{1}{2} a \frac{gt^2}{a} + bt\sqrt{\frac{g}{a}} = \frac{1}{2}gt^2 + bt\sqrt{\frac{g}{a}}$: and these equations contain the whole theory of projectiles moving without resistance: they show that the horizontal velocity is uniform, and that the velocity in a vertical direction is the same as if the body fell in a right line.

[SCHOLIUM. It seems to be an unnecessary departure from the simple order of investigation to examine a very complicated and intricate case in order to deduce from it a very simple one: and yet it may be said that unless this were done, we should have frequent repetitions from considering the same case in its simple form, and then as an inference from a more general law. But for a student, it is better to have such repetitions, than to be without a clear conception of the shortest path by which he may arrive at an elementary conclusion. It seems, therefore, not altogether superfluous to insert here a few illustrations of the motions of projectiles, demonstrated in the most natural and simple manner.

274. THEOREM. The velocity of a projectile may be resolved into two parts, its horizontal and its vertical velocity: the horizontal motion will not be affected by the action of gravitation perpendicular to it, and will therefore continue uniform; and the vertical motion will be the same as if it had no horizontal motion.

For gravitation, being considered as a uniformly accelerating force, must act, by the definition of such a force, equally on a body in motion and at rest, so that the vertical motion will not be affected by the horizontal motion; and the diagonal motion, resulting from the combination, will terminate in the same vertical line as the simple horizontal motion would have done; and consequently the horizontal motion must remain unaltered.

SCHOLIUM. Thus if we let fall, from the head of the mast of a ship, sailing uniformly along in smooth water, a weight, which partakes of its progressive motion, the weight will descend by the side of the mast in the same manner, and in the same time, as if neither the ship nor the weight had any horizontal motion.

275. THEOREM. The greatest height, to which a projectile will rise, may be determined by finding the height from which a body must fall, in order to gain a velocity equal to its vertical velocity; and the horizontal range may be found, by calculating the distance described by its horizontal velocity, in twice the time of rising to its greatest height.

This is evident from the equality of the velocity of ascending and descending bodies at equal heights, and from the independence of the vertical and horizontal motions of the projectile

SCHOLIUM. For example, suppose a musket to be so elevated, that the muzzle is higher than the but end by half of the length, that is, at an angle of 30°; and let the ball be discharged with a velocity of 1000 feet in a second; then its vertical velocity will be half as great, or 500 feet in a second: now the square of one eighth of 500 is $\frac{250000}{8 \times 8} = 3906$, consequently the height, to which the ball would rise, if unresisted by the air, is 3906 feet, or three quarters of a mile. But in fact a musket ball, actually shot directly upwards, with a velocity of 1670 feet in a second, which would rise six or seven miles in a vacuum, is so retarded by the air, that it does not attain the height of a single mile. The time, in which the velocity of 500 feet would be destroyed, is found by dividing it by 32, or twice the time if we divide by 16: we have, therefore, 31 seconds for the time of the whole range; and the horizontal velocity, being $1000 \times \sqrt{(1 - \frac{1}{4})} = 886$ feet, the ball would describe in 31 seconds, with this velocity, nearly 28000 feet, or above five miles. But the resistance of the air will reduce this distance also to less than one mile.

276. THEOREM. With a given velocity, the horizontal range is proportional to the sine of twice the angle of elevation.

The time of ascent being as the vertical velocity, that is as the sine of the angle of elevation, when the oblique velocity is given, the range must be as the product of the horizontal and vertical velocities, or as the product of the sine and cosine; that is, as the sine of twice the angle (140).

SCHOLIUM. Hence it follows, that the greatest horizontal range will be when the elevation is half a right angle; supposing the body to move in a vacuum. But the resistance of the air increases with the length of the path, and the same cause also makes the angle of descent much greater than the angle of ascent, as we may observe in the track of a bomb. For both these reasons, the best elevation is somewhat less than 45°, and sometimes, when the velocity is very great, as little as 30°. But it usually happens in the operations of natural causes, that near the point where any quantity is greatest or least, its variation is slower than elsewhere: a small difference, therefore, in the angle of elevation, is of little consequence to the extent of the range, provided that it continue between the limits of 45° and 35°; and for the same reason, the angular adjustment requires less accuracy in this position than in any other, which, besides the economy of powder, makes it in all respects the best elevation for practice, where the object is to carry a ball or shell to the greatest possible distance.]

[277. LEMMA A. If the equation $a+bx+cx^2+dx^3+\ldots=0$ be true for all values of x, it will follow that each coefficient must be separately $=0$.

For, putting $x=0$, we have $a=0$, therefore $bx+cx^2+\ldots=0$; then, dividing by x, $b+cx+dx^2+\ldots=0$; consequently $b=0$; and in the same manner all the coefficients may be made to vanish in succession.

278. LEMMA B. The binomial or dinomial theorem (244) is true for all powers, whether entire or fractional.

Its truth may be the most easily shown from the principles of fluxions, and the Taylorian theorem combined. For since $d(x^n)=nx^{n-1}dx$, making dx constant, we have $d^2(x^n)=n(n-1)x^{n-2}dx^2$, and $d^3(x^n)=n(n-1)(n-2)x^{n-3}dx^3$; whence, taking $u=(x+h)^n$, we have $\Delta u' = h\frac{du'}{dx}+\frac{h^2}{1.2}$

$\frac{d^2u'}{dx^2} + \ldots$, u' being $= x^n$, the initial value of u; and Δx^n $= hnx^{n-1} + h^2 . \frac{n(n-1)}{1.2} x^{n-2} \ldots$; consequently $(x+h)^n =$ $x^n + nx^{n-1}h + \frac{n(n-1)}{1.2} x^{n-2}h^2 + \frac{n(n-1)(n-2)}{1.2.3} x^{n-3}h^3 + \ldots$, which is the theorem in question, without limitation.

279. Lemma C. The fluent $\int \sin^{\text{M}} z dz =$ $-\frac{1}{\text{M}} \sin^{\text{M}-1} z \cos z + \int \frac{\text{M}-1}{\text{M}} \sin^{\text{M}-2} z dz$; and ${}^0\!\int_{\frac{\pi}{2}}$ $\sin^{\text{M}} z dz = \frac{\text{M}-1}{\text{M}} {}^0\!\int_{\frac{\pi}{2}} \sin^{\text{M}-2} z dz$.

The fluxion $d(\sin^{\text{M}-1} z \cos z) = \sin^{\text{M}-1} z d \cos z + \cos z d \sin^{\text{M}-1} z = -\sin^{\text{M}} z dz + \cos^2 z (\text{M}-1) \sin^{\text{M}-2} z dz =$ $\sin^{\text{M}} z dz + (1 - \sin^2 z)(\text{M}-1) \sin^{\text{M}-2} z dz = (\text{M}-1) \sin^{\text{M}-2} z dz - \text{M} \sin^{\text{M}} z dz$; consequently $\sin^{\text{M}} z dz = \frac{\text{M}-1}{\text{M}} \sin^{\text{M}-2} z dz - \frac{1}{\text{M}} d(\sin^{\text{M}-1} z \cos z)$; and in the case of $z = \frac{\pi}{2}$, or a quadrant, the cosine vanishing, the first term of the fluent vanishes.

Corollary. When $\text{M} = 2$, the particular fluent becomes $\int' \sin^2 z\, dz = \frac{1}{2} \int' dz = \frac{1}{2} z = \frac{1}{4} \pi$; when $\text{M} = 3$, $\int'$ $\sin^3 dz = \frac{2}{3} \int' \sin z dz = -\frac{2}{3} \cos z = 0$; if $\text{M} = 4$, $\frac{\text{M}-1}{\text{M}} = \frac{3}{4}$, and the fluent is $\frac{1.3}{2.4} . \frac{\pi}{2}$; in the same manner for $\text{M} = 6$, we have $\frac{1.3.5}{2.4.6} . \frac{\pi}{2}$; and the series may be continued at pleasure for all the even values of M.]

280. THEOREM. The oscillations of a gravitating body, moving freely on a spherical surface, of which the radius is r, are performed in a time $T=\pi\sqrt{\frac{r}{g}}.\ \gamma\sqrt{\frac{2r(a+b)}{a^2-b^2}}\left\{1+\left(\frac{1}{2}\right)^2\gamma^2+\left(\frac{1.3}{2.4}\right)^2\gamma^4+\left(\frac{1.3.5}{2.4.6}\right)^2\gamma^6+\ldots\right.$; π being the semicircumference to the radius 1, a the greatest and b the least distance of the body below the centre, g the space described by a heavy body in the unit of time, and $\gamma^2=\frac{a^2-b^2}{a^2+2ab+r^2}$.

In this case we obtain from the equation $\Sigma S\delta s+R\delta r=0$, compared with $P=\frac{ddx}{dt^2}$, $Q=\frac{ddy}{dt^2}$, and $R=\frac{ddz}{dt^2}$ (264), the three equations $0=\frac{ddx}{dt^2}+\Lambda\frac{x}{r}$, $0=\frac{ddy}{dt^2}+\Lambda\frac{y}{r}$, and $0=\frac{ddz}{dt^2}+\Lambda\frac{z}{r}-g$, Λ being the pressure on the surface; for since $r^2=x^2+y^2+z^2$, we have $\frac{\delta' r}{\delta x}=\frac{x}{r}$, $\frac{\delta' r}{\delta y}=\frac{y}{r}$, and $\frac{\delta' r}{\delta z}=\frac{z}{r}$. Now since $v^2=\frac{dx^2+dy^2+dz^2}{dt^2}=\int(Pdx+Qdy+Rdz)$, P, Q, and R being the accelerating forces concerned (264 Cor.), the fluent here becomes $v^2=2\int g dz=c+2gz$, consequently the pressure derived from the centrifugal force will be simply $\frac{c+2gz}{r}$ (272 Schol.) to which adding the force of gravity, reduced in the ratio $\frac{\delta' r}{\delta z}$, or $\frac{z}{r}$, that is $\frac{gz}{r}$, we obtain $\frac{c+3gz}{r}$, for the whole pressure on the surface.

If we multiply the first equation by $-y$, and the second by x, and add them together, we have $\frac{y\mathrm{dd}x}{\mathrm{d}t^2} - \Lambda\frac{xy}{r} + \frac{x\mathrm{dd}y}{\mathrm{d}t^2} + \Lambda\frac{xy}{r} - 0 = \frac{x\mathrm{dd}y - y\mathrm{dd}x}{\mathrm{d}t^2}$ · but $\mathrm{d}(x\mathrm{d}y) = x\mathrm{d}^2y + \mathrm{d}x\mathrm{d}y$, and $\mathrm{d}(y\mathrm{d}x) = y\mathrm{d}^2x + \mathrm{d}x\mathrm{d}y$, consequently $\mathrm{d}(x\mathrm{d}y - y\mathrm{d}x) = x\mathrm{d}^2y - y\mathrm{d}^2x$, and $0 + \frac{c'}{\mathrm{d}t} = \frac{x\mathrm{d}y - y\mathrm{d}x}{\mathrm{d}t^2}$, $\mathrm{d}t$ being constant. Now the equation of the surface gives us $x\mathrm{d}x + y\mathrm{d}y + z\mathrm{d}z = 0$: we have therefore the three equations $x\mathrm{d}x + y\mathrm{d}y = -z\mathrm{d}z$, $x\mathrm{d}y - y\mathrm{d}x = c'\mathrm{d}t$, and $\frac{\mathrm{d}x^2 + \mathrm{d}y^2 + \mathrm{d}z^2}{\mathrm{d}t^2} = c + 2gz$. Adding together the squares of the two first, we obtain $x^2\mathrm{d}x^2 + y^2\mathrm{d}y^2 + x^2\mathrm{d}y^2 + y^2\mathrm{d}x^2 = z^2\mathrm{d}z^2 + c'^2\mathrm{d}t^2 = (x^2 + y^2)(\mathrm{d}x^2 + \mathrm{d}y^2) = (r^2 - z^2)\,\mathrm{d}x^2 + \mathrm{d}y^2 = (r^2 - z^2)\left\{(c + 2gz)\,\mathrm{d}t^2 - \mathrm{d}z^2\right\}$: consequently $(r^2 - z^2)(c + 2gz)\mathrm{d}t^2 - c'^2\mathrm{d}t^2 = (r^2 - z^2)\,\mathrm{d}z^2 + z^2\mathrm{d}z^2 = r^2\mathrm{d}z^2$, and $\mathrm{d}t = \frac{\pm r\mathrm{d}z}{\sqrt{\left\{(r^2 - z^2)(c + 2gz) - c'^2\right\}}}$.

But it is most convenient to substitute for the denominator $\sqrt{\left\{(a - z)(z - b)(2gz + f)\right\}}$; for which we find, by actual multiplication, $cr^2 + 2gr^2z - cz^2 - 2gz^3 - c'^2 = (az - ab - z^2 + bz)(2gz + f) = 2agz^2 - 2abgz - 2gz^3 + 2bgz^2 + afz - abf - fz^2 + bfz$; then, by equating the coefficients of z (277), $2gr^2 = -2abg + af + bf$, consequently $f = 2g.\frac{r^2 + ab}{a + b}$ · we have next, from z^2, $-c = 2ag + 2bg - f$, $c = f - 2g(a + b) = 2g\left(\frac{r^2 + ab}{a + b} - (a + b)\right) = 2g.\frac{r^2 - a^2 - ab - b^2}{a + b}$; and lastly $cr^2 - c'^2 = -abf$, whence $c'^2 = cr^2 + ab \times 2g.\frac{r^2 + ab}{a + b} =$

$$2g\left(\frac{r^4-a^2r^2-abr^2-b^2r^2}{a+b}+\frac{a\,b\,r^2+a^2b^2}{a+b}\right)=2\,g.$$

$\frac{r^4-a^2r^2-b^2r^2+a^2b_2}{a+b}$ It must be observed that the quantities a and b will be the greatest and least values of z; [for otherwise the fluxion $r$dz would not vanish, as it must do, when the curve becomes horizontal].

Making now $\sin\theta=\sqrt{\frac{a-z}{a-b}}$, we have $d\sin\theta=\cos\theta d\theta-$

$\frac{-dz}{2\sqrt{(a-z)}\sqrt{(a-b)}}$, and since $\cos\theta=\sqrt{\frac{z-b}{a-b}}$, $d\,\theta$ $-$

$$\frac{-dz}{2\sqrt{(a-z)}\sqrt{(a-b)}}.\sqrt{\frac{a-b}{z-b}}=\frac{-dz}{2\sqrt{(a-z)}\sqrt{(z-b)}}:$$

consequently, in the ascent of the body,

$$\frac{-rdz}{\sqrt{(a-z)}\sqrt{(z-b)}\sqrt{(2gz+f)}}=dt=\frac{2rd\theta}{\sqrt{(2gz+f)}}.$$ Now

since $\sin^2\theta=\frac{a-z}{a-b}$, $(a-b)\sin^2\theta=a-z, z=a-(a-b)\sin^2\theta$,

and f being $-2g.\frac{r^2+ab}{a+b}$, $2gz+f=2g\,(a-(a-b)\sin^2\theta+$

$\left.\frac{r^2+ab}{a+b}\right)=2\,g\,\frac{a^2+ab+r^2+ab-(a^2-b^2)\sin^2\theta}{a+b}$, and

making $\frac{a^2-b^2}{a^2+r^2+2ab}=\gamma^2$, we have $dt=\sqrt{\frac{r}{g}}.\sqrt{}$

$$\frac{2r(a+b)}{a^2+r^2+2ab}\cdot\frac{d\theta}{\sqrt{(1-\gamma^2\sin^2\theta)}}.$$

Since $\sin^2\theta=\frac{a-z}{a-b}$, and $\cos^2\theta=\frac{z-b}{a-b}$, we have $a\cos^2$

$\theta+b\sin^2\theta=\frac{az-ab+ab-bz}{a-b}-z$, and $\frac{z}{r}$ will be the cosine of the inclination of the radius to the vertical diameter.

If ϖ be the angle made by the revolving vertical plane of r and z, with the plane of x and z, we have $\tan\varpi=\frac{y}{x}$ and $d\tan\varpi=$

$(1+\mathrm{ta}^2\varpi)\,\mathrm{d}\varpi = \mathrm{d}\frac{y}{x} = (1+\frac{yy}{xx})\mathrm{d}\varpi - \frac{x\mathrm{d}y - y\mathrm{d}x}{xx} = \frac{xx+yy}{xx}$ $\mathrm{d}\varpi$, and $x\mathrm{d}y - y\mathrm{d}x = (x^2+y^2)\mathrm{d}\varpi = (r^2 - z^2)\,\mathrm{d}\varpi$: and since it has been shown that $x\mathrm{d}y - y\mathrm{d}x = c'\mathrm{d}t$, we have $\mathrm{d}\varpi = \frac{c'\mathrm{d}t}{r^2 - z^2}$; hence, substituting for z and $\mathrm{d}t$ their values in terms of θ, we shall have the relation of ϖ and θ, which is sufficient for determining the place of the moving body.

If we call the time occupied by the body, in its passage from the highest to the lowest point of its motion, a semi-oscillation, or $\frac{1}{2}\,T$, we may determine it by finding the fluent of the value of $\mathrm{d}t$, taken from $\theta = 0$ to $\theta = \frac{1}{2}\pi = 90^0$; first resolving $\frac{1}{\sqrt{(1-\gamma^2\sin^2\theta)}}$ into a series by means of the dinomial theorem, which gives us $\frac{1}{\sqrt{(1-x^2)}} = 1 + \frac{1}{2}x^2 + \frac{1.3}{2.4}x^4 + \frac{1.3.5}{2.4.6}x^6 + \ldots$, and then taking the particular fluents of $\mathrm{d}\theta$ multiplied by the powers of $\gamma^2\sin^2\theta$, by means of the formula ${}^0\!\!\int_{\frac{\pi}{2}} \sin^{2\mathrm{M}} z\mathrm{d}z = \frac{1.3.5..(2\mathrm{M}-1)}{2.4..2\mathrm{M}}\frac{\pi}{2}$; whence we obtain $T = \pi\sqrt{\frac{r}{g}}\sqrt{\frac{2r(a+b)}{a^2+r^2+2ab}}\left\{1+\left(\frac{1}{2}\right)^2\gamma^2 + \left(\frac{1.3}{2.4}\right)^2\gamma^4 + \left(\frac{1.3.5}{2.4.6}\right)^2\gamma^6 + \quad\right\}$.

COROLLARY 1. Supposing the point to be suspended by a thread, without weight or inertia, and fixed at its upper extremity, its length being r, the motion will be exactly the same as if it rested on a spherical surface; and the greatest deviation of the thread, from the vertical direction, will be the angle of which the cosine is $\frac{b}{r}$. If the velocity, in this situation, be supposed to vanish, the oscillation will be in a vertical plane: we shall then have $a=r$, $\gamma^2=\frac{r^2-b^2}{r^2+2rb+r^2}=\frac{r^2-b^2}{2r^2+2rb}=\frac{r-b}{2r}$, γ being the sine of half the greatest angle that the thread forms with the vertical line, and its square half the verse sine of that angle. The time of the oscillation will then be $T=\pi\sqrt{\frac{r}{g}}$

$$\left\{1+\left(\frac{1}{2}\right)^2\cdot\frac{r-b}{2r}+\left(\frac{1.3}{2.4}\right)^2\cdot\left(\frac{r-b}{2r}\right)^2+\left(\frac{1.3.5}{2.4.6}\right)^2\cdot\left(\frac{r-b}{2r}\right)^3+\ldots\right\}.$$

COROLLARY 2. If the oscillation is very small, $\frac{r-b}{2r}$, being a very minute fraction, may be neglected in comparison with unity: we may therefore call, in this case, $T=\pi\sqrt{\frac{r}{g}}$, and we may consider the small vibrations as isochronous, whatever their comparative extent may be.

COROLLARY 3. We may, therefore, employ experiments on the length of a pendulum, vibrating in a given time, for the determination of the variations of the intensity of gravitation in different parts of the earth. If z be the height through which a body would fall in the time T, we have $z=\frac{1}{2}gT^2$ (232); consequently since $T^2=\pi^2\frac{r}{g}$, $z=\frac{1}{2}\pi^2r$; hence we may determine the space described by a gravitating body with the greatest precision by means of the pendulum.

SCHOLIUM. It has been found, by very accurate experiments, first made by Newton, that the length of the pendulum vibrating in a given time is the same, whatever is the nature of the substances composing it: whence it follows, that gravitation acts equally on all bodies, producing in them the same velocity in the same time; that is, in the absence of a resisting medium.

281. The equation of the tautochronous curve, in a resisting medium, is $ngdz=kds(1-e^{-ns})$; g being the force of gravity, z the vertical ordinate, s the length of the curve from the lowest point, and k a constant quantity: the resistance being expressed by $m\frac{ds}{dt}+n\frac{ds^2}{dt^2}$.

The forces acting on the moving point are, first, the force of gravity reduced to the direction of the curve, which is expressed by $g\frac{dz}{ds}$; and secondly r, the resistance of the medium, which depends in general on the velocity $\frac{ds}{dt}$: and it follows from the definition of an accelerating force, that the fluxion of the velocity is its measure, (228, 229), consequently, in the ascent of the body, $-dv=g\frac{dz}{ds}+r$, and $0=d\frac{ds}{dt}+g\frac{dz}{ds}+r$, or, making dt constant, $0=\frac{dds}{dt^2}+g\frac{dz}{ds}+r$, which is more circuitously expressed in the original notation $0=\frac{dds}{dt^2}+g\frac{dz}{ds}+\varphi\left(\frac{ds}{dt}\right)$, ($i$), r being called a "function" of $\frac{ds}{dt}$. The notation is, however, immediately exchanged for the more convenient supposition of a resistance proportional to the sum of two powers of the velocity, $\varphi\left(\frac{ds}{dt}\right)$ being $m\frac{ds}{dt}+n\frac{ds^2}{dt^2}$. We must now assume a variable quantity u, dependent on x, and making $p=$

$\frac{ds}{du}$, and $q = \frac{dp}{du}$, we shall have $\frac{ds}{dt} - p\frac{du}{dt}$, and $\frac{dds}{dt^2} - p\frac{ddu}{dt^2} + dp\frac{du}{dt^2} = p\frac{ddu}{dt^2} + q\frac{du^2}{dt^2}$: and the equation ($i$) will become $0 = p\frac{ddu}{dt^2} + q\frac{du^2}{dt^2} + g\frac{dz}{pdu} + mp\frac{du}{dt} + np^2\frac{du^2}{dt^2}$, or, dividing by p, $0 - \frac{ddu}{dt^2} + m\frac{du}{dt} + \frac{q+np^2}{p}.\frac{du^2}{dt^2} + \frac{gdz}{p^2du}$, which is expressed, in the original notation, by $0 = \frac{dds'}{dt^2} + m\frac{ds'}{dt} + \frac{ds'^2}{dt^2}.\frac{\psi''s' + n(\psi's')^2}{\psi's'} + \frac{gdz}{ds'(\psi's')^2}$, ($l$). In this equation we may destroy the coefficient of $\frac{du^2}{dt^2}$, by making $q + np^2 = 0$, that is, $\frac{dp}{du} + np^2 = 0$, and $\frac{dp}{np^2} + du = 0$, whence $\frac{1}{np} = u + c$, $\frac{1}{p} - n(u+c)$ and $pdu = \frac{du}{n(u+c)} - ds$, consequently $s = \frac{1}{n}\text{hl}(u+c) + c'$, or $= \text{hl}\left\{h(u+c)\right\}^{\frac{1}{n}}$, h and c being constant quantities: and supposing u to begin with s, we have $hc^{\frac{1}{n}} = 1$; and it will be simplest to make $h = 1$, and $c = 1$, so that s becomes $= \text{hl}\,(u+1)^{\frac{1}{n}}$, $ns = \text{hl}(u+1)$, and $u + 1 = e^{ns}$, if hl $e = 1$, whence $u = e^{ns} - 1$, and $p = \frac{1}{n(u+1)} - \frac{1}{ne^{ns}} - \frac{1}{n}e^{-ns}$. We thus reduce the equation to $0 = \frac{ddu}{dt^2} + m\frac{du}{dt} + \frac{gdz}{p^2du}$; and supposing u to be small, the last term is capable of being developed in the form of a series ascending according to its powers, which will be of this form, $ku + lu^i + \ldots$, i being greater than unity, so that the equation will become $0 = \frac{ddu}{dt^2} + m\frac{du}{dt} + ku + lu^i + \ldots$ In order to obtain the fluent of this equation, which in its present form cannot be integrated for want of the relation between u and dt, we may multiply it by $e^{\frac{mt}{2}}(\cos\gamma t+\sqrt{-1}\sin\gamma t)\,dt$, which we may call $e^{\frac{mt}{2}}\Gamma\,dt$, observing that $d\Gamma=\Gamma\sqrt{-1}\gamma dt$, and $d(e^{\frac{mt}{2}}\Gamma)=e^{\frac{mt}{2}}\Gamma\left(\frac{m}{2}+\sqrt{-1}\gamma\right)dt$. Now, beginning with the first term $\frac{ddu}{dt^2}$, and taking the fluxion of its fluent multiplied by $e^{\frac{mt}{2}}\Gamma$, we obtain $\int e^{\frac{mt}{2}}\Gamma\frac{ddu}{dt}= e^{\frac{mt}{2}}\Gamma\frac{du}{dt}-\int e^{\frac{mt}{2}}\Gamma\left(\frac{m}{2}+\sqrt{-1}\gamma\right)du$: the next step must, therefore, be with $mdu-\left(\frac{m}{2}+\sqrt{-1}\gamma\right)du$, or $\left(\frac{m}{2}-\sqrt{-1}\gamma\right)du$; and we have $\int e^{\frac{mt}{2}}\Gamma\left(\frac{m}{2}-\sqrt{-1}\gamma\right)du=e^{\frac{mt}{2}}\Gamma\left(\frac{m}{2}-\sqrt{-1}\gamma\right)u-\int e^{\frac{mt}{2}}\Gamma\left(\frac{m}{2}-\sqrt{-1}\gamma\right)\left(\frac{m}{2}+\sqrt{-1}\gamma\right)du$: and this last term, that is $\int e^{\frac{mt}{2}}\Gamma\left(\frac{mm}{4}+\gamma^2\right)u$, together with ku, may be made to disappear by putting $\frac{mm}{4}+\gamma^2=k$, and $\gamma=\sqrt{\left(k-\frac{mm}{4}\right)}$: so that the whole equation will become $e^{\frac{mt}{2}}(\cos\gamma t+\sqrt{-1}\sin\gamma t)\left(\frac{du}{dt}+\frac{m}{2}-\sqrt{-1}\gamma)u\right)=-l\int e^{\frac{mt}{2}}(\cos\gamma t+\sqrt{-1}\sin\gamma t)u'dt$—

If we compare the real and imaginary parts of this equation separately, which, as is well known, must always be allowable, because imaginary quantities can never be equated with real ones, unless they are compensated by some other imaginary quantities, we shall obtain two equations for finding the value of $\frac{du}{dt}$: but it will be sufficient at present to consider that part which is multiplied by $\sqrt{-1}$, and which affords us the equation $e^{\frac{mt}{2}} \sin \gamma t \frac{du}{dt} + e^{\frac{mt}{2}} \left(\frac{m}{2} \sin \gamma t - \gamma \cos \gamma t\right) u = -l \int e^{\frac{mt}{2}} \sin \gamma t \ u^i dt - \ldots$; the flowing quantities in the second member being supposed to begin with t. Now, at the end of the ascent, putting the time T, the fluxion ds vanishes, and with it du, which is $=(nu+1)ds$; at this moment, then, we have $e^{\frac{mT}{2}} u \left(\frac{m}{2} \sin \gamma T - \gamma \cos \gamma T\right) = -l \int e^{\frac{mt}{2}} \sin \gamma t \ u^i dt - \ldots$; which being universally true, it must be true also when the whole value of u is evanescent, and since in this case u^i is infinitely small in comparison with u, the whole of the fluents in the second member of the equation, which depend on the powers of u, must vanish in comparison with the first member, and we shall have $0 = \frac{m}{2} \sin \gamma T - \gamma \cos T$, and $\frac{m}{2} \cdot \frac{\sin}{\cos} \gamma T = \gamma$, or tang $\gamma T = \frac{2\gamma}{m}$, T being the whole time of describing the arc s, whatever its length may be, since, by the conditions of the problem, this time must always be the same, so that the equation $0 = \frac{m}{2} \sin \gamma T - \gamma \cos \gamma T$ will be true in all cases, whence in general $-l. {}^0\!\int_T e^{\frac{mt}{2}} \sin \gamma t \ u^i dt - .. = 0$; but when s and u are small, the first term is the only one that remains considerable, the others vanishing in comparison with it, consequently this term must also vanish, which can only happen if $l = 0$, since none of the quantities concerned change their values from positive to negative within the limits of $t=0$ and $t=T$. We must therefore make ku alone equal to $\frac{g\,\mathrm{d}z}{p^2\mathrm{d}u}=\frac{g\mathrm{d}z}{p\mathrm{d}s}$, and $k\,(e^{ns}-1)-\frac{g\mathrm{d}z}{p\mathrm{d}s}$, whence $k\mathrm{d}s\,(e^{ns}-1)=g\mathrm{d}z.\ \frac{1}{p}=g\mathrm{d}z.ne^{ns}$, and $ng\mathrm{d}z=k\mathrm{d}s\,(1-e^{-ns})$.

282. Corollary. When the resistance either disappears, or is proportional to the velocity only, $n=0$, and the equation becomes $g\mathrm{d}z=ks\mathrm{d}s$, which belongs to the cycloid.

For since $e^{-ns}=1-ns+\frac{n^2s^2}{2}+\ldots$ (247, Cor. 3), when h vanishes, $1-e^{-ns}=ns$, and $ng\mathrm{d}z=nks\mathrm{d}s$. [This equation is shown to belong to the cycloid in article 287.]

Scholium 1. It is remarkable that the coefficient n of the part of the resistance proportional to the square of the velocity does not enter into the expression of the time T; and it is obvious, from the steps of the analysis, that the expression would be the same, if we added to the preceding law of the resistance terms proportional to the higher powers of the velocity $\frac{\mathrm{d}s^3}{\mathrm{d}t^3}, \frac{\mathrm{d}s^4}{\mathrm{d}t^4}\ldots$ [That k is independent of n, appears from making s very small, when $ng\mathrm{d}z$ $-nks\mathrm{d}s$, and $k-\frac{s\mathrm{d}s}{g\mathrm{d}z}$, whether n be greater or smaller.]

Scholium 2. "In general, if the retarding force in the curve be R, we shall have $0=\frac{\mathrm{dd}s}{\mathrm{d}t^2}+R$, the space s being a function of the time t and of the whole arc described, which is of course a function of t and s; and by taking the fluxion of this last function, we may obtain an equation of the form $\frac{\mathrm{d}s}{\mathrm{d}t}=V$, the velocity being thus represented by its relation to t and s, and this function vanishing, according to the conditions of the problem, when t has a determinate value, independent of the arc described. Supposing V, for example, represented by ST, S being a function of s alone, and T of t alone, we shall have $\frac{dds}{dt^2}-d\frac{(ST)}{dt}=T\frac{dS}{ds}\cdot\frac{ds}{dt}+S\frac{dT}{dt}$;" which indeed might be written $T\frac{dS}{dt}+S\frac{dT}{dt}$, since S can only vary with s, and this expression could cause no ambiguity. " But, since $T=\frac{V}{S},=\frac{ds}{Sdt}, T\frac{dS}{ds}\cdot\frac{ds}{dt}=\frac{dS}{sds}\cdot\frac{ds^2}{dt^2}$. Now since t is a function of T, or of $\frac{ds}{Sdt}$, we may also suppose $\frac{dT}{dt}$ to be a function of $\frac{ds}{Sdt}$, and we may call it $\frac{ds^2}{S^2dt^2}\psi\left(\frac{ds}{Sdt}\right)$, and we shall have $\frac{dds}{dt^2}-\frac{ds^2}{Sdt^2}\left\{\frac{dS}{ds}+\psi\left(\frac{ds}{Sdt}\right)\right\}=-R$. Such is the expression for the resistance derived from the differential equation $\frac{ds}{dt}=ST$; which comprehends the case of a resistance proportional to the two first powers of the resistance, multiplied by constant coefficients: but other differential equations representing $\frac{ds}{dt}$ would give different forms to the expression of the resistance."

[SCHOLIUM 3. Instead of attempting to show the utility of this very general formula, which is certainly not extremely obvious in its present state, it will probably be more useful to insert here a more elementary view of the properties of the pendulum, remarking first that this proposition is only demonstrated with respect to the ascent of a body in the curve to be investigated, and that the descent will require some of the signs to be changed, the re-

and counteracting it in the other. Since however the steps of the demonstration do not depend on the positive character of the symbols m and n, we may simply make m negative, and we shall have tang $\gamma T - \frac{-2\gamma}{m}$, implying that the time is as much greater in the descent, as it is less in the ascent, than when the body moves without resistance: so that the whole time of the oscillation can never be sensibly affected by any small resistance of this kind: a conclusion which is of the more importance, as the resistances acting on pendulums, vibrating in common circumstances, appears to vary very nearly in the simple ratio of the velocity, the arcs decreasing proportionally in equal intervals of time.]

[283. THEOREM. "255." When a body descends along an inclined plane, without friction, the force in the direction of the plane is to the whole force of gravity as the height of the plane is to its length.

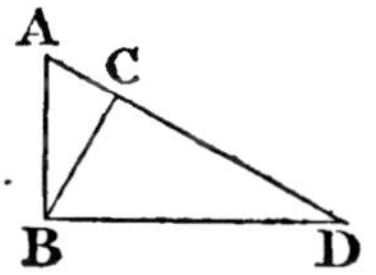

For if AB represent the motion which would be produced by gravity in a given time, this motion may be resolved into AC and CB; by means of AC the body arrives at the line CB in the same time as if it were at liberty; but the motion CB is destroyed by the resistance of the plane; and as AB to AC, so is AD to AB (121). But forces are measured by the spaces described in the same time (230).

SCHOLIUM. Hence, by employing a plane differing but little from a horizontal direction, we may lessen the velocity of descent, so as to make some illustrative experiments on the effects of accelerating forces, without the inconvenience of too great a velocity: although, if the weights employed roll down the plane, some force will be lost in the production of rotatory motion; and if they slide, they will be retarded by friction.

284. Theorem. "256." When bodies descend on any inclined planes of equal height, their times of descent are as the lengths of the planes, and the final velocities are equal.

Since $t=\sqrt{\left(\frac{2x}{a}\right)}$ (233), and here $a=\frac{1}{x}$, $t=\sqrt{(2x^2)}=\sqrt{2}x$; and the times vary as the spaces: but the times being greater in the same proportion as the forces are less, the velocities acquired are equal (230).

Scholium. Thus a body will acquire a velocity of 32 feet in a second, after having descended 16 feet, either in a vertical line or in an oblique direction; but the time of descent will be as much greater than a second, as the oblique length of the path is greater than 16 feet: and if we suffer three balls to descend together along three grooves of the same height, but of the lengths of 1, 2, and 3 feet respectively, we may estimate by the ear the equality of the intervals at which they reach the bottom.

285. Theorem. "257." The times of falling through all chords drawn to the lowest point of a circle are equal.

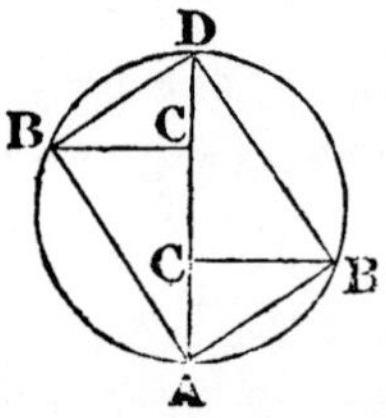

The accelerating force in any chord A B is to that of gravity as A C to A B, or as A B to A D (121), therefore the forces being as the distances, the times are equal;

for their squares are as the spaces directly and the forces inversely (233).

SCHOLIUM. This elegant proposition may be illustrated by an easy experiment: if we place two bodies at different points of a circle, fixed in a vertical situation, and suffer them to descend at the same instant along two planes, which meet in the lowest point of the circle, they will arrive there at the same time.

286. THEOREM "258." When a body is retained in any curve by its attachment to a thread, or descends along any perfectly smooth surface of continued curvature, its velocity is the same, at the same height, as if it fell freely.

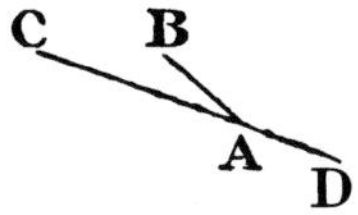

Since the velocity is the same at A, whether the body has descended an equal vertical distance from B or from C, it will proceed in A D with the same velocity in both cases, provided that no motion be lost in the change of its direction, and therefore its velocity will be the same, after passing any number of surfaces, as if it had fallen perpendicularly from the same height. But where the curvature is continued, no velocity is lost in the change of direction; for let A B be the thread, or its evolved portion, the body B, if no longer actuated by gravity, would proceed in the circular arc with uniform motion (263); consequently no velocity is destroyed by the resistance of the thread, nor by that of the surface BC, which can only act in the same direction, perpendicular to the direction of the moving body.

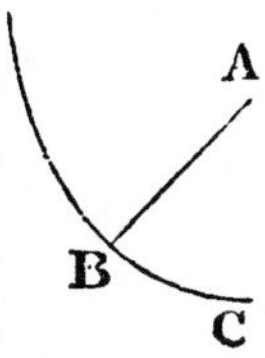

SCHOLIUM. We may easily show, by an experiment on a suspended ball, that its velocity is the same, when it descends from the same height, whatever may be the form of its path; and this we prove by observing the height to which it rises on the opposite side of the lowest point, whether in the same curve, or in different ones. We may alter the form of its path both in descending and in ascending, by placing pins at different points, so as to interfere with the thread that supports the ball, and to form, in succession, temporary centres of motion; and we shall find, in all cases, that the body ascends to a height equal to that from which it has descended, with a small deduction on account of friction. Thus, the same ball, descending from equal heights at A, B, or C, by different paths, will rise to the same height at D on the opposite side of E, and the reverse.

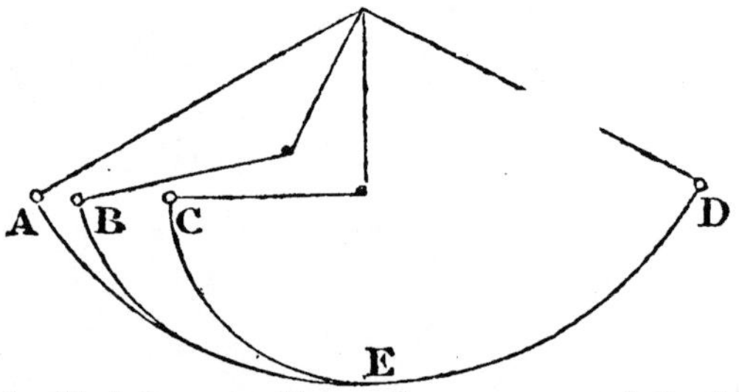

287. THEOREM. "259" If a body be suspended by a thread between two cycloidal cheeks, it will describe an equal cycloid by the evolution of the thread (208); and the time of descent will be equal, in whatever part of the curve the motion may begin, and will be to the time of falling through one half of the length of the thread, as half the circumference of a circle is to its diameter: and the space described in the cycloid will be always equal

to the verse sine of an arc which increases uniformly.

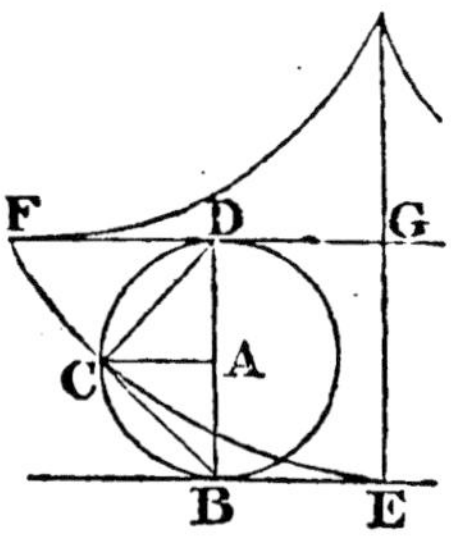

For since the accelerating force, in the direction of the curve, is always to the force of gravity as AB to BC, or as BC to the constant quantity BD, it varies as BC, or as its double, CE, the arc to be described, and CE being called s, the force $g\frac{dz}{ds}$ must vary as s (208). If therefore any two arcs be supposed to be equally divided into an equal number of evanescent spaces, the force will be every where as the space to be described; and it may be considered, for each space, as equable, and the increments of the times, and consequently the whole times, will be equal. Supposing the generating circle to move uniformly, the velocity of the describing point C will always be as CD (209), or since AD : CD :: CD : BD, and CD $= \surd(AD.BD)$ as $\surd AD$; but the velocity of a body falling in DA, or descending in FC, varies in the same ratio (232, 230, 286); therefore if the velocity at E be equal to that which a body acquires by falling through GE, the describing point C will always coincide with the place of a heavy body descending in FCE; and the velocity of the point of contact D is half that of C at E (209), it would therefore describe a space equal to GE in the time that a body would fall through GE, and will describe FG in a time which is to that time as FG to GE, or as half the circumference of a circle to its diameter, and this will be the time of descent in a cycloidal arc. And since FC$=$2DB $-$2BC, FC is equal to the verse sine of the angle CBD, to the radius 2BD: but the angle CAD increasing uniformly, its half CBD must also increase uniformly. And if the motion begin at any other point of the curve, it follows, from the former part of the demonstration, that the velocity will be in a constant ratio to the velocity in similar points of the whole cycloid. It is also obvious that the arc of ascent will be equal to the arc of descent, and described in an equal time, supposing the motion without friction.

288. THEOREM. "260." The times of vibration of different cycloidal pendulums are as the square roots of their lengths.

For the times of falling through half their lengths are in the ratio of the square roots of these halves, or of the wholes.

SCHOLIUM. Major Kater has ascertained, by a great number of very accurate experiments, performed with an apparatus of his own invention, that the length of the pendulum vibrating in a second in London, on the level of the Thames, and in a vacuum, is 39·14 inches, very nearly. Hence the time of falling through 19·57 inches will be $\frac{1''}{\pi}$, and the space described in a second $19{\cdot}57 \times \pi^2$. Now Log $3{\cdot}141592^2 = {\cdot}9943$ and Log $19{\cdot}57 = 1{\cdot}2916$; their sum 2·2859 is the logarithm of 193·15 inches, or 16·096 feet, the space described by a heavy body in the first second of its descent. More accurately the numbers are 39.1387 and 16.095.

289. THEOREM. "261." The cycloid is the curve of swiftest descent between any two points not in the same vertical line.

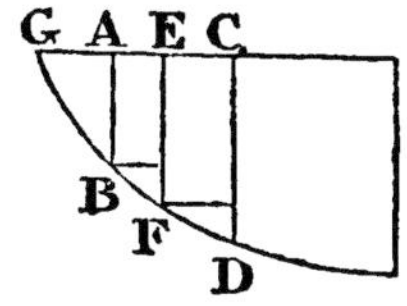

Let AB and CD be two parallel vertical ordinates at a constant evanescent distance, in any part of the curve of swiftest descent, and let a third, EF, be interposed, which is always in length an arithmetical mean between them, and which, as it approaches more or less to AB, will vary the curvature of the element BFD. Call AB, a, EF, b; $b-a$, c; AE, u; and EC, v: then BF$=\surd(u^2+c^2)$, and since CD$-$EF$=$EF$-$AB, FD$=\surd(v^2+c^2)$. But the velocities at B and F are as $\surd a$ and $\surd b$, and the elements BF, FD being supposed to be described with their velocities, the time of describing BD is $\surd\left(\frac{uu+cc}{a}\right)+\surd\left(\frac{vv+cc}{b}\right)$; which must be a minimum, and its fluxion must vanish: or $\frac{2u\mathrm{d}u}{2\surd(a\{uu+cc\})}+\frac{2v\mathrm{d}v}{2\surd(b\{vv+cc\})}=0$; but since AC, or $u+v$, is constant, $\mathrm{d}u+\mathrm{d}v=0$, or $\mathrm{d}u=-\mathrm{d}v$; therefore $\frac{u}{\surd a\surd(uu+cc)}=\frac{v}{\surd b\surd(vv+cc)}$. Let the variable absciss GA be now called x, the ordinate AB, y, and the arc GB, z, then u and v are increments of x, and BF and FD of z, when y becomes equal to a and b respectively; we have, therefore, $\frac{x'}{\surd yz'}$, the same in both cases, so that it may be called $\frac{1}{a}$, and $\frac{x'}{z'}$, or $\frac{\mathrm{d}x}{\mathrm{d}z}=\frac{\surd y}{a}$. Now in the cycloid the chord of the generating circle must be always a mean proportional between the verse sine y and the radius, since, in article 287, CD$=\surd$(AD.BD) and the arc z being perpendicular to that chord, its fluxion, by similar triangles, is to that of the absciss x, as the diameter to $\surd y$: therefore the cycloid answers the condition in every part, and consequently in the whole curve.

SCHOLIUM 1. The demonstration implies that the origin of the curve must coincide with the uppermost given point: now only one cycloid can fulfil this condition and pass through the other point, and it will often happen that the curve must descend below the second point, and rise again.

SCHOLIUM 2. The method of independent variations may be applied with great elegance and simplicity to problems of this kind, although it has too commonly been made complicated and perplexed by unnecessary abstraction. An example of its application has already occurred in the investigation of the properties of $\int v\mathrm{d}s$ (266), but it will not be superfluous to enter into some further illustration of the method on this occasion.

Let it be required, for example, to determine the equation of the line which gives the shortest distance between two points, from the property of maximums and minimums which are unaltered by any slight variation of their elements. We have, therefore, $\delta s=0$; but $\delta s=\int \mathrm{d}\delta s$, the characteristic $\int$ relating to the fluxional variation expressed by d; and $\int \mathrm{d}\delta s=\int \delta \mathrm{d}s$ (265). Now, x and y being the ordinates, and s the curve, we have $\mathrm{d}s^2=\mathrm{d}x^2+\mathrm{d}y^2$, and $\delta \mathrm{d}s=\frac{\mathrm{d}x\delta \mathrm{d}x+\mathrm{d}y\delta \mathrm{d}y}{\mathrm{d}s}$; and, for the sake of simplicity, we may make $\delta \mathrm{d}x=0$, supposing the curve to pass into a neighbouring form by the variation of $\mathrm{d}y$ only: we have, then, $\delta \mathrm{d}s=\frac{\mathrm{d}y}{\mathrm{d}s}\delta \mathrm{d}y$, of which we must find the fluent. Now $\mathrm{d}\left(\frac{\mathrm{d}y}{\mathrm{d}s}\delta y\right)=\frac{\mathrm{d}y}{\mathrm{d}s}\,\mathrm{d}\delta y+\mathrm{d}\,\frac{\mathrm{d}y}{\mathrm{d}s}\,\delta y=\frac{\mathrm{d}y}{\mathrm{d}s}\,\delta \mathrm{d}y+\mathrm{d}\,\frac{\mathrm{d}y}{\mathrm{d}s}\,\delta y$; consequently $\int \delta \mathrm{d}s=\frac{\mathrm{d}y}{\mathrm{d}s}\,\delta y-\int \mathrm{d}\,\frac{\mathrm{d}y}{\mathrm{d}s}\,\delta y=\delta s$. This expression implies, when geometrically considered, that the variation of the length of the curve, δs, is expressed by the variation of the ordinate $\dot{y}$ at any given point, reduced to the direction of the curve, and lessened by the length of a minute curve of equal angular extent to the curve in question, and of which the radius of curvature is equal to the variation δy reduced to a direction perpendicular to the curve. Now, in order to determine the shortest distance, we must put $\delta s = 0$, and $\frac{dy}{ds}\delta y = \int d\frac{dy}{ds}\delta y$. But at the beginning and at the end of the line in question δy must be $=0$, both the points being fixed; consequently the fluent $\int d\frac{dy}{ds}\delta y$ $=0$, which can only happen when $d\frac{dy}{ds}=0$, since δy is not $=0$, and the fluent cannot have different values, destroying each other, in different parts of the line, because the value must vanish equally for all parts of the line, which must be always the shortest distance between their extremities: and the sine of the inclination $\frac{dy}{ds}$ being constant, the curve must become a right line.

In the case of the present problem, we have $0=\delta t - \int \delta dt = \int \delta \frac{dz}{a\sqrt{y}} = \int \frac{dx\delta dx + dy\delta dy}{a\sqrt{y}dz} - \int \frac{dz\delta y}{2ay\sqrt{y}}$, which we may simplify by making $\delta dy=0$ and $\delta y=0$, confining the variation to dx, according to the spirit of the preceding demonstration of the theorem; consequently $\delta dt = \frac{dx\delta dx}{a\sqrt{y}dz}$ $=\frac{dx}{a\sqrt{y}dz}d\delta x$; and comparing this equation with the $\frac{dy}{ds}d\delta y$ of the former example, we have in a similar manner

$\delta t = \int \frac{dx}{a\sqrt{ydz}} d\delta x = \frac{dx}{a\sqrt{ydz}} \delta x - d\frac{dx}{a\sqrt{ydz}} \delta x = 0$. Hence $\frac{dx}{a\sqrt{ydz}} \delta x = d\frac{dx}{a\sqrt{ydz}} \delta x$, which vanishing for the whole curve and for all its parts, as $d\frac{dy}{ds}$ was shown to vanish before, it follows that $\frac{dx}{a\sqrt{ydz}}$ must be a constant quantity; which is the property of the cycloid.

290. THEOREM. "262." The time of vibration of a simple circular pendulum in a small arc, is ultimately the same as that of a cycloidal pendulum of the same length; "but in larger arcs the times are greater." (280).

In small cycloidal arcs the radius of curvature is very nearly constant; but at greater distances from the lowest point, the circular arc falls without the cycloidal, and is less inclined to the horizon, so that the force is smaller, and consequently the velocity is smaller.

291. THEOREM. "263." If a body suspended by a thread revolve freely round the vertical line, the times of revolution will be the same, when the height of the point of suspension above the plane of revolution is the same, whatever be the length of the thread.

For, by the resolution of forces, the force urging the body towards the vertical line is to that of gravity as the distance from that line to the vertical height; the other part of the force being counteracted by the tension of the thread; and when the forces are as the distances, the times must be equal. (261).

SCHOLIUM. Thus, if a number of balls are fixed to threads, or rather wires, connected to the same point of an axis, which is made to revolve by means of the whirling table, they will so arrange themselves, as to remain very nearly in the same horizontal plane.

292. THEOREM. "264." The time of a revolution of a body, suspended by a thread, is equal to the time occupied by a cycloidal pendulum, of which the length is equal to the height of the point of suspension above the plane of revolution, in vibrating once forwards and once backwards to the point at which its motion began; and if the revolutions be small, and the thread nearly vertical, they will be very nearly isochronous, whatever be their extent.

For, supposing the distance to be equal to the height, the centrifugal force will be equal to the force of gravity, and while the body describes a distance equal to the radius, another body, actuated by the same force, would describe half that radius, (259) and the whole time of revolution is, therefore, to this time, as the circumference to the radius, and is consequently equal to the time of four semivibrations of a cycloidal pendulum, of which the length is equal to the given height (287). And since the time varies, in the same revolving pendulum, only as the square root of the cosine of the angle of inclination, it will be nearly constant for all small revolutions.

SCHOLIUM. The near approach of these revolutions to isochronism has sometimes been applied to the measurement of time, but more frequently, and more successfully, to the regulation of the motions of machines. Thus, in Mr. Watt's steam-engines, two balls are fixed at the ends of rods in continual revolution, and as soon as the motion becomes a little too rapid, the balls rise considerably, and turn a cock, which regulates the quantity of of steam admitted.

293. THEOREM. "265." The vibrations of a cycloidal pendulum will be performed in the same time, whether they be without resistance, or retarded by a uniform force.

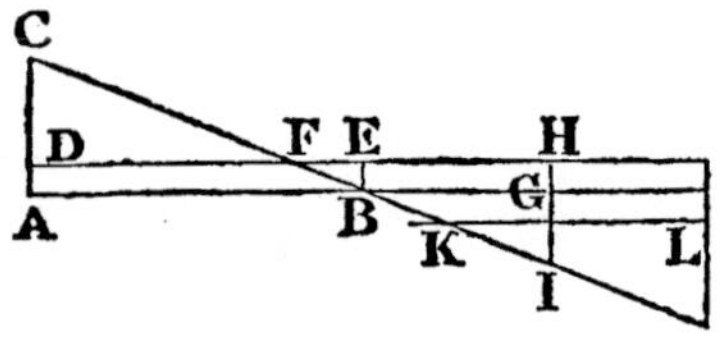

Let the relative force of gravity, at the distance AB in the curve from its lowest point, be always represented by the ordinate AC; then CB will be a right line: now the resistance may always be represented by the equal ordinates AD, BE; and DC will express the remaining force, which becomes neutral at F, and then negative: therefore the force is always the same, at equal distances on each side of F, as in the simple pendulum on each side of B, and the vibration will be perfectly similar to the vibration of the simple pendulum in a smaller arc, but it will extend only to G, where the ordinate HI is equal to DC, and FH=FD. In the return of the body from G, the neutral point will be determined by the intersection of KL, parallel to AB, and as much below it, as DE was above it; this vibration will terminate in a point as far on one side of K as I is on the other: so that the extent of each vibration will be less than that of the preceding one, by twice the length of FE, until the whole force is exhausted, the time of each complete vibration remaining unaltered.

294. Theorem. "A." (Nich. Journ. 1813.) If the point of suspension (A) of a pendulum (AB) be made to vibrate in a regular manner, that is, according to the law of cycloidal vibrations, the pendulum itself may also vibrate regularly in the same time, provided that the extent of its vibrations (BC) be to that of the vibrations of the point of suspension (AD) as the length of the thread (AE) supposed to carry this point as a pendulum, is to the difference of the lengths of the two threads.

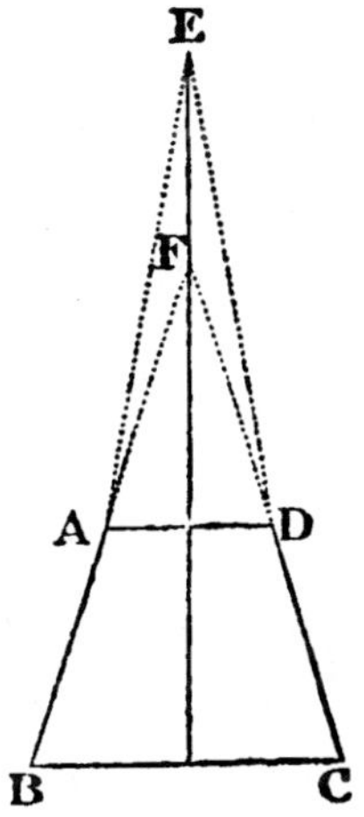

In representing the vibrations, we may disregard the curvature of the paths, considering them as of evanescent extent, the forces being however still supposed to depend on the inclination of the threads, which must be exaggerated in the figures employed. Let F be the intersection of AB with the vertical line EF; then, upon the conditions of the theorem, BF will be equal to AE; that is, if BC : AD = AE : AE ∾ AB, since

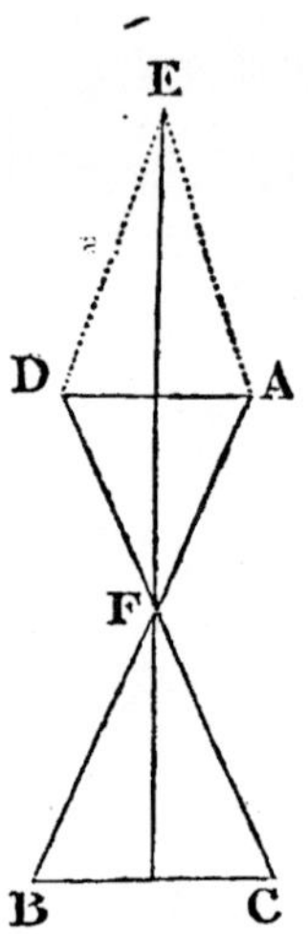

by similar triangles BC : AD=BF : BF∾AB, it follows that AE=BF. Consequently the inclination of the thread AB will always be the same as if F were its fixed point of suspension, and the body B will begin and continue its vibrations like a simple pendulum attached to that point, the true point of suspension accompanying it with a proportional velocity, so as to be always in the right line passing through it and through F. It is obvious, that when the thread supposed to suspend the moveable point of suspension is the longer of the two, the vibrations will be in the same direction; when the shorter, in contrary directions.

SCHOLIUM 1. The truth of this proposition may easily be illustrated, by holding any pendulous body in the hand, and causing it to vibrate more or less rapidly, by moving the hand regularly backwards or forwards, in a longer or in a shorter time than that of the spontaneous vibrations.

SCHOLIUM. 2. The same mode of reasoning is applicable to oscillations of any other kinds, which are governed by forces proportional to the distances of the bodies concerned, from a point of which the situation, either in a quiescent space, or with respect to another moveable point, varies according to the law of the cycloidal pendulum, or may be expressed by the sines of arcs varying with the time: such forces always producing periodical variations, of which the extent is to that of the excursions of the supposed point of suspension in the ratio of n to $n-1$, n being to 1 as the square of the time of the forced to that of the time of the spontaneous vibration; and

and when $n-1$ is negative, the displacement being in a direction opposite to that of the supposed point of suspension. Consequently, when a body is performing oscillations by the operation of any force, and is subjected to the action of any other periodical forces, we have only to inquire at what distance a moveable point must be situated before or behind it, in order to represent the actual magnitude of the periodical force by the relative situation, according to the law of the primary force concerned, and to find an expression for this distance in terms of the sines of arcs increasing equably, in order to obtain the situation and velocity of the body at any time, provided that we suppose it to have attained a permanent state of vibration.

SCHOLIUM 3. We may easily express this reasoning in a form more strictly algebraical: thus the time, with respect to the forced vibration of the centre of suspension, being called t, the place of the vertical line passing through that point will be indicated by $\sin t$, supposing t to begin from the middle of a vibration: now the force acting on the moving body will always be as its distance from this moveable vertical line, considered with relation to the length of the true pendulum m; that is, it will be expressed by $f = \frac{s - \sin t}{m}$, the unit of m being the length of the imaginary pendulum carrying the point of suspension, since when $s=0$ and $\sin t=1$, the force must be $=1$ or $=g$. Now we may satisfy this equation by the particular solution $s - \sin t = a \sin t$, which represents a vibration either corresponding in its direction with the former, or in an opposite direction, accordingly as a is positive or negative; and s, the space actually described, will be the sum or difference of the spaces belonging to the separate vibrations so combined: then since $v=-\int f dt$, and $s=\int v dt$, we have $v=-\int\frac{a\sin t}{m}dt=\frac{a}{m}\cos\ t+c$, and $s=\frac{a}{m}\sin t+ct=a$ $\sin t+\sin t$, and $c=0, \frac{a}{m}-a+1, a=1:\left(\frac{1}{m}-1\right)-$ $\frac{m}{1-m}$, or if $n=\frac{1}{m}$, $\frac{1}{n-1}$, as before.

SCHOLIUM 4. If the oscillating body be initially in any other condition, its subsequent motion may be determined, by considering it as performing a secondary vibration with respect to a point vibrating in the manner here supposed, which will consequently represent its mean place; but if there be no resistance, the body will have no tendency to assume the form of a regular simple vibration, rather than any other. Supposing, for example, that the point had been initially at rest in the middle vertical line, when the centre of suspension passed that line; it will then agree in situation with the point representing its mean place, but not in velocity; and it will return to its mean place after every interval equal to a complete single spontaneous vibration of the true pendulum; and when this coincidence happens in the middle vertical line as at first, the whole cycle of motions will begin again, after a period depending on the comparative lengths of the supposed pendulums: and at some intermediate time the coincidence will in most cases occur near the extremity of the vibration representing the mean place, and the excursion will be much greater than that of this vibration, while at another part of the cycle it may be almost obliterated. Such a succession of cycles may be often observed in the actual vibrations of elastic bodies of irregular forms, the excursions being alternately greater and smaller without any interference of external causes.

SCHOLIUM 5. A more general analytical solution of the problem may be obtained by making $s = b \sin t + c \sin (et + h)$ whence $v = -\int\int f dt = -\int n \left\{ (b-1) \sin t + c \sin (et + h) \right\} dt = n \left\{ (b-1) \cos t + \frac{c}{e} \cos (et+h) \right\} + i$, since $d \cos (et+h) = -\sin (et+h) \, e dt$; and $s = \int v dt = n \left\{ (b-1) \sin t + \frac{c}{ee} \sin (et+h) \right\} + it + k = b \sin t + c \sin (et+h)$; whence $n(b-1) = b$, $\frac{nc}{ee} = c$, $i=0$ and $k=0$; consequently $n = \frac{b}{b-1}$, and $b = \frac{1}{n-1}$, $\frac{n}{ee} = 1$, and $e = \sqrt{n}$, h and c remaining altogether undetermined. We may, therefore, accommodate this expression to any relative values of the supposed vibrations, or of the forces belonging to them, and to any conditions of motion or rest in the initial state of the moving body. Thus, if we suppose it initially at rest, so that $s=0$ and $v=0$ when $t=0$, the length n being given, we have $s = b \sin t + c \sin (et+h) = 0$, and consequently $h=0$, and $v = n(b-1) \cos t + \frac{c}{e} \cos et = b + \frac{c}{e} = 0$, and $\frac{c}{e} = -b = \frac{-1}{n-1}$ whence $c = \frac{-e}{n-1} = \frac{\sqrt{n}}{1-n}$, and we have $s = \frac{\sin t}{n-1} + \frac{\sqrt{n}}{n-1} \sin \sqrt{n} t$.

295. THEOREM. "B." If the resistance be simply proportional to the velocity, a pendulum with a vibrating point of suspension may perform regular vibrations, isochronous with those of the point of suspension, provided

that, at the middle of a vibration, the point of suspension (A) be so situated, as to cause a propelling force equal to the actual resistance, the extent of the vibrations being reduced, in the ratio of the whole excursion of the point of suspension (BC) to its distance from the middle, at the beginning of the motion of the pendulous body (DC): and it will ultimately acquire this mode of vibration, whatever may have been its initial condition.

Let FG be the supposed length of the thread carrying the point of suspension, and draw FE passing through D instead of B; then if HC, =EG, be the extent of the vibration, it will be maintained according to the law of the cycloidal pendulum. Draw the concentric circles BI, DK, HL: then the initial force may be represented by

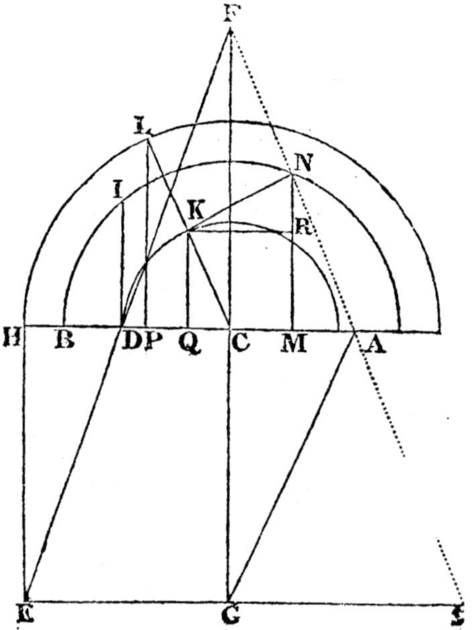

HD, which determines the greatest inclination of the thread; and at any subsequent part of the vibration, if the point of suspension be advanced from D to M, the time elapsed will be expressed by the arc IN, DI and MN being perpendicular to AB, and taking HL similar to IN, the perpendicular LP will show the place of the pendulous

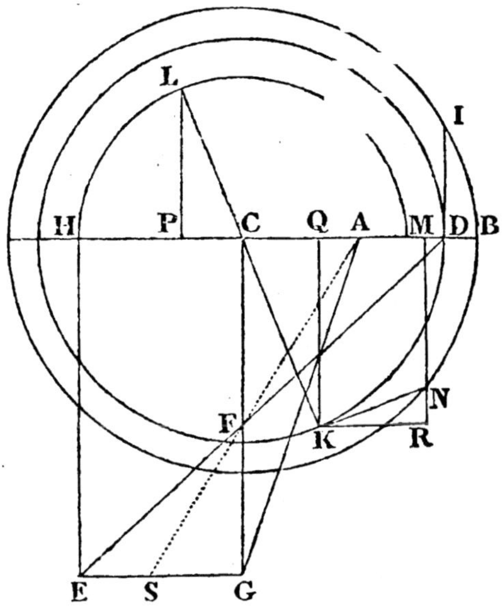

body, and PM the force, which may be divided or resolved into two parts, PQ and QM. But PQ is to LK, or HD, as PC to LC, or HC; consequently this part of the force will always be employed in generating the regular velocity; and QM is equal to KR, which is the sine of the angle KNR or BCL to the radius KN=DI=AC, each of these lines being equal to the sine of BI; the line QM therefore varies as the velocity, and will always be equivalent to the friction, provided that it be once equivalent to it, as it is supposed to be at A; the ratio of the forces concerned, in any two succeeding instants, being always such as to maintain a regular vibration.

If the pendulum be initially in any other situation than that which is here supposed, its subsequent motion may be determined by comparison with that of a point so vibrating, and its progress may be estimated, with tolerable accuracy, while this deviation exists, by supposing it to perform small vibrations with respect to its mean place, in which the immediate effect of resistance may be neglected: but since these vibrations are not supported by any new sustaining force, they will evidently be rendered by degrees smaller and smaller, so that the pendulum will ultimately approach infinitely near to the regular state of vibration here described, which may, therefore, be considered as affording a stable equilibrium of motion.

SCHOLIUM 1. Supposing the relation of the resistance to the velocity to be altered, the relation of the sine AC to the cosine CD must be similarly altered, the force equivalent to the resistance varying as the sine, and the extent of the vibrations, and consequently the velocity, as the cosine of the displacement BI: but the relation of the sine to the cosine is that of the tangent to the radius: so that the tangent of the displacement will be as the mean resistance: and the sine of the displacement, AC, is to the radius BC, as the greatest resistance is to the greatest force which would operate on the pendulous body if it remained at rest at G: the displacement at the extremity of the vibration having the same angular measure, but becoming, with respect to the place of the body, the verse sine only, instead of the sine.

SCHOLIUM 2. It is obvious, from the figures, that the body G will always be behind the place S, which it would have occupied without the resistance, when the vibration is direct, but before it when inverted.

SCHOLIUM 3. When the resistance is very small, a simple pendulum with a similar resistance may be conceived to vibrate nearly in a similar manner: and if we neglect the diminution of the velocity in the consideration of the resistance, and call $r = mv = m \cos t$, we have $v = -\int f \mathrm{d}t = -\int (\sin t + m \cos t)\mathrm{d}t = \cos t - m \sin t$, and $s = \int v \mathrm{d}t = \sin t + m \cos t - a = \surd(1+m^2) \sin (t+b) - a$, b being the angle of which the tangent is m (216), and $a = \surd(1+m^2) \sin b =$

$$\surd(1+m^2)\frac{m}{\surd(1+mm)} = m,$$ consequently $s = \surd(1+m^2) \sin (t+b) - m$, which implies a vibration observing the period of t, but beginning at a point at the distance b further back in the circle, so that the time of ascent will be diminished and that of descent increased very nearly in an eqnal de-

gree, as may be inferred from Laplace's formula (282) tang $\gamma T = \frac{2\gamma}{m}$, whence cot $\gamma T = \frac{m}{2\gamma}$, γ^2 being $1 - \frac{mm}{4}$; and ultimately cot $T = \frac{m}{2}$: the value of m in this scholium being equal to $\frac{m}{2}$ of article 282, since here the greatest velocity in the pendulum, due to a height equal to half its length, is made the unit of v and of r, instead of a more direct comparison with the value of g the force of gravity.]

CHAPTER III.

OF THE EQUILIBRIUM OF A SYSTEM OF BODIES.

---

§. 13. [*Introduction*]. *Conditions of the equilibrium of two systems of points, meeting each other, with velocities directly contrary. Definition of the quantity of motion, and of similar moveable points.* **P. 36.**

[296. Definition. "266." A moveable body is to be imagined as a point, composed of single points or particles equally moveable, which, as they differ in number, constitute the proportionally different mass or bulk of the body.

297. Definition. "267." A reciprocal action between two bodies is an action which affects the single particles of both equally, increasing or diminishing their distance.

298. Definition. "268." The centre of inertia of two bodies is that point, in the right line joining them, which divides it reciprocally in the ratio of their magnitudes.

299. THEOREM. "269" The centre of inertia of two bodies, initially at rest in any space, remains at rest, notwithstanding any reciprocal action of the bodies.

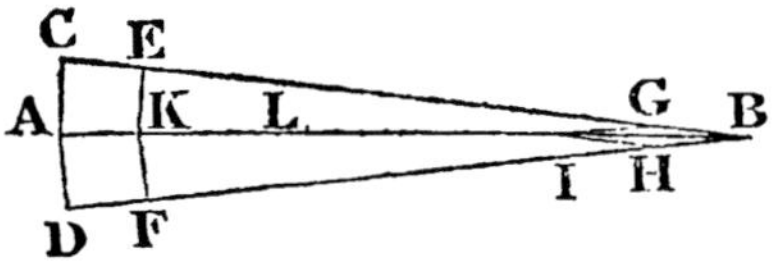

Suppose the bodies equal, and consisting each of a single particle, then it is obvious that both will be equally moved by any reciprocal action, and the centre of inertia will still bisect their distance (217). Again, let one body A be double the other B, and suppose A to be divided into two points placed very near each other, as C, D. Join BC, BD, take any equal distances CE, DF, BG, BH, and they will represent the mutual actions of B on C and D, and of C and D on B, and the motions produced by these equal actions; complete the parallelogram BGIH, and the diagonal BI will be the joint result of the motions of B; which, when C and D coincide in A and K, becomes equal to 2BG, 2CE, or 2AK; but L being the centre of inertia, BL=2AL (298) therefore IL remains equal to 2KL (15), and L is still the centre of inertia. And in the same manner the theorem may be proved when the bodies are in any other proportion.

SCHOLIUM. This important theorem is capable of an easy experimental illustration; first observing, that all known forces are reciprocal, and among the rest the action of a spring: we place two unequal bodies so as to be separated when a spring is set at liberty, and we find that they describe, in any given interval of time, distances which are inversely as their weights; and that consequently the place of the centre of inertia remains unaltered. They may either be made to float on water, or may be suspended by long threads: the spring may be detached by burning a thread that confines it, and it may be observed whether or no they strike at the same instant two obstacles, placed at such distances as the theory requires; or, if they are suspended as pendulums, the arcs which they describe may be measured, the velocities being always nearly proportional to these arcs, and accurately so to the chords, which are as the square roots of the verse sines, representing the heights of ascent.

300. DEFINITION. "270." The joint ratio of the masses and velocities of any two bodies is the ratio of their momenta.

301. THEOREM. "271." The momentum of any body is the true measure of the quan tity of its motion.

For the same reciprocal action produces in a double body half the velocity, the common centre of inertia remaining at rest; and, the cause being the same, the effects must be considered as equal: and when the reciprocal force varies, the velocity of both bodies varies in the same ratio.

SCHOLIUM 1. We may also demonstrate experimentally, by means of Mr. Atwood's machine, that the same momentum is generated, in a given time, by the same preponderating force, whatever may be the quantity of matter moved. Thus if the preponderating weight be one sixteenth of the whole weight of the boxes, it will fall one foot in a second, instead of 16, and a velocity of two feet will be acquired by the whole mass, instead of a velocity of 32 feet, which the preponderating weight alone would have acquired. And when we compare the centrifugal forces of bodies revolving in the same time, at different distances from the centre of motion, we find that a greater quantity of matter compensates for a smaller force; so that two balls, connected by a wire, with liberty to slide either way, will retain each other in their respective situations, when their common centre of inertia coincides with the centre of motion; the centrifugal force of each particle of the one being as much greater than that of an equal particle of the other, as its weight, or the number of the particles, is smaller.

302. SCHOLIUM 2, A.] The simplest case of the equilibrium of several bodies is that of two material points meeting each other with equal and directly contrary velocities; their mutual impenetrability must evidently annihilate their motion, and reduce them to a state of rest.

[B.] Let us now suppose a number m of contiguous material points, arranged in a right line, and moving in its direction with the velocity u: and again another number m' of contiguous points, disposed in the same line, and moving with the velocity u' in a contrary direction, so that the two systems meet each other; there must exist a relation between u and u', such that the systems may both remain at rest after the shock.

[C.] In order to determine this condition, we may observe that the system m, moving with the velocity u, would destroy the motion of a single point, moving with the velocity mu, for every point in the system would destroy, in this last point, a velocity equal to u, and consequently the m points would destroy the whole velocity mu: we may therefore substitute for this system a single point, moving with the velocity mu. In the same manner we may substitute for the system m' a single point moving with the velocity $m'u'$: but the two systems being supposed capable of destroying each other's motion, the two points, possessing respectively equal quantities of motion, must remain at rest after meeting, consequently their velocities must be equal (A); we have therefore, for the condition of the equilibrium of the two systems, $mu = m'u'$.

[D.] The mass of a body consists in the number of its material points, and the product of the mass by the velocity is called the quantity of motion of a body: and this product is also [sometimes] considered as the force of a body in motion. In order that two bodies meeting may destroy each other's motion, the quantities of motion in opposite directions must be equal, and consequently the velocities must be inversely as the masses.

[E.] The density of a body depends on the number of material points which it contains within a given volume or bulk. In order to ascertain its absolute density, it would be necessary to compare it with a body having no pores: but since we know of no such body, we can only compare any given substance with some other as a standard with respect to density. It is obvious that the mass of a body is in the joint proportion of the volume and the density, so that calling the mass M, the bulk U, and the density D, we have in general $M = DU$; the quantities M, D, and U, relating to different units, each of its own species.

[F.] In this reasoning we suppose that bodies are formed of similar material points, and that they only differ in the relative situation of the atoms composing them. But the intimate nature of matter being unknown, this assumption is at least hypothetical, and it is perfectly possible that there may be a difference in the elementary particles of matter. Fortunately, however, the truth of the hypothesis is of no consequence to the science of mechanics, and we may adopt it without any danger of error, provided that, by similar material points, we understand points, which, when they meet with equal velocities, destroy each other's motion, whatever their nature may be.

§. 14. *Of the reciprocal action of material points. Reaction is always equal and contrary to action. Equation of the equilibrium of a system of bodies, giving the law of virtual velocities. Method of determining the pressure of bodies on the surfaces or the curves to which they are confined.* P. 37.

303. THEOREM. Action and reaction are always equal and contrary.

Two material points, of which the masses are m and m', can only act on each other in the direction of the right line joining them. If, indeed, they are united by a thread passing over a pulley, their reciprocal action may be otherwise directed: but in this case the fixed pulley may be considered as having at its centre a body of infinite density, which reacts on the two bodies m and m', so as to make their mutual action indirect only.

If the action of m on m', exerted by means of an inflexible line, without inertia, uniting them, be called p, and if it be met by a contrary force, expressed by $-p$, this force will destroy in the body m a force equal to p, and the force p in the right line will be communicated entirely to m'. This loss of force in m, occasioned by its action on m', is called the reaction of m'; so that, in the communication of motion, "reaction is always equal and contrary to action." And it is found by observation that this principle holds good with respect to all forces in nature.

[SCHOLIUM 1. All the forces in nature, with which we are acquainted, act reciprocally between different masses of matter, so that any two bodies, repelling or attracting each other, are made to recede or approach with equal momenta. This circumstance is generally expressed by the third law of motion, that action and reaction are equal. There would be something peculiar, and almost inconceivable, in a force which could affect unequally the similar particles of matter; or in the particles themselves, if they could be possessed of such different degrees of mobility, as to be equally moveable with respect to one force, and unequally with respect to another. For instance, a magnet and a piece of iron, each weighing a pound, will remain in equilibrium when their weights are opposed to each other by means of a balance; they will be separated with equal velocities, if impelled by the unbending of a spring placed between them; and it is difficult to conceive that they could approach each other with unequal velocities in consequence of magnetic attraction, or of any other natural force. The reciprocality of force is, therefore, a necessary law in the mathematical consideration of mechanics, and it is also perfectly warranted by experience. The contrary supposition is so highly improbable, that the principle may almost as justly be termed a necessary axiom, as a phenomenon collected from observation.

SCHOLIUM 2. Sir Isaac Newton observes, in his third law of motion, that "reaction is always contrary and equal to action, or, that the mutual actions of two bodies are always equal, and directed contrary ways." He proceeds, "if any body draws or presses another, it is itself as much drawn or pressed. If any one presses a stone with his finger, his finger is also pressed by the stone. If a horse is drawing a weight tied to a rope, the horse is also equally drawn backwards towards the weight; for the rope, being distended throughout its length, will, in the same endeavour to contract, urge the horse towards the weight, and the weight towards the horse, and will impede the progress of the one as much as it promotes the advance of the other." Now, although Newton has always applied this law in the most unexceptionable manner, yet it must be confessed that the illustrations here quoted are clothed in such language as to have too much the appearance of paradox. When we say that a thing presses another, we commonly mean, that the thing pressing has a tendency to move forwards into the place of the thing pressed: but the stone would not sensibly advance into the place of the finger, if it were removed; and in the same manner we understand, that a thing pulling another has a tendency to recede further from the thing pulled, and to draw this after it: but it is obvious that the weight, which the horse is drawing, would not return towards its first situation, with the horse in its train, although the exertion of the horse should entirely cease; in these senses, therefore, we cannot say, that the stone presses, or that the weight pulls; and we have no reason to offend the natural prejudices of a beginner, by introducing paradoxical expressions without necessity. Yet it is true in both cases, that if all friction, and all connexion with the surrounding bodies, could be instantaneously destroyed, the point of the finger and the stone would recede from each other, and the horse and the weight would approach each other, with equal quantities of motion. And this is what we mean by the reciprocality of forces, or the equality of action and reaction.

304. THEOREM. "285" If two gravitating bodies be suspended at constant distances from each other and from a given point, they will be at rest when their centre of inertia is in the vertical line passing through the point of suspension: and the equilibrium will be stable when the centre of inertia would ascend in quitting the vertical line, tottering when it would descend, and neutral when it cannot quit it.

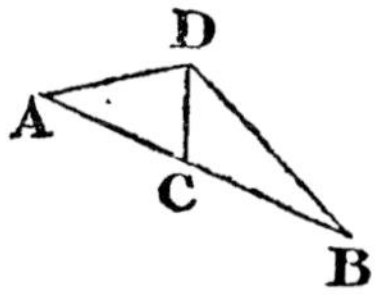

Suppose the bodies A and B, of which C is the centre of inertia, to be suspended from D by the threads AD, BD, and to be retained at the distance AB by the rod AB, and let C be in the vertical line DC. Let the force of gravity be represented by DC, then AD will represent the action of the thread, and AC the pressure exerted by A on any obstacle at C (241); and in the same manner BC will represent the pressure of B in the direction BC, supposing the weights A and B equal, and each represented by DC; but since they are unequal, the ratio of their masses must be compounded with that of the relative forces, and A.AC will represent the actual force of A, and B.BC that of B; but these products, by the supposition, are equal, since A : B = BC : AC (298);

therefore the pressures are equal, and the bodies will remain in equilibrium. If now the centre of inertia ascended towards either weight, as A, the segment AC, which determines the action of A, would be increased, and BC lessened; therefore the weight of A would prevail, and the centre would return to the vertical line. But, supposing C above D, the rod and threads must change places, and the same demonstration will hold good; and since in this case the weights pull against each other, the prevalence of A, if the centre of inertia descended towards its place, would draw it still further from the vertical line, and the equilibrium would be lost.

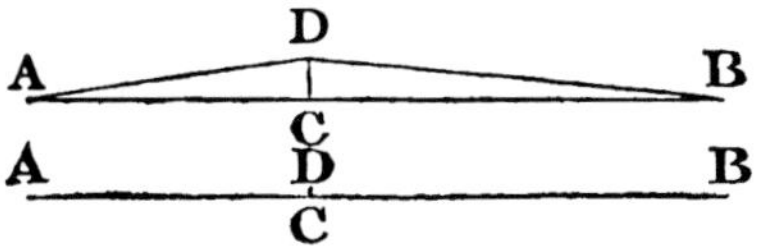

Now the magnitude of the distance of C above or below D is of uo consequence to the existence of the equilibrium; therefore when that distance vanishes, and the thread and rod are united into one inflexible right line or lever, those points will coincide, and there will still be an equilibrium; which may properly be termed neutral, since no change of the position of the bodies will create a tendency either to return to their places, or to proceed further from them. But the case of an inflexible right line is perfectly out of the reach of experiment, since the strength, necessary for the inflexibility of a mathematical line, becomes infinite, and that, in an infinitely small quantity of matter.

SCHOLIUM. The demonstrations of the fundamental property of the lever have been very various. Archimedes himself has given us two. Huygens, Newton, Maclaurin, Dr. Hamilton, and Mr. Vince, have elucidated the same subject by different methods of considering it. The demonstration of Archimedes, as improved by Mr. Vince, is ingenious and elegant; but it is neither so general and natural as one of Dr. Hamilton's, which is here adopted, nor so simple and convincing as Maclaurin's, which it may also be worth our while to notice. Supposing two equal weights, of an ounce each, to be fixed at the ends of the equal arms of a lever; in this case it is obvious that there will be an equilibrium, since there is no reason why either weight should preponderate. It is also evident, that the fulcrum supports the whole weight of two ounces, neglecting that of the lever; consequently we may substitute for the fulcrum a force equivalent to two ounces, drawing the lever upwards; and instead of one of the weights, we may place the end of the lever under a firm obstacle, and this equilibrium will still remain, the lever being now of the description which is called the second kind, the fixed point being at one end. Here, therefore, the weight remaining at the other end of the lever counterbalances a force of two ounces, acting at half the distance from the new fulcrum; and we may substitute for this force a weight of two ounces, acting at an equal distance on the other side of that fulcrum, supposing the lever to be sufficiently lengthened; and there will still be an equilibrium. In this case the fulcrum will sustain a weight of three ounces; and we may substitute for it a force of three ounces, acting upwards, and proceed as before. In a similar manner the demonstration may be extended to any commensurable proportion of the arms; and it is easy to show that the same law must be true of all ratios whatever, even if they happen to be incommensurable (120, Sch.); the forces remaining always in equilibrium, when they are to each other inversely as the distances at which they are applied. Lagrange, in his Mécanique Analytique, has entered very fully and clearly into the history of this proposition.

305. THEOREM. If a system of bodies be in equilibrium, the sum of the products of the forces, acting on the several bodies, into the infinitely small variations of their places, in the directions of the forces, the variations being so taken as to be subjected to the conditions of the system, must be equal to nothing. Or, if p be the force acting on each body, and δf the variation of the place of the body in its direction, $0 = \Sigma p \delta f$; which is the Law of virtual velo cities.

Let us first suppose two heavy bodies, m and m', fixed to the extremities of a horizontal line, supposed to be inflexible and without weight, being at liberty to turn round a fixed point within its length. In order to conceive the action of these bodies on each other when they are in equilibrium, we must suppose the right line to be infinitely little bent at the fixed point, so as to be formed of two right lines, making at that point an angle which differs but infinitely little from two right angles; and this difference we may call ω. Let f and f' be the distances of m and m' from the fixed point; if we decompose the weight of m into two parts, the one acting on the fixed point, in the direction of the bent line, the other directed towards m', this last will be $\frac{mg(f+f')}{\omega f'}$, mg being the weight of the body: [for since AB : sin ADB = DB : sin DAC, (P. 175) we have $\sin \mathrm{DAC} = \frac{\mathrm{DB}.\sin \mathrm{ADB}}{\mathrm{AB}} = \frac{f'\omega}{f+f'}$ and $\mathrm{DC} = \sin$ DAC.AD: but if DC represent the weight mg, AC or AD will be the pressure in the direction AB, which will be $mg.\frac{\mathrm{AD}}{\mathrm{DC}} = \frac{mg}{\sin \mathrm{DAC}} = mg.\frac{f+f'}{f'\omega}$.] For the same reason the action of m' on m will be $m'g.\frac{f+f'}{\omega f}$, and since these two forces must be equal, in the case of equilibrium, we shall have $mf = m'f'$, which is the well known law of the action of a lever, and which explains how two forces, acting in a parallel direction, may cause reciprocal effects, and balance each other [that is, by calling into action a third force equal to their sum, and acting in a contrary direction].

We may next consider the equilibrium of a system of points, m, m', m''..., actuated by any number of forces, and reacting on each other. Let f be the distance of m from m', f' that of m from m'', and f'' the distance of m' from m''; let p be the reciprocal action of m on m', p' that of m on m'', p'' that of m' on m''; and lastly, let $m\,S$, $m'\,S'$, $m''\,S''$..., be the forces acting on m, m', and m'', and s, s', s'', the distances of any fixed points, in the directions of those forces, from the bodies to which they belong. We may consider the point m either as being perfectly at liberty, but held in equilibrium by means of its own force mS, and the action of the other bodies m', m''..., or as subject, besides these forces, to the reaction of a surface or a curve to which it may be confined. Now, if δs be the variation of s, and $\delta_{\prime\prime} f$ that of f taken with regard to this variation only, supposing m' to be fixed; and if $\delta_{\prime} f'$ be the variation of f', supposing m'' to be fixed; R and R' being the reaction of the two surfaces, forming, by their intersection, the curve to which the motion of m is confined, and r, r' the lines perpendicular to these surfaces, we shall have, from the equation $0=\Sigma S\delta s+R\delta r+R'\delta r'$ (d) (253), $0=mS\delta s+p\delta_{,}f+p'\delta_{,}f'+\ldots+R\delta r+R'\delta r'$. In the same manner m' may be considered as a point held in equilibrium by means of the force $m'S'$, together with the actions of the bodies m, m'', . . . , and the reactions of the surfaces, which may be called R'' and R'''. If, then, the variation of s' be called $\delta s'$, that of f, taken with regard to this variation, and supposing m to be fixed, $\delta_{,,}f$, that of f'', supposing m'' fixed, $\delta_{,}f''$, and the variations in the directions of R'' and R''' be $\delta r''$ and $\delta r'''$, we shall have, for the equilibrium of m', $0=m'S'\delta s'+p\delta_{,,}f+p''\delta_{,}f''+\ldots+R''\delta r''+R'''\delta r'''$: and the rest of the points will afford similar variations, which we may add together, observing that for the total variations, $\delta f=\delta_{,}f+\delta_{,,}f$, $\delta f'=\delta_{,}f'+\delta_{,,}f'$.,..; each distance being liable to two partial variations, one at each end. We shall thus obtain

$$0=\Sigma mS\delta s+\Sigma p\delta f+\Sigma R\delta r. \quad (k)$$

In estimating the forces acting on each point m, m'..., it is obvious that we may either consider any number of different forces separately multiplied by the respective variations of their distances, or consider the whole as combined, for each body, into a single result, by the equation (a) $V\delta u=\Sigma S\delta s$ (250).

If the bodies are united at fixed distances from each other, the lines f, f', f''..., becoming constant, this condition may be expressed by making $\delta f=0$, $\delta f'=0$, $\delta f''=0$. The variations of the coordinates, comprehended in the equation (k), may be subjected to this condition, and then the forces p, expressing the reciprocal actions of the bodies, will no longer be concerned in it: we may also omit the terms $R\delta r$, $R'\delta r' \ldots$, if we limit the variations to the surfaces in which the bodies are compelled to move. The equation (k) will then become

$$0 = \Sigma m S \delta s. \qquad (l)$$

Hence it follows that, in the case of equilibrium, the sum of the products of the forces, into the elementary variations of their directions, will be equal to nothing, provided that the conditions of the connexion of the system be observed in those variations.

It may be further shown that this theorem, which is here demonstrated upon the supposition that the bodies are united at invariable distances, is true in general, for all conditions of the connexion of the different parts of the system. In order to prove this, it will be sufficient to show, that, observing these conditions, we have, in the equation (k), $0 = \Sigma p \delta f + \Sigma R \delta r$, since it will then follow that $\Sigma m S \delta s = 0$ also. But it is clear that δr, $\delta r' \ldots$ will necessarily vanish when the variations are confined to the given surfaces, and we have only to show that $\Sigma p \delta f = 0$ under the same circumstances.

Let us, therefore, conceive the system to be subjected only to the forces $p, p', p'', \ldots$, and suppose the bodies to be at liberty to move in obedience to them upon the given surfaces: these forces may be resolved into others, some of which $q, q', q'', \ldots$, will act in the directions of $f, f', f'', \ldots$; which will destroy each other [as the forces p in the former supposition, in virtue of the equality of action and reaction], without producing any motion in the curves in question; others $T, T', T'', \ldots$, will be perpendicular to the curves described; and others again will be in the directions of the tangents of those curves, and capable separately of giving motion to the system: but it is easy to see that the sum of these last forces must be equal to nothing, since the system is at liberty to move in the respective directions, [unless each point were held at rest by equal and opposite forces, so that the sums of the opposite forces must be equal for all the points, and all these forces will vanish,] producing neither pressure on the given curves, nor reaction between the bodies, so that they may be excluded from the equation, and the forces p, p', p'' must be in equilibrium without them, or in other words $-p, -p', -p'', \ldots$ together with $q, q', q'', \ldots$, must afford an equilibrium among themselves. Now, if $\delta i, \delta i', \ldots$ be the variations of the lines of direction of the forces $T, T', T'', \ldots$, we shall have, from the equation $(k), 0=\Sigma(q-p)\,\delta f+\Sigma T\delta i$; but the system being supposed to remain at rest in consequence of the forces $q, q', \ldots$, without any action upon the curves or surfaces, the equation (k) gives us also $0=\Sigma q\delta f$: consequently $0=\Sigma p\delta f-\Sigma T\delta i$. But in the conditions of the problem $\delta i=0$, $\delta i'=0, \ldots$, the variations being confined to the curves, so that we have finally $0=\Sigma p\delta f$, whence it follows, that with the conditions of the connexion of the system, $\Sigma mS\delta s=0$, as before.

[SCHOLIUM. The object of the second part of the demonstration is to prove that if $p, p', p'', \ldots$, represent not the reciprocal actions, but the total forces exerted on each body, exclusive of the pressure of the surfaces, these forces may be decomposed so as to afford forces equivalent to the reciprocal actions of the respective bodies, and that the remaining portions of the forces, as well as these reciprocal actions, will balance each other, in the case of equilibrium, according to the terms of the proposition].

306. COROLLARY The converse of this proposition is equally true, and whenever the

law of virtual velocities is observed, the system must remain in equilibrium.

For if it were otherwise, and the points m, m', ..., acquired the increments of velocity v, v', ..., while $\Sigma mS\delta s$ remained$=0$, the system would be held in equilibrium by the forces mS, $m'S'$, diminished by the forces expended on the velocities, which may be called mv, $m'v'$, ... [making the increment of time unity]; and if we call the variations in the directions of these forces δv, $\delta v'$, ..., we shall have, by the proposition, $0=\Sigma mS\delta s-\Sigma mv\delta v$: and since $\Sigma mS\delta s=0$, we have also $0=\Sigma mv\delta v$. But as the variations δv, $\delta v'$, must be subject to the conditions of the system, we may suppose them equal to vdt, or to v, and we have then $0=\Sigma mv^2$, which can only happen when $v=0$, $v'=0$, since all squares are positive: it follows, therefore, that the system must remain at rest in consequence of the forces mS, $m'S'$, ..., alone.

SCHOLIUM. The conditions of the connexion of the different parts of a system with each other may always be reduced to equations between the coordinates of the different bodies concerned. Suppose these equations to be $u=0$, $u'=0$, $u''=0$, we may always add to the equation $0=\Sigma mS\delta s$ (l) the quantity $\Sigma\lambda\delta u$, the functions $\lambda\delta u$, $\lambda'\delta u'$,... of which it is the sum, being dependent on the coordinates, [and of such a nature as to substitute an expression derived from them for the variations of the perpendiculars to the surfaces and for those of the distances of the bodies (245, Sch. 3)]; the equation will then become $0=\Sigma mS\delta s+\Sigma\lambda\delta u$. In this case the variations of all the coordinates will be arbitrary, and their coefficients may be separately made equal to nothing, which will give as many different equations for the determination of λ and λ'. If we compare this equation with the equation (k), we shall have $\Sigma\lambda\delta u = \Sigma p\delta f + \Sigma R\delta r$; whence it will be easy to infer the reciprocal actions of the bodies $m, m', \ldots$, as well as the pressures $-R, -R', \ldots$, which they exert on the surfaces to which they are confined.

§ 15. *Conditions of equilibrium for a system, of which all the points are united in an invariable manner. Centre of gravity: mode of determining its position with respect to three planes or three given points.* P. 42.

307. THEOREM. The forces acting on any system of bodies in equilibrium being referred to three orthogonal directions, the sum of all the forces acting in each direction must vanish, as well as the sum of the rotatory pressures with respect to axes in each of the three directions.

If all the bodies of a given system be invariably united to each other, its position will be determined by that of any three points belonging to it, which are not in a right line: now the position of each of these points depends on three coordinates, so that nine different distances are comprehended in their equations: but since the three distances of the points are given, they reduce the number of independent quantities to six, which will afford as many arbitrary variations: and by supposing the coefficients of these to vanish, we shall obtain six equations, which will include all the conditions of the equilibrium

For this purpose, we may suppose x, y, z, to be the coordinates of m; x', y', z', those of m', and x'', y'', z'', those of m'', ...; we shall then have

$$f=\sqrt{\{(x'-x)^2+(y'-y)^2+(z'-z)^2\}}$$
$$f'=\sqrt{\{(x''-x)^2+(y''-y)^2+(z''-z)^2\}}$$
$$f''=\sqrt{\{(x''-x')^2+(y''-y')^2+(z''-z')^2\}}\ldots;$$

and if we suppose $\delta x=\delta x'=\delta x''=\ldots$, $\delta y=\delta y'=\delta y''=\ldots$, and $\delta z=\delta z'=\delta z''=\ldots$, we shall have $\delta f=0$, $\delta f'=0$, $\delta f''=0,\ldots$; and the distances will be invariable, according to the conditions of the system. We may then infer, from the equation $0=\Sigma mS\delta s$, (l),

$$0=\Sigma mS\frac{\delta s}{\delta x};\ 0=\Sigma mS\frac{\delta s}{\delta y};\ 0=\Sigma mS\frac{\delta s}{\delta z}. \qquad (m)$$

For since $\delta x=\delta x'=\ldots$ the quantity $\Sigma mS\delta s$, which is the sum of the partial differences with respect to x, x', ..., must be divisible by δx; and the same is true with respect to y and z. It is obvious that these equations constitute the first part of the proposition.

It will still be consistent with the conditions $\delta f=0$, $\delta f'=0,\ldots$, to suppose z, z', z'', ..., invariable, and to make $\delta x=y\delta\varpi$, $\delta x'=y'\delta\varpi,\ldots$; $\delta y=-x\delta\varpi$, $\delta y'=x'\delta\varpi$, $\delta\varpi$ being any variation at pleasure [for example, that of an angle described round an axis parallel to z]: and substituting their values in two of the equations $0=\Sigma mS\delta s$, we have, since $\Sigma mS\frac{\delta s}{\delta x}\delta x=\Sigma mS\frac{\delta s}{\delta x}y\delta\varpi$, and $\Sigma mS\frac{\delta s}{\delta y}\delta y=\Sigma mS\frac{\delta s}{\delta y}(-x\delta\varpi)$, adding these together, and dividing them by $\delta\varpi$, $0=\Sigma mS\left(y\frac{\delta s}{\delta x}-x\frac{\delta s}{\delta y}\right)$; [the third equation disappearing, because δz is supposed to vanish, as when the variation takes place in a circle described on the axis parallel to z.] For the same reasons, we may obtain similar equations for x and z, omitting y, and for y and z, omitting x, so that

$$0=\Sigma mS\left(y\frac{\delta' s}{\delta x}-x\frac{\delta' s}{\delta y}\right);\ 0=\Sigma mS\left(z\frac{\delta' s}{\delta x}-x\frac{\delta' s}{\delta z}\right);$$

$$0=\Sigma mS\left(y\frac{\delta' s}{\delta z}-z\frac{\delta' s}{\delta y}\right). \qquad (n)$$

Now the quantity $\Sigma mSy\frac{\delta' s}{\delta x}$ is the rotatory pressure of all the forces reduced to a direction parallel to x, with regard to an axis parallel to z (256, 304). In the same manner the quantity $\Sigma mSx\frac{\delta' s}{\delta y}$ is the sum of the rotatory pressures of all the forces parallel to y, tending to turn the system round the axis of z, but in a direction contrary to the former: it follows therefore from the first of the equations (n), that the whole rotatory pressure must vanish with respect to the axis parallel to z. The second and third equations indicate, in a similar manner, that the sum of the rotatory pressures is nothing with respect to axes parallel to y and to x: and these six equations complete the conditions of equilibrium expressed in the proposition.

308. Corollary. If any point in the system, invariably connected with the whole, be permanently at rest, it must be in consequence of a force equal and opposite to the result of the three forces acting in the three given directions; and the conditions of equilibrium will then be reduced to the equality of the rotatory pressures with respect to the three orthogonal axes.

Supposing the bodies m, m', m'', to be subject to the force of gravitation only, its action and direction being the same with respect to the whole system, we shall have $S=S'=S''$, ..., $\frac{\delta s}{\delta x}=\frac{\delta s}{\delta x_{,}}=\frac{\delta s}{\delta x_{,,}}=\ldots$; $\frac{\delta s}{\delta y}=\frac{\delta s}{\delta y_{,}}=\frac{\delta s}{\delta y_{,,}}=\ldots$; $\frac{\delta s}{\delta z}=\frac{\delta s}{\delta z_{,}}=\frac{\delta s}{\delta z_{,,}}$, ..., and the equation $0=\Sigma mS\left(y\frac{\delta s}{\delta x}-x\frac{\delta s}{\delta y}\right)$ (n), becomes $S\left(y\frac{\delta s}{\delta x}\Sigma my-\frac{\delta s}{\delta y}\Sigma mx\right)$, since the quantity $\frac{\delta s}{\delta x}$ is the same for all the bodies concerned, as well as the force S: and the conditions of the equations, thus transformed, may be fulfilled, by putting

$$\Sigma mx=0,\ \Sigma my=0,\ \text{and } \Sigma mz=0. \qquad (o)$$

The three forces $\Sigma mS\frac{\delta s}{\delta x}$, $\Sigma mS\frac{\delta s}{\delta y}$, and $\Sigma mS,\frac{\delta s}{\delta z}$ parallel the three axes, which are destroyed by the reaction of the fixed point, become, for a similar reason, $S\frac{\delta s}{\delta x}\Sigma m$, $S\frac{\delta s}{\delta y}\Sigma m$, and $S\frac{\delta s}{\delta z}\Sigma m$; and these forces compose a force $S\Sigma m$, which is equal to the weight of the body; since $\left(\frac{\delta s}{\delta y}\right)^2+\left(\frac{\delta s}{\delta y}\right)^2+\left(\frac{\delta s}{\delta z}\right)^2$ are always$=1$, and the resulting force is expressed by the diagonal of the parallelepiped.

SCHOLIUM 1. The origin of the coordinates, thus considered as the fixed point of the system, is very remarkable for the property of affording an equilibrium of the weight of the whole system, whenever it is simply supported, whatever the angular situation of the system may be. Hence it is called the *centre of gravity* of the system. Its place is determined by the property, that if we suppose any plane to pass through this point, the sum of the products of all the separate bodies, into their distances from this plane, is equal to nothing: for the distances must be in some given proportion to all the coordinates x, y, and z, [depending on the properties of similar triangles (117) and therefore " linear functions", not involving their squares; for example nx, $n'y$, or $n''z$: but when $\Sigma mx=0$, it is obvious that $\Sigma mnx=0$, since n is constant;] whence the property of the plane passing through the centre of gravity is evident.

In order to determine the position of the centre of gravity of any body, we may suppose X, Y, and Z to be its coordinates with respect to any given origin, x, y, and z being those of m, x', y', and z' of m', . . ., with respect to the same point. We shall then have, from the equations (o), $0=\Sigma m\,(x-X)$ [the x of those equations being supposed to begin at the centre of gravity, and therefore answering to $x-X$ here]; now $\Sigma mX=X\Sigma m$, Σm being the mass of the system; we have therefore $X=\dfrac{\Sigma mx}{\Sigma m}$; and in the same manner $Y=\dfrac{\Sigma my}{\Sigma m}$, and $Z=\dfrac{\Sigma mz}{\Sigma m}$. It is also evident that the coordinates X, Y, and Z, being thus completely determined by the magnitude and position of the separate bodies of the system, they can only belong to a single point for any one system of bodies at the same time. For the direct distance of the centre of gravity we have the equation $X^2+Y^2+Z^2=\dfrac{(\Sigma mx)^2+(\Sigma my)^2+(\Sigma mz)^2}{\Sigma m^2}$; which may be transformed into

$$X^2+Y^2+Z^2=\frac{\Sigma m(x^2+y^2+z^2)}{\Sigma m}-\frac{\Sigma mm'\left\{(x'-x)^2+(y'-y)^2+(z'-z)^2\right\}}{\Sigma m^2}$$

The finite integral being understood as comprehending all the combinations of the different bodies in pairs. [Thus for two bodies, m and m', Σm being $m+m'$, $\Sigma mx = mx + m'x'$, and $\Sigma mm' = mm'$, we have $(\Sigma mx)^2 = m^2x^2 + m'^2x'^2 + 2mm'xx' = \Sigma mx^2 . \Sigma m - mm'(x'-x)^2 = (mx^2 + m'x'^2)(m+m') - mm'(x'^2 + x^2 - 2x'x) = m^2x^2$ " $+ mm'x'^2 + m'mx^2$" $+ m'^2x'^2$ " $- mm'x'^2 - mm' x^2$" $+ 2mm'xx'$: and adding a third body, if Σmx be $mx + m'x' + m''x''$, we have $(\Sigma mx)^2 = m^2x^2 + m'^2x'^2 + m''^2x'' + 2mm' xx' + 2mm''xx'' + 2m'm''x'x'' = (mx^2 + m'x'^2 + m''x''^2)(m + m' + m'') - mm'(x'-x)^2 - mm''(x''-x)^2 - m'm''(x''-x')^2$; and a similar proof may be extended to any number of bodies.]

By this mode of computation, we may determine the distance of the centre of gravity from any fixed point, when we know the distances of the different bodies of the system from this point and from each other: and when the distance of the centre of gravity from any three points is thus found, its situation is in all respects completely ascertained.

SCHOLIUM 2. The denomination of "centre of gravity" has [sometimes] been extended to any system of bodies with or without weight, as determined by the three coordinates X, Y, and Z, thus computed [,but it is more correct to employ, in this sense, the term "centre of inertia" (298)].

§ 16. *Conditions of the equilibrium of a solid of any figure whatever.* P. 46.

309. THEOREM. For a single solid body, whatever its figure may be, we have the same conditions of equilibrium as for a system of bodies, substituting fluxions nnd fluents for single bodies and finite integrals: that is

$0 = \int P\,dm$, $0 = \int Q\,dm$, $0 = \int R\,dm$; $0 = \int (Py - Qx)\,dm$, $0 = \int (Pz - Rx)\,dm$, $0 = \int (Ry - Qz)\,dm$.

In fact we have only to conceive the solid as a system of an infinite number of points, united in an invariable manner. If, then, we suppose Δm to be an infinitely small point or atom of the body, of which x, y, and z are the orthogonal coordinates, and P, Q, R, the forces acting on the particle in the directions of x, y, and z, the equations (m) and (n) will only require the substitution of P for $S \frac{\delta' s}{\delta x}$, Q for $S \frac{\delta' s}{\delta y}$, and R for $S \frac{\delta' s}{\delta z}$, to which they are respectively equal, and we shall have $\Sigma P \Delta m = 0$, . . ., and consequently $\int P\,dm = 0$; [for since the fluxions are always in a constant ratio to the evanescent increments, whenever $\Sigma P \Delta m = 0$, we may make $\int P\,dm = 0$ also; and in the same manner the substitutions in all the six equations may be shown to be admissible: the character of integration $\int$ being understood as extending to the whole solid, in all its dimensions.

SCHOLIUM. If the body is only at liberty to move round a given point, at which the coordinates begin, the latter three equations are sufficient to determine the conditions of its equilibrium.

CHAPTER IV.

OF THE EQUILIBRIUM OF FLUIDS.

---

§ 17. [*Introduction*]. *General equations of this equilibrium. Application to the equilibrium of a homogeneous fluid, of which the surface is at liberty, and which covers a solid nucleus of any figure.* P. 47.

[310. Definition "367." A fluid is a collection of particles considered as infinitely small spheres, moving freely on each other without friction.

311. Theorem. "368." The surface of a gravitating fluid, at rest, is horizontal.

If the surface were in the least inclined to the horizon, the particles found in it could not remain in equilibrium, but would descend, in virtue of their power of perfect freedom of motion, until the level were restored. But it is more satisfactory to consider the immediate action of the particles concerned: and we may suppose two minute straight tubes, differently inclined to the horizon, and joined at the bottom by a curved portion, to be filled with evanescent spherules: then the relative force of gravity is inversely as the length, when the height is the same

(283), and the number of particles is directly as the length: consequently the absolute pressure will be equal, and there will be an equilibrium; and if the fluid in either arm be higher, it will preponderate. The pressure on the tube at any part is only the effect of the particle immediately in contact with it, and acts in the direction perpendicular to the tube, therefore if another similar row of particles in equilibrium were placed on the first, this pressure, acting in the same direction, would not disturb the equilibrium of the particles among themselves, however they might be situated with respect to the first. And conceiving any fluid to be divided into an infinite number of tubes, bent or straight, in which the particles form a continuous series, there can be no force to preserve the equilibrium in each of them, unless the height of each portion be equal.

312. THEOREM. "370." The pressure of a fluid on every particle of the vessel containing it, or of any other surface, real or imaginary, in contact with it, is equal to the weight of a column of the fluid, of which the base is equal to that particle, and the height to its depth below the surface of the fluid.

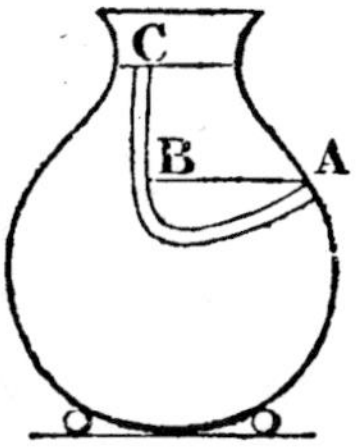

Imagine an equable tube to be so bent, that one of its arms may be vertical, and the other perpendicular to the given surface: then drawing a horizontal line AB, the fluid in the portion of the tube AB will remain in equilibrium, and will only transmit the pressure of BC to the surface at A, and this will be true whatever be the position of the imaginary tube; and since some particles of the fluid may be so arranged as to be no more disturbed in their initial tendency to motion than the fluid in such a tube would be, the equilibrium can never be permanent, unless the pressures be such as are here assigned.

SCHOLIUM 1. If therefore any portion of the superior part of a fluid be replaced by a part of the vessel, the pressure against this from below will be the same which before supported the weight of the fluid removed, and, every part remaining in equilibrium, the pressure on the bottom will be the same as if the horizontal section of the vessel were every where of equal dimensions. In this manner the smallest given quantity of a fluid may be made to produce a pressure capable of sustaining a weight of any magnitude, either by diminishing the diameter of the column and increasing its height, or by increasing the surface which supports the weight: a property which has been called the hydrostatic paradox, and which is the foundation of the construction of Bramah's powerful presses.

SCHOLIUM 2. These properties may be still further illustrated by imagining a vessel to be made of ice, and to be immersed in a larger reservoir of water, and then thawed: in this case the water will make a part of the general contents of the reservoir, and consequently will remain at rest, if its surfaces are level with that of the reservoir: and it is obvious that the vessel has acquired no new power of supporting the pressure from being thawed: consequently the water will stand at the same height in every part of the vessel of ice as if it had remained water; exerting the same pressure on the sides of the vessel, as if it had to react against the weight of a fluid column imme-

tubes or branches belonging to it, the water will stand at the same height in all.

313. LEMMA A. The partial variations $\delta_y(\delta_x u)$ and $\delta_x(\delta_y u)$ are equal.

For, when the variation of u is taken with respect to x, the quantities depending on y remain unaltered, and the process leads to the same result, when the variation is afterwards taken with respect to y, as if it had been inverted. For example, if $u=x^m y^n$, $\delta_x u=mx^{m-1}\,\delta x.y^n$, and $\delta_y(\delta_x u)=mnx^{m-1}y^{n-1}\delta x\delta y=\delta_x(\delta_y u)$: again, if $u=ax^2+by^2$, $\delta_x u=2ax\delta x$, $\delta_y(\delta_x u)=0$; $\delta_y u=2by\delta y$, $\delta_x(\delta_y u)=0$: and if $u=x^m y^n z^p$, the same results will be obtained, for the variations with respect to x and y, as if z were a constant quantity.

314. LEMMA B. If $\delta u=N\delta x+M\delta y+L\delta z$, we have $\frac{\delta' N}{\delta y}=\frac{\delta' M}{\delta x}$, $\frac{\delta' N}{\delta z}=\frac{\delta' L}{\delta x}$, and $\frac{\delta' M}{\delta z}=\frac{\delta' L}{\delta y}$.

For $\delta_x u=N\delta x$, and $\delta_y u=M\delta y$, and $\delta_y(\delta_x u)=\delta_y(N\delta x)=\delta_x(\delta_y u)$ (313) $=\delta_x(M\delta y)-\frac{\delta' M}{\delta x}\,\delta x\delta y=\frac{\delta' N}{\delta y}\,\delta y\,\delta x$; consequently $\frac{\delta' M}{\delta x}=\frac{\delta' N}{\delta y}$: and in the same manner the other equations are obtained, by comparing the variations in pairs.

315. COROLLARY. An exact variation containing two or more variable quantities, must always be conformable to the condition of this proposition.

SCHOLIUM. This condition of integrability was first laid down by Nicolas Bernoulli, in 1728.]

316. THEOREM. The surfaces, dividing the different strata of a fluid of different densities, must be perpendicular to the results of the forces acting on them.

If we wished to determine the laws of the equilibrium and motion of the separate particles of fluids, it would be necessary that we should ascertain their precise form, which is totally unknown to us: but in fact we have only occasion to obtain such laws as are applicable to fluids considered as masses, or assemblages of particles, and for this purpose the knowledge of the figures of the particles is superfluous. Whatever these figures may be, and whatever may be the affections of the separate particles as depending on them, all fluids, taken as aggregates, must afford the same phenomena in their equilibrium and their motions, so that the observation of the phenomena can lead us to no conclusions respecting the forms of the particles. These general phenomena depend on the perfect mobility of the particles, which may be displaced by the slightest force: and it is by this mobility that fluids are distinguished from solids. It is the necessary consequence of this mobility, that every particle of a fluid must be held in equilibrium by means of the forces acting on it, together with the pressures to which it is subjected, and which are transmitted by the surrounding particles. We must now examine the equations which may be deduced from this constitution of a fluid.

We may, therefore, consider a system of elementary particles, forming an infinitely small rectangular parallelepiped; and we may suppose the coordinates, x, y, and z, to belong to the angle nearest to their common origin. Let the infinitely small differences Δx, Δy, Δz, be the sides of the parallelepiped: let p be the mean pressure on the different points of the surface $\Delta y \Delta z$, which is perpendicular to x, and p' the same quantity belonging to its opposite surface: the parallelepiped will be urged in the direction of x by a force equal fo $(p-p') \Delta y \Delta z$. Now $(p'-p)$ is the difference of p, taken on the supposition that x alone is variable; for though p' is supposed to act in the direction contrary to that of p, yet the pressure, that a point of a fluid undergoes, being the same in all directions, we may consider $p'-p$ as the difference of the two forces, acting in the same direction, at an infinitely small distance from each other: so that we have $p'-p=\Delta_x p$, and $(p-p')\Delta y \Delta z$ $=-\Delta_x p \Delta y \Delta z - \frac{\Delta' p}{\Delta x} \Delta x \Delta y \Delta z$. Let P, Q, and R be the three accelerating forces which act on the fluid particles, independently of their connexions, in directions parallel to x, y, and z: if we call the density of the parallelepiped ϱ, its mass will be $\varrho \Delta x \Delta y \Delta z$, and the product of the force P by this mass will represent the whole motive force derived from it; consequently the whole force, acting in the direction of x, will be $\left(\varrho P - \frac{\Delta' p}{\Delta x}\right) \Delta x \Delta y \Delta z$. For similar reasons, the elementary system will be solicited, in directions parallel to y and z, by the forces $\left(\varrho Q - \frac{\Delta' p}{\Delta y}\right) \Delta x \Delta y \Delta z$, and $\left(\varrho R - \frac{\Delta' p}{\Delta z}\right) \Delta x \Delta y \Delta z$. We shall therefore have, for the conditions of equilibrium (*b*) (251)

$$0 = \left(\varrho P - \frac{\Delta' p}{\Delta x}\right) \delta x + \left(\varrho Q - \frac{\Delta' p}{\Delta y}\right) \delta y + \left(\varrho R - \frac{\Delta' p}{\Delta z}\right) \delta z; \text{ or}$$

$\delta p = \varrho(P\delta x + Q\delta y + R\delta z)$ [: since $\frac{\delta' p}{\delta x} = \frac{\Delta' p}{\Delta x}$, and $\delta p = \frac{\delta' p}{\delta x} \delta x$ +

$\frac{\delta p}{\delta y}\delta y+\frac{\delta p}{\delta z}\delta z$]. Now p being a possible and consistent quantity, its variations, and consequently its fluxion, must be exact (315): we have therefore (314), $\frac{d'(\varrho P)}{dy}=\frac{d'(\varrho Q)}{dx}$; $\frac{d'(\varrho P)}{dz}=\frac{d'(\varrho R)}{dx}$; $\frac{d'(\varrho Q)}{dz}=\frac{d'(\varrho R)}{dy}$; consequently [, since $d'(\varrho P)=\varrho d'P+Pd'\varrho$, ..., we have, by combining the three last equations, multiplied by P, Q, and R, $0=P\left(\varrho\frac{d'Q}{dz}+Q\frac{d'\varrho}{dz}\right.$ $\varrho\frac{d'R}{dy}-R\left.\frac{d'\varrho}{dy}\right)-Q\left(\varrho\frac{d'P}{dz}+P\frac{d'\varrho}{dz}-\varrho\frac{d'R}{dx}-R\frac{d'\varrho}{dx}\right)+R\left(\varrho\frac{d'P}{dy}+P\frac{d'\varrho}{dy}-\varrho\frac{d'Q}{dx}-Q\frac{d'\varrho}{dx}\right)$: and since the terms containing $d'\varrho$ obviously destroy each other, we obtain, from those which are multiplied by ϱ, the equation]

$$0=P\frac{d'Q}{dz}-Q\frac{d'P}{dz}+R\frac{d'P}{dy}-P\frac{d'R}{dy}+Q\frac{d'R}{dx}-R\frac{d'Q}{dx}.$$

And this equation expresses the relation between the forces P, Q, and R, which is required in order that the equilibrium may be possible.

If the surface of the fluid, or any part of the surface, is at liberty, the value of p must be evanescent at that point, since there is no pressure that could be measured by p; we have therefore for the direction of the surface $\delta p=0$, the variations δx, δy, δz, being so related as to belong to it. The independent forces must therefore balance each other with respect to any motion in the direction of the surface, and $0=P\delta x+Q\delta y+R\delta z$: but this can only happen when the result of these forces is perpendicular to the surface, the general eqnation $\Sigma S\delta s+$ "R" $\delta r=0$ (c) (252) becoming here $P\delta x+Q\delta y+R\delta z+$ "R" $\delta r=0$, and $P\delta x+Q\delta y+R\delta z$ $=-$ "R" δr, indicating a result in the direction of r, the perpendicular to the surface.

Supposing the variation $P\delta x+Q\delta y+R\delta z$ to be exact, which must be the case whenever it arises from any attractive forces that can be combined in nature, and calling this variation δf, we shall have $\delta p=\varrho\delta f$: consequently ϱ must depend on p and f; and since the fluent of this equation gives us f in terms of p, we shall have p determinable from ϱ, so that the pressure p must be the same wherever the density ϱ is the same, and dp or Δp must vanish with respect to those strata of the fluid, in the direction of which the density is constant: we have therefore, with regard to these surfaces, $0=P\delta x+Q\delta y+R\delta z$, consequently the result of the forces, acting at any such surface, must be perpendicular to it: and such strata are called level strata [,at least with respect to the force of gravity]. This condition is always satisfied throughout the fluid, when it is homogeneous and incompressible, since then the strata, to which the result is perpendicular, are always of the same density.

For the equilibrium of a homogeneous fluid, of which the upper surface is at liberty, it is necessary, and it is sufficient, first that the quantity $P\delta x+Q\delta y+R\delta z$ be an exact variation, and secondly, that the result of these forces, at the exterior surface, be directed perpendicularly towards that surface.

CHAPTER V

GENERAL FRINCIFLES OF THE MOTION OF A SYSTEM OF BODIES.

---

§ 18. *General equation of the motion of a system.* **P. 50.**

317. Theorem. If we have any number of bodies, m, m', m'' ..., the places of which are denoted by the coordinates x, y, z, x', y', z', , and which are subject to the forces P, Q, R, P', Q', R', ..., respectively, we shall have, supposing dt constant, $0=\Sigma\left\{m\delta x\left(\frac{ddx}{dt^2}-P\right)+m\delta y\left(\frac{ddy}{dt^2}-Q\right)+m\delta z\left(\frac{ddz}{dt^2}-R\right)\right\}$; the characteristic Σ implying the sum of all the quantities of the same form, belonging to each of the bodies respectively.

The laws of the motion of a point have been compared with those of its equilibrium, by [conceiving the motion created or destroyed in each instant to form an equilibrium with the force or forces producing the change, or, in other words, by] decomposing its momentary motion into two parts, one of which it retains in the next instant, while the other is destroyed by the effect of the forces to which it is subjected. The same method may be employed in order to determine the motion of a system of bodies, m, m', m'',

... Thus, let mP, mQ, mR, be the motive forces which impel the body m in directions parallel to the orthogonal coordinates x, y, z; let $m'P'$, $m'Q$, $m'R'$, be the forces belonging to m'; and let the time be t. The momentum of m, reduced to the respective directions, will be $m\frac{dx}{dt}$, $m\frac{dy}{dt}$, and $m\frac{dz}{dt}$: to this the force P, so far as it is not otherwise compensated, will add a momentum, which may be expressed by m ' P ' Δt, and which is obviously equal to $m\,\Delta\frac{dx}{dt}$, since in the time Δt the momentum becomes $m\frac{dx}{dt}+m\Delta\frac{dx}{dt}$; and m ' P ' $dt=md\frac{dx}{dt}$: consequently the uncompensated force in the direction of x will be $m\,P\,dt - \frac{ddx}{dt}$ [or more properly $m\,P-\frac{ddx}{dt^2}$; for it is unnecessary to combine the idea of time with that of force in estimating its comparative magnitude]; and the same may be shown with respect to the other forces concerned. We have, therefore, from the principle of virtual velocities, that is $0=\Sigma mS\delta s$ (l) (305), $0=m\delta x\left(\frac{ddx}{dt^2}-P\right)+m\,\delta\,y$

$$\left(\frac{ddy}{dt^2}-Q\right)+m\delta z\left(\frac{ddz}{dt^2}-R\right)+m'\delta x'\left(\frac{ddx'}{dt^2}-P'\right)+m'\delta y'\left(\frac{ddy'}{dt^2}-Q'\right)+m'\,\delta z'\left(\frac{ddz'}{dt^2}-R'\right) \qquad (P)$$

From this general equation we may eliminate, by means of the particular conditions of the system, as many of the variations as there are of these conditions; and then by making the coefficients of the remaining variations vanish separately, we shall obtain all the equations necessary for determining the motion of the different bodies of the system.

§ 19. *Of the principle of living force. It is only true where the motions change by imperceptible degrees. Mode of estimating the alteration of the living force in the abrupt changes of the motions of a system.* P. 51.

[318. DEFINITION. The product of the mass of any body, into the square of its velocity, is called its impetus or energy.

319. THEOREM. The joint impetus of any system of bodies is equally increased or diminished by the action of any combination of forces, provided that the initial and final places of the system are the same, whatever may have been the intermediate paths described by the different bodies.]

We may derive from the equation (*P*) of the last proposition several general principles of motion, which it will be proper to examine in detail. The variations δx, δy, δz, $\delta x'$, . . ., will obviously be subjected to all the conditions of the connexion of the system, if they be supposed proportional to the fluxions dx, dy, dz, dx', . . . , which represent the actual motion; we may, therefore, make this substitution in the equation (*P*) and it will then become $0 = \Sigma \left\{ m dx \left(\frac{ddx}{dt^2} - P \right) + m dy \left(\frac{ddy}{dt^2} - Q \right) + m\, dz \left(\frac{ddz}{dt^2} - R \right) \right\}$; whence we have $0 = \Sigma m \frac{dx^2 + dy^2 + dz^2}{2dt^2} - \Sigma \int m\, (P dx + Q dy + R dz)$ and $\Sigma m \frac{dx^2 + dy^2 + dz^2}{dt^2} = C + 2 \Sigma m \int (P dx + Q dy + R dz)$, C being a constant quantity. (*Q*)

If the forces P, Q, R, are the results of attractions directed to fixed points, and of attractions of the bodies to each other, the quantity $\Sigma m\,(P\mathrm{d}x+Q\mathrm{d}y+R\mathrm{d}z)$ is an exact fluxion. For the part which depends on the attraction to fixed points is an exact fluxion, because the forces in the three directions are obtained by the resolution of single forces acting in given lines, each of which must afford a true or exact variation when resolved, so that their sum, however combined, must still be an exact variation. And with respect to the parts depending on the mutual attractions of the bodies of the system, if we call the distance of m from m', f, and the attraction of m' for m, $m'F$, the part of $m\,(P\mathrm{d}x+Q\mathrm{d}y+R\mathrm{d}z)$ that relates to this attraction will be $mm'F\mathrm{d}'f$, the fluxion $\mathrm{d}'f$ relating to the change of the coordinates of m only; but since reaction is alway equal and contrary to action, the part of $m'\,(P'\mathrm{d}x'+Q'\mathrm{d}y'+R'\mathrm{d}z')$ depending on the action of m or m' is equal to $-mm'F\mathrm{d}'f$, supposing $\mathrm{d}'f$ to relate to the change of the coordinates of m' only: consequently the whole effect of the reciprocal action of m and m' is represented by the product $-mm'F\mathrm{d}f$, $\mathrm{d}f$ being the total variation of f: and $F\mathrm{d}f$ is an exact fluxion whenever F is a function of f, or when the attraction is dependent on the distance, as we suppose to be the case with respect to attractive forces in general. Consequently the sum of all such actions must be expressed by an exact fluxion, whenever the forces concerned depend on the attraction of the bodies of the system for each other, or for any fixed points. If then we suppose this fluxion to be $\mathrm{d}\varphi$, and if we call the velocity of m, v, that of m', v', . . . , we shall have

$$\Sigma mv^2 = c + 2\varphi \qquad (R)$$

This equation is analogous to the simpler equation $v^2 = c + 2\varphi(g)$ (264), and expresses algebraically the law of living forces [or energies. Dr. Wollaston has given to this function of a moving body the very appropriate name of impetus; a short time before, the term energy had been proposed, and either or both of these words may be employed with advantage: energy is perhaps more likely to be misconstrued in a moral sense, but it is more convenient when a plural is wanted].

320. SCHOLIUM 1. This principle is, however, only applicable when the motions of the bodies concerned are changed by imperceptible degrees.

For if the motions undergo abrupt changes, the impetus is diminished in a manner which may be thus determined. We may employ, in this case, the character Δ (317) as denoting a finite variation of the velocity, and we shall have for the part of the force P not accelerating m, $m\left(P - \Delta\frac{dx}{dt}\right)$, and the equation ($P$) will become $0 = \Sigma m$

$$\left\{\frac{\delta x}{dt}\Delta\frac{dx}{dt} + \frac{\delta y}{dt}\Delta\frac{dy}{dt} + \frac{\delta z}{dt}\Delta\frac{dz}{dt}\right\} - \Sigma m(P\delta x + Q\delta y + R\delta z).$$

In this equation we may substitute for δx, $dx + \Delta dx$, for δy, $dy + \Delta dy$, and for δz, $dz + \Delta dz$, since it is perfectly consistent with the conditions of the system, to make the arbitrary variations such as actually happen, the variations preserving the proportions of these fluxions though they remain infinitely small. The equation will then become

$$0 = \Sigma m\left\{\left(\frac{dx}{dt} + \Delta\frac{dx}{dt}\right)\Delta\frac{dx}{dt} + \left(\frac{dy}{dt} + \Delta\frac{dy}{dt}\right)\Delta\frac{dy}{dt} + \left(\frac{dz}{dt} + \Delta\right.\right.$$

$\left.\frac{dz}{dt}\right) \Delta \frac{dz}{dt} \Big\} - \Sigma m \Big\{ P(dx + \Delta dx) + Q(dy + \Delta dy) + R(dz +$

$\Delta dz) \Big\}$.

The sum or integral of this expression, considered with regard to the finite differences, may be denoted by $\Sigma_{,}$, the sum of the similar expressions, derived from the separate bodies of the system, being still distinguished by Σ. Now $\Sigma_{,} m P(dx + \Delta dx)$ is evidently equal to $\int mPdx$: and we have $0 = \Sigma m \frac{dx^2 + dy^2 + dz^2}{dt^2} + \Sigma_{,} \Sigma m \Big\{ \left(\Delta \frac{dx}{dt}\right)^2 + \left(\Delta \frac{dy}{dt}\right)^2$

$+ \left(\Delta \frac{dz}{dt}\right)^2 \Big\} - 2\Sigma \int m(Pdx + Qdy + Rdz)$: [for, if Δu be the finite difference of u, $\Delta(u^2) = (u + \Delta u)^2 - u^2 = 2u\Delta u + \Delta u^2$, and $\Delta(u^2) + \Delta u^2 = 2u\Delta u + 2\Delta u^2$, consequently $u^2 + \Sigma_{,} \Delta u^2 = 2\Sigma_{,}(u\Delta u + \Delta u^2)$, and, in the present case $dx^2 + \Sigma_{,}(\Delta dx)^2 = 2\Sigma_{,}(dx + \Delta dx)\Delta dx$: and with respect to the integral of $mP(dx + \Delta dx)$ it is evident that the expression being only of one dimension, the product $mPdx$ will remain unaltered, whether it be supposed to vary by finite or by infinitely small differences, provided that the same value of P be always attributed to the same value of x, so that the difference of the values of $\int mPdx$ for any two values of x will be equal to the difference of the values of $\Sigma_{,} m(Pdx + \Delta dx)$; that fluent may, therefore, be considered as the integral represented by the character $\Sigma_{,}$.] If, therefore, we denote by v, v', v'', .., the velocities of m, m', m'', . . ., we shall have $\Sigma mv^2 = C - \Sigma_{,} \Sigma m \Big\{ \left(\Delta \frac{dx}{dt}\right)^2 + \left(\Delta \frac{dy}{dt}\right)^2 +$

$\left(\Delta \frac{dz}{dt}\right)^2 + 2\Sigma \int m(Pdx + Qdy + Rdx)$. Now the quantity under the sign $\Sigma_{,}$ being necessarily positive, we see that the impetus of the system is diminished by the mutual action of the bodies concerned, whenever, in the course of the motion, any of the variations $\Delta\frac{dx}{dt}, \Delta\frac{dy}{dt}, \ldots$, are finite: and the preceding equation affords a very easy method of determining this diminution.

At every abrupt variation of the motion of the system, we may conceive the velocity of m to be divided into two portions, the one v, which it retains, the other V, destroyed by the actions of the other bodies [, for, even if the velocity be increased, we have only to suppose that a negative portion of it has been destroyed, in order to justify this expression of Dalembert, which is so often used by Laplace]: now the velocity of m being $\sqrt{\frac{dx^2+dy^2+dz^2}{dt^2}}$, before this decomposition, and afterwards $\sqrt{\frac{(dx+\Delta dx)^2+(dy+\Delta dy)^2+(dz+\Delta dz)^2}{dt^2}}$, it is easy to see that $V^2 = \left(\Delta\frac{dx}{dt}\right)^2 + \left(\Delta\frac{dy}{dt}\right)^2 + \left(\Delta\frac{dz}{dt}\right)^2$ [since the diagonal of a parallelepiped, of which the square is equal to the sum of the squares of its sides, may be divided into two portions of which the squares must be respectively equal to the sums of the squares of the parts of those sides: in fact $+V$ must be simply equal to the square root of this quantity; since the sum of the squares of the finite differences of the velocities, in the three orthogonal directions, must necessarily give the square of the difference of the actual velocity:] and the preceding equation may be expressed in this form, $\Sigma mv^2 = C - \Sigma_{,} \Sigma m V^2 + 2\Sigma\int m(Pdx + Qdy + Rdz)$.

[SCHOLIUM 2. It is very doubtful whether an abrupt change of velocity ever takes place in nature, though the loss of force by friction, and by the change of the form of aggregation may sometimes produce almost the same phenomena: but the investigation of such cases scarcely requires to be conducted in a very general manner, or in great detail. It may be of more utility to insert here a geometrical demonstration, subservient to the illustration of the principle of the preservation of impetus or living force, though it might, without impropriety, have been introduced somewhat earlier, since it relates to a single moving point only.

321. COROLLARY. "245." Two bodies being attracted towards a given centre, with equal forces at equal distances, if their velocities be once equal at equal distances, they will always remain equal at equal distances, whatever their direction may be.

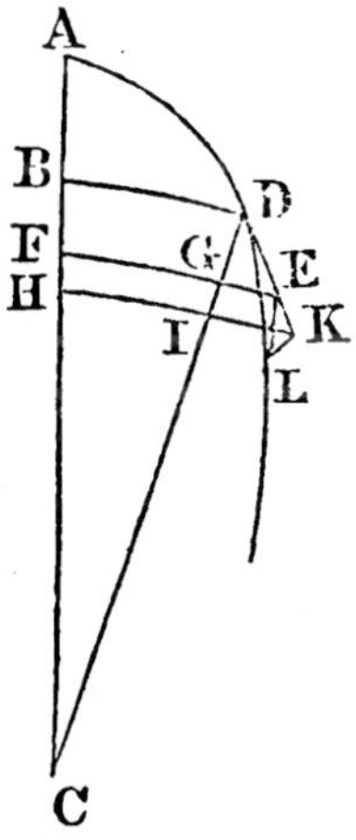

Let one of the bodies descend in the right line AB towards C, and let the other describe the curve AD, and let the velocities at B and D be equal; let DE, in the tangent of AD, be the space which would be described in an evanescent portion of time by the velocity at D, FG the arc of a circle of which the centre is C, and GE its tangent; and while BF would be described by the velocity at B, let FH be added to it by the attractive force; draw the arc HI and its tangent IK, and EL parallel to DC, and KL perpendicular to DK, then DG : DE :: GI : EK :: EK : EL, by similar triangles; therefore GI is to EL in the duplicate ratio of DG to DE, or as the square of DG to the square of **DE** (194): consequently **EL** will be the space described by the attractive force, while **DE** would have been described by the velocity at **D**; for the force may be considered as uniform during the evanescent increments, and the spaces described by such a force are as the squares of the times: hence the joint result will be **DL**, which is ultimately equal to **DK**, and the whole velocity will be increased in the ratio of **DK** to **DE**, or **DI** to **DG**, or **BH** to **BF**; consequently, since **H**, **I**, and **K** are ultimately equidistant from **C**, the velocities in **AB** and **AD**, being always equally increased at equal distances, will therefore always remain equal at equal distances.

SCHOLIUM 3. We may observe that every known force in nature acts in conformity with this condition, and operates always equally at equal distances from its origin: as Laplace has himself remarked in this article, asserting that F is always a function of f: and if the case were otherwise, with respect to gravitation or magnetism, for example, we might easily obtain a source of perpetual motion, by causing a body to describe, in its descent, a path in which the force is greater, and to ascend by one in which it is smaller at the same distance. There is indeed a supposed exception, in the hypothesis, which Laplace has elsewhere adopted, respecting the extraordinary refraction of crystallized bodies: but the exception is by far too paradoxical, to be admitted by any person, not previously determined to deduce the motions of light from the laws of attractive and repulsive forces: for here it is assumed that the force depends, not on the distance of the attracting substance, but on the direction of the motion, with which it varies perpetually. The Newtonian demonstration of the laws of ordinary refraction had the advantage, on the other hand, of simplifying their supposed cause, since it shows that the phenomena may be deduced from the operation of a constant force, acting equally upon the moving body, whatever its direction might be, and fulfilling the condition, that "l'attraction est comme une fonction de la distance, ainsi que nous le supposerons *toujours*." P. 58.]

§ 20. *Of the principle of the preservation of the motion of the centre of gravity: which is true even when the bodies exert abrupt actions on each other.* P. 54.

322. THEOREM. The centre of gravity of any system of bodies perseveres in its state of rest or uniform rectilinear motion, notwithstanding any reciprocal action between the bodies.

If we substitute, for the variations of the places of all the bodies m', m'', . . ., the variations of the place of m augmented by the difference of the variations, and make

$$\delta x' = \delta x + \delta x_{\prime}', \quad \delta y' = \delta y + \delta y_{\prime}', \quad \delta z' = \delta z + \delta z_{\prime}',$$
$$\delta x'' = \delta x + \delta x_{\prime}'', \quad \delta y'' = \delta y + \delta y_{\prime}'', \quad \delta z'' = \delta z + \delta z_{\prime}'',$$

substituting these values in the expressions for the variations of f, f', . . ., the distances of the bodies (307); it is obvious that δx, δy, δz will disappear from these expressions[; thus $\delta_x f = \frac{2(x'-x)(\delta x' - \delta x)}{f} = \frac{2(x'-x)}{f} \delta x_{\prime}'$, (307)].

Now if the system is at liberty, none of its parts being connected with any foreign bodies, the conditions, relating to their mutual connexion, depending only on their distances from each other, the variations δx, δy, δz, which relate to a quiescent point, will be independent of these conditions; whence it follows, that if we substitute these values of the variations in the equation (P) (317), we may suppose either δx, δy, or δz to subsist alone, so that its coefficients will vanish: we have thus the three equations $0=\Sigma m\left(\frac{\mathrm{dd}x}{\mathrm{d}t^2}-P\right)$, $0=\Sigma m\left(\frac{\mathrm{dd}y}{\mathrm{d}t^2}-Q\right)$, $0=\Sigma m\left(\frac{\mathrm{dd}z}{\mathrm{d}t^2}-R\right)$. Now supposing X, Y, and Z to be the three coordinates of the centre of gravity of the system, we have $X=\frac{\Sigma mx}{\Sigma m}$; $Y=\frac{\Sigma my}{\Sigma m}$; $Z=\frac{\Sigma mz}{\Sigma m}$: consequently, since $\mathrm{dd}X=\frac{\Sigma m\mathrm{dd}x}{\Sigma m}$, we have $0=\frac{\mathrm{dd}X}{\mathrm{d}t^2}-\frac{\Sigma mP}{\Sigma m}$; $0=\frac{\mathrm{dd}Y}{\mathrm{d}t^2}-\frac{\Sigma mQ}{\Sigma m}$, and $0=\frac{\mathrm{dd}Z}{\mathrm{d}t^2}-\frac{\Sigma mR}{\Sigma m}$; so that the motion of the centre of gravity of the system is the same, as if all the bodies, and all the forces acting on them, were united in it. (264).

If the system is only subjected to the mutual actions of the bodies composing it, we shall have

$$0=\Sigma mP;\ 0=\Sigma mQ;\ 0=\Sigma mR;$$

For if we express the mutual action of m and m' by p, and their distance by f, we shall have, as far as this action alone is concerned,

$$mP=\frac{p(x-x')}{f};\ mQ=\frac{p(y-y')}{f};\ mR=\frac{p(z-z')}{f}$$

$$m'P'=\frac{p(x'-x)}{f};\ m'Q'=\frac{p(y'-y)}{f};\ m'R'=\frac{p(z'-z)}{f}$$

Hence $mP+m'P'=0$; $mQ+m'Q'=0$; $mR+m'R'=0$: the mutual actions of the bodies in the respective directions obviously destroying each other: and it is manifest that these equations would be equally true if p represented any finite and instantaneous action. We have also, in the absence of any foreign force,

$0=\frac{\mathrm{dd}X}{\mathrm{d}t^2}$, $0=\frac{\mathrm{dd}Y}{\mathrm{d}t^2}$, $0=\frac{\mathrm{dd}Z}{\mathrm{d}t^2}$; and by taking the fluent twice, $X=a+bt$, $Y=a'+b't$, and $Z=a''+b''t$, the *a*s and *b*s being constant quantities. These equations will give us linear relations between X, Y, and Z, if we exterminate t; whence it follows that the motion of the centre of gravity is rectilinear: and its velocity being equal to $\surd\left\{\left(\frac{dX}{dt}\right)^2+\left(\frac{dY}{dt}\right)^2+\left(\frac{dZ}{dt}\right)^2\right\}$, or to $\surd(b^2+b'^2+b''^2)$, it is always constant, and the motion is uniform.

SCHOLIUM. It is obvious from this analysis that the invariability of the motion of the centre of gravity of a system of bodies, whatever their mutual actions may be, holds good even in the case of an instantaneous loss of a finite quantity of motion in the separate bodies, by means of their mutual action.

§ 21. *Of the principle of the constancy of areas. It subsists notwithstanding the abruptness of any changes in the system. Determination of a system of coordinates, for which the sum of the areas described by the projections of the revolving radii vanishes for two of the planes of the ordinates, the sum being a maximum on the third, and vanishing for every plane perpendicular to it.* P. 56. [*General properties of projections.*]

323. THEOREM. The sum of the areas described by the projections of the revolving radii of any system of bodies, upon any given plane, multiplied respectively by their masses, is proportional to the time, supposing the bodies subject only to their reciprocal actions, and to a force directed to the origin of the radii.

We may obtain from the equation (P) (317) the particular value $0=\Sigma m\,\frac{x\mathrm{dd}y-y\mathrm{dd}x}{\mathrm{d}t^2}+\Sigma m\,(Py-Qx)$, if we cause the variation δx to disappear from the expression $\delta f=\delta\sqrt{\{(x'-x)^2+(y'-y)^2+(z'-z)^2\}}$ by making $\delta x'=\frac{y'\delta x}{y}+\delta x'_{,}$; $\delta x''=\frac{y''\delta x}{y}+\delta x''_{,}$; $\delta y=\frac{-x\delta x}{y}+\delta y_{,}$; $\delta y'=\frac{-x'\delta x}{y}+\delta y'_{,}$; $\delta y''=\frac{-x''\delta x}{y}+\delta y''_{,}$; ...; [the part of each of these expressions, that involves δx, belonging to a supposed revolution of the body round the axis parallel to z: for if the distance of m from this axis be s, and that of m', s', the elementary arc described by m will be $\frac{s}{y}\,\delta x$, and the arc described by m', $\frac{s'}{s}.\frac{s}{y}\,\delta x=\frac{s'}{y}\,\delta x$, whence the variation of x' will be $\frac{y'}{s'}.\frac{s'}{y}\,\delta x=\frac{y'}{y}\,\delta x$]. This substitution gives us the value of δf, $\delta f'$, $\delta f''$, ..., independently of δx, [as it must necessarily do from the agreement of the variations substituted with a rotatory motion]: we are therefore at liberty to assign any value to δx at pleasure, while we observe these conditions, and its coefficients may be made to vanish, [as they must obviously do if δx be infinitely greater than the other variations concerned]. In making this substitution for $\delta x'$, ..., in the equation (P) (317), that is, $0=m\delta x\left(\frac{\mathrm{dd}x}{\mathrm{d}t^2}-P\right)\ldots+$, $m'\delta x'\left(\frac{\mathrm{dd}x'}{\mathrm{d}t^2}-P'\right)$ . . . we are only required to employ for $\delta x'$, $\frac{y'\delta x}{y}$, since $\delta x'_{,}$ is supposed to vanish in comparison with

δx, and we have $0 = m\left(\frac{ddx}{dt^2} - P\right) + m'.\frac{y'}{y}\left(\frac{ddx'}{dt^2} - P'\right)$, or $0 = m\left(\frac{yddx}{dt} - Py\right) + m'\left(\frac{y'ddx'}{dt^2} - P'y'\right)\ldots = \Sigma m\left(\frac{yddx}{dt^2} - Py\right)$ and in the same manner the substitution of $\frac{-x\delta x}{y}, \ldots,$ for $\delta y, \delta y', \ldots,$ gives us, for $m\delta y\left(\frac{ddy}{dt^2} - Q\right) + m'\delta y'\left(\frac{ddy'}{dt^2} - Q'\right)\ldots, -\Sigma m\left(\frac{xddy}{dt^2} - Qx\right)$: whence we obtain $\Sigma m\,\frac{xddy - yddx}{dt^2} + \Sigma m\,(Py - Qx) = 0$: and by taking the fluent, we have $c = \Sigma m\,\frac{xdy - ydx}{dt} + \Sigma\!\int\! m\,(Py - Qx)\,dt$, [since $d(xdy) = xd^2y + dxdy$, and $d(ydx) = yd^2x + dxdy$]; c being a constant quantity. By employing the same mode of reasoning with respect to the variations of x and z, and of y and z, compared together, we obtain two other similar equations; consequently

$$c = \Sigma m\,\frac{xdy - ydx}{dt} + \Sigma\!\int\! m\,(Py - Qx)\,dt,$$

$$c' = \Sigma m\,\frac{xdz - zdx}{dt} + \Sigma\!\int\! m\,(Pz - Rx)\,dt, \text{ and}$$

$$c'' = \Sigma m\,\frac{ydz - zdy}{dt} + \Sigma\!\int\! m\,(Qz - Ry)\,dt.$$

Let us now suppose that the different bodies are only subjected to each other's reciprocal actions, and to a force directed to the origin of the coordinates. Calling the reciprocal action of m and m', p, we shall have, as far as this action is concerned, $0 = m(Py - Qx) + m'(P'y' - Q'x')$; [for $mP = \frac{p(x - x')}{f}$, $m'P' = \frac{p(x' - x)}{f}$, $mQ = \frac{p(y - y')}{f}$, $m'Q' = \frac{p(y' - y)}{f}$, as in article 322, and $mPy + m'P'y' = \frac{p(x - x')}{f}\,y$

$+\frac{p(x'-x)}{f}y'$, and $mQx+m'Q'x'=\frac{p(y-y')}{f}x-\frac{p(y'-y)}{f}$ x'; but these two sums being equal, their difference $\Sigma m(Py-Qx)$ vanishes:] and the same is, therefore, true respecting all the other reciprocal actions of the system, and with respect to all these the sum $\Sigma m\,(Py-Qx)$ vanishes. Again, if S be the force which urges m towards the origin of the coordinates, we shall have, as far as this force alone is concerned, $P=\frac{-Sx}{\surd(xx+yy+zz)}$, and $Q=\frac{-Sy}{\surd(xx+yy+zz)}$: consequently $Py=Qx$, and their difference vanishes. When, therefore, the bodies are only subjected to their mutual action, and to the forces directed to the origin of the coordinates, we have

$$c=\Sigma m\frac{x\mathrm{d}y-y\mathrm{d}x}{\mathrm{d}t}\,;\quad c'=\Sigma m\frac{x\mathrm{d}z-z\mathrm{d}x}{\mathrm{d}t}\,;$$

$$c''=\Sigma m\frac{y\mathrm{d}z-z\mathrm{d}y}{\mathrm{d}t}. \qquad (Z)$$

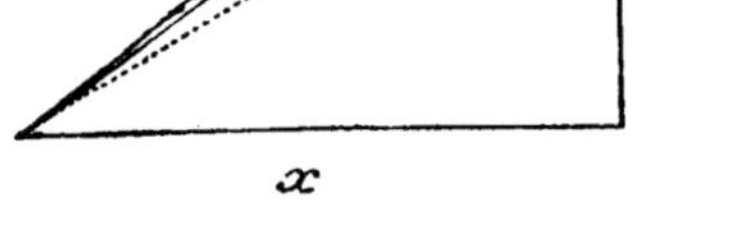

If we suppose the place of the body m to be projected on the common plane of x and y, the fluxion $\frac{1}{2}(x\mathrm{d}y-y\mathrm{d}x)$ will represent the area traced by the radius drawn, from the origin of the coordinates, to the projection of m: it follows, therefore, that the sum of the areas described by the radii, belonging to the different bodies of the system, multiplied by their masses, is proportional to the fluxion of the time, and, for any finite interval, proportional to the time itself. This constitutes the principle of the constancy of areas, which is obviously true for any plane whatever, since the motion of the bodies bears no determinate relation to x, y, and z; and, if the attractive force vanishes, the principle is also true with respect to any point whatever: nor is the demonstration limited to changes produced by insensible degrees.

324. LEMMA. If we have two systems of orthogonal coordinates, x, y, z, and $x_{,,,}$, $y_{,,,}$, $z_{,,,}$, originating from the same point: if θ be the inclination of the plane of $x_{,,,}$ and $y_{,,,}$ to that of x and y, [its positive values implying that $z_{,,,}$ inclines towards the same side of x with $+y$], and if ψ be the angular distance of x from the intersection of these planes, and φ that of $x_{,,,}$, the equations between the coor dinates will be

$$x = x_{,,,}(\cos\theta\sin\psi\sin\varphi + \cos\psi\cos\varphi)$$
$$+ y_{,,,}(\cos\theta\sin\psi\cos\varphi - \cos\psi\sin\varphi)$$
$$+ z_{,,,}\sin\theta\sin\psi$$
$$y = x_{,,,}(\cos\theta\cos\psi\sin\varphi - \sin\psi\cos\varphi)$$
$$+ y_{,,,}(\cos\theta\cos\psi\cos\varphi + \sin\psi\sin\varphi)$$
$$+ z_{,,,}\sin\theta\cos\psi$$
$$z = z_{,,,}\cos\theta - y_{,,,}\sin\theta\cos\varphi - x_{,,,}\sin\theta\sin\varphi.$$

In order to assist the imagination, we may suppose the origin of the coordinates to be at the centre of the earth, the plane of x and y to be the ecliptic, and z to be directed to its north pole [x being considered as positive when it tends more or less to approach the vernal equinox ♈, and y when it tends towards the sign ♋, and negative on the opposite side of the centre]: then if the plane of $x_{\prime\prime\prime}$ and $y_{\prime\prime\prime}$ be that of the equator, we shall have $z_{\prime\prime\prime}$ parallel to the earth's axis, pointing to the north pole [, and inclining towards the sign ♋, towards which y is positive]; the obliquity of the ecliptic will then be [+] θ, and ψ will be the longitude of the axis x with respect to the vernal equinox, which is the intersection of the two planes on the side of $+x$; the distance of $x_{\prime\prime\prime}$ and $y_{\prime\prime\prime}$ from the same line will be φ and $\varphi+\frac{\pi}{2}$ respectively, these angles varying with the rotation of the earth.

Now if $x_{\prime}$, $y_{\prime}$, and $z_{\prime}$, be an intermediate system of orthogonal coordinates, $x_{\prime}$ being the line of the vernal equinox, $y_{\prime}$ the projection of the earth's axis on the plane of the ecliptic, and $z_{\prime}$ coinciding with the axis of the ecliptic z; the ordinates x, y, $x_{\prime}$, and $y_{\prime}$ being in the same plane, we have

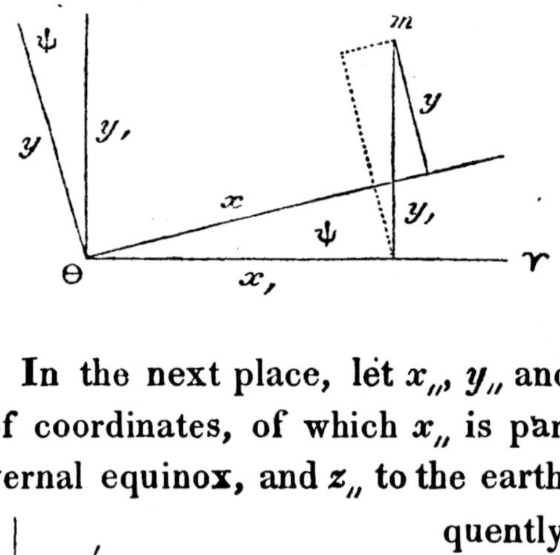

$x = x_{\prime}\cos\psi + y_{\prime}\sin\psi$;

$y = y_{\prime}\cos\psi - x_{\prime}\sin\psi$;

$z = z_{\prime}$.

In the next place, let $x_{\prime\prime}$, $y_{\prime\prime}$ and $z_{\prime\prime}$ be another system of coordinates, of which $x_{\prime\prime}$ is parallel to the line of the vernal equinox, and $z_{\prime\prime}$ to the earth's axis, $y_{\prime\prime}$ being consequently in the plane of the equator: we have then $y_{\prime\prime}$ and $z_{\prime\prime}$ in the plane passing through $y_{\prime}$ and $z_{\prime}$, while $x_{\prime}$ and $x_{\prime\prime}$ coincide: consequently

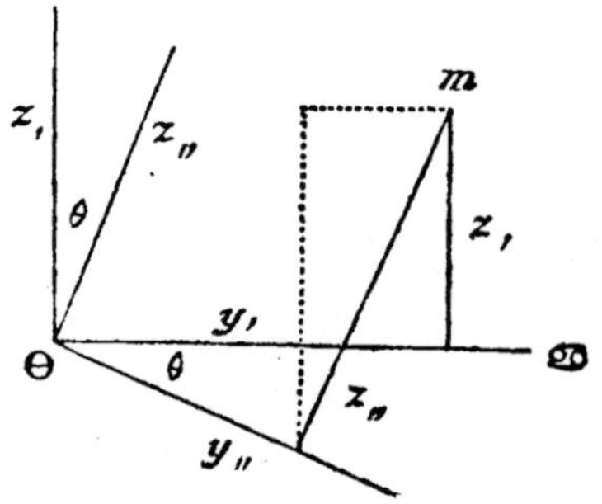

$x_{\prime} = x_{\prime\prime}$;

$y_{\prime} = y_{\prime\prime}\cos\theta + z_{\prime\prime}\sin\theta$;

$z_{\prime} = z_{\prime\prime}\cos\theta - y_{\prime\prime}\sin\theta$.

Lastly, while $z_{\prime\prime\prime}$ is substituted for its equal $z_{\prime\prime}$, with which it is identical, we shall have $x_{\prime\prime\prime}$ and $y_{\prime\prime\prime}$ in the same plane with $x_{\prime\prime}$ and $y_{\prime\prime}$, which is that of the equator: we have thus

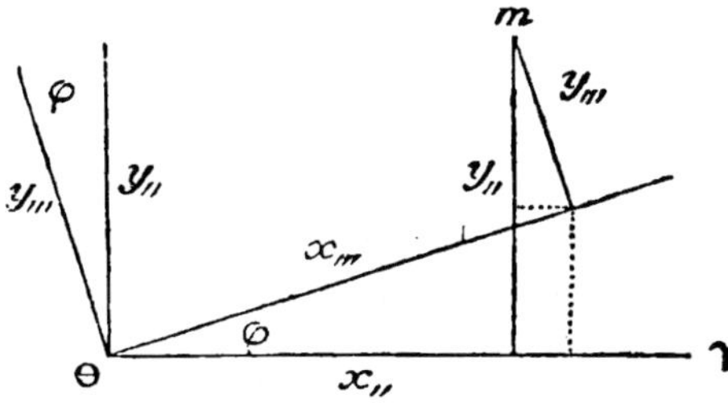

$x_{\prime\prime}=x_{\prime\prime\prime}\cos\varphi - y_{\prime\prime\prime}\sin\varphi;$
$y_{\prime\prime}=y_{\prime\prime\prime}\cos\varphi + x_{\prime\prime\prime}\sin\varphi;$
$z_{\prime\prime}=z_{\prime\prime\prime}.$

[The second sign in the value of $x_{\prime\prime}$ is here negative, because the axis $x_{\prime\prime}$ is not between $x_{\prime\prime\prime}$ and $y_{\prime\prime\prime}$, while $y_{\prime\prime}$ is between $y_{\prime\prime\prime}$ and $x_{\prime\prime\prime}$.] By substituting successively the values thus obtained, we have [first

$x_{\prime}=x_{\prime\prime\prime}\cos\varphi - y_{\prime\prime\prime}\sin\varphi;$
$y_{\prime}=y_{\prime\prime\prime}\cos\varphi\cos\theta + x_{\prime\prime\prime}\sin\varphi\cos\theta + z_{\prime\prime\prime}\sin\theta;$
$z_{\prime}=z_{\prime\prime\prime}\cos\theta - y_{\prime\prime\prime}\cos\varphi\sin\theta - x_{\prime\prime\prime}\sin\varphi\sin\theta;$ then
$x=x_{\prime\prime\prime}\cos\varphi\cos\psi - y_{\prime\prime\prime}\sin\varphi\cos\psi + y_{\prime\prime\prime}\cos\varphi\cos\theta\sin\psi + x_{\prime\prime\prime}\sin\varphi\cos\theta\sin\psi + z_{\prime\prime\prime}\sin\theta\sin\psi;$
$y=y_{\prime\prime\prime}\cos\varphi\cos\theta\cos\psi + x_{\prime\prime\prime}\sin\varphi\cos\theta\cos\psi + z_{\prime\prime\prime}\sin\theta\cos\psi - x_{\prime\prime\prime}\cos\varphi\sin\psi + y_{\prime\prime\prime}\sin\varphi\sin\psi;$
$z=z_{\prime\prime\prime}\cos\theta - y_{\prime\prime\prime}\cos\varphi\sin\theta - x_{\prime\prime\prime}\sin\varphi\sin\theta;$ or, collecting the coefficients]

$x=x_{\prime\prime\prime}(\cos\varphi\cos\psi + \sin\varphi\cos\theta\sin\psi) + y_{\prime\prime\prime}(\cos\varphi\cos\theta\sin\psi - \sin\varphi\cos\psi) + z_{\prime\prime\prime}\sin\theta\sin\psi;$
$y=x_{\prime\prime\prime}(\sin\varphi\cos\theta\cos\psi - \cos\varphi\sin\psi) + y_{\prime\prime\prime}(\cos\varphi\cos\theta\cos\psi + \sin\varphi\sin\psi) + z_{\prime\prime\prime}\sin\theta\cos\psi;$
$z=-x_{\prime\prime\prime}\sin\varphi\sin\theta - y_{\prime\prime\prime}\cos\varphi\sin\theta + z_{\prime\prime\prime}\cos\theta.$

COROLLARY 1. We find also

$x_{\prime\prime\prime}=x(\cos\theta\sin\psi\sin\varphi + \cos\psi\cos\varphi) + y(\cos\theta\cos\psi\sin\varphi - \sin\psi\cos\varphi) - z\sin\theta\sin\varphi;$
$y_{\prime\prime\prime}=x(\cos\theta\sin\psi\cos\varphi - \cos\psi\sin\varphi) + y(\cos\theta\cos\psi\cos\varphi + \sin\psi\sin\varphi) - z\sin\theta\cos\varphi;$
$z_{\prime\prime\prime}=x\sin\theta\sin\psi + y\sin\theta\cos\psi + z\cos\theta.$

These values are obtained by multiplying each of the former equations by the respective coefficients of $x_{,,,}$, and adding the three products together; and by repeating the operation for $y_{,,,}$ and $z_{,,,}$ in the same manner [: or, much more simply, by merely substituting $-\theta$, φ, and ψ for θ, ψ, and φ, $x_{,,,}$, $y_{,,,}$, and $z_{,,,}$, for x, y, and z, and the reverse, according to the terms of the proposition].

SCHOLIUM. These different transformations of the coordinates will be very useful hereafter. We may distinguish those which belong to the bodies m', m'', . . ., by adding accents above the respective characters, as x', x'', . . ., $x_{,,,}'$, $x_{,,,}''$, . . .

[COROLLARY 2. Putting $y_{,,,}=0$, and $z_{,,,}=0$, we have $x=x_{,,,}(\cos\theta\sin\psi\sin\varphi+\cos\psi\cos\varphi)$ and in this case $\frac{x}{x_{,,,}}$ is the cosine of the angle formed by x and $x_{,,,}$, or of the arc intercepted between them: while θ is the spherical angle opposite to that arc or side, and ψ and φ the two sides including it. We have also $\frac{z}{x_{,,,}}=-\sin\theta\sin\varphi$, for the cosine of the angle formed by z and $x_{,,,}$, which is equivalent to sin ☉ Lat=sin Obl Ecl × sin ☉ Long.]

[325. LEMMA A. If a perpendicular be let fall from the vertex of a triangle on the base, the difference of the segments will be a fourth proportional to the base and to the sum and difference of the two sides.

The segments of the base being a' and a'', the difference of their squares is $a'^2-a''^2$; but the difference of their squares is equal to the difference of the squares of the two sides, since the perpendicular is the same for both the right angled triangles formed by the division of the base: we have therefore $a'^2-a''^2=b^2-c^2$: but $a'^2-a''^2=(a'-a'')\ (a'+a'')=(a'-a'')\ a$, and $b^2-c^2=(b+c)(b-c)$: consequently $a'-a''=\frac{(b+c)\,(b-c)}{a}$.

326. LEMMA B. If an angle of a rectangular parallelepiped be cut off by a plane passing through three of its diagonals, the three planes perpendicular to the section, and passing through the edges meeting in the angle, will be perpendicular to the opposite sides of the section.

For the perpendicular falling from the solid angle on the diagonal between the sides or edges a and b will divide that diagonal into two segments, of which the difference is equal to $\frac{aa-bb}{\surd(aa+bb)}$, (325), and the perpendicular from the opposite angle of the section will fall on the same point, for in this case the difference of the squares of the sides is $a^2+c^2-(b^2+c^2)$, which is equal to a^2-b^2, and the diagonal is common to both triangles: but both the perpendiculars being perpendicular to the same line, the plane in which they lie will be perpendicular to this line and to the section; and this plane passes through the edge in question.

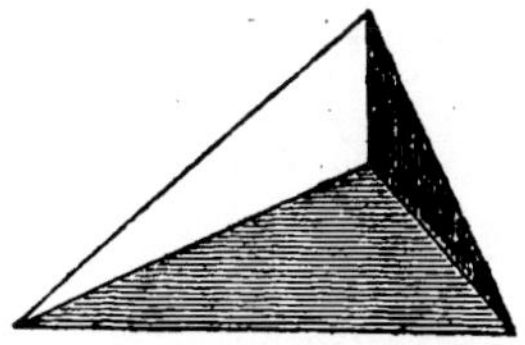

327. LEMMA C. If an angle of a parallelepiped be cut off by a plane, the square of the area of the section will be equal to the sum of

the squares of the areas of the three triangular faces of the solid angle.

The area of the face between a and b is $\frac{1}{2}ab$, and the perpendicular falling on its base from the solid angle is $\frac{ab}{\sqrt{(aa+bb)}}$: but this perpendicular must be perpendicular to the third side c, and the square of the hypotenuse of the triangle lying between them must be $c^2+\frac{aabb}{aa+bb}$, which multiplied by the square of the side to which it is perpendicular, or a^2+b^2, must be the square of twice the area; consequently the square of the area is $\frac{1}{4}\left\{(a^2+b^2)c^2+a^2b^2\right\}$ $=\frac{1}{4}a^2b^2+\frac{1}{4}a^2c^2+\frac{1}{4}b^2c^2$, which is the sum of the squares of the areas of the three faces.

328. Lemma D. The sum of the squares of the projections of any area, on three orthogonal planes, is equal to the square of the area itself.

For the projection of the area on each plane is to the original in the same proportion, as the whole face of the parallelepiped is to the whole oblique section: the proportion of the areas being determined by the inclination of the planes, whatever the form of the area projected may be.

329. Lemma E. The cosine of the inclination of the section to either of the faces will be expressed by the area of that face divided by the area of the section.

For if the area be resolved into elementary rectangles, the breadth of each, parallel to the common section of the planes, being the same in the projection as in the original, the length of the projection will be to that of the original as the cosine of the inclination to the radius; and the whole areas will be in the same constant ratio as their elements.

Corollary. Hence the sum of the squares of the sines of the angles, formed by the three faces of the parallelepiped with the section, is equal to the square of the radius, or unity.]

330. Theorem. For every independent system of bodies, a fixed plane may be determined, with respect to which the sum of the projections of all the areas, described by the revolving radii, multiplied by the masses of the respective bodies, is the greatest possible; and for every plane perpendicular to which, the sum of the projections vanishes.

" By taking the fluxions of the equations for the values of $x_{,,,}$, $y_{,,,}$, and $z_{,,,}$" [, the angles remaining constant], " and substituting c, c′, and c″, for

$$\Sigma m \frac{x\mathrm{d}y - y\mathrm{d}x}{\mathrm{d}t}, \ \Sigma m \frac{x\mathrm{d}z - z\mathrm{d}x}{\mathrm{d}t}, \text{ and } \Sigma m \frac{y\mathrm{d}z - z\mathrm{d}y}{\mathrm{d}t},$$ we obtain

$$\Sigma m \frac{x_{,,,}\mathrm{d}y_{,,,} - y_{,,,}\mathrm{d}x_{,,,}}{\mathrm{d}t} = c \cos\theta - c' \sin\theta \cos\psi + c'' \sin\theta \sin\psi;$$

$$\Sigma m \frac{x_{,,,}\mathrm{d}z_{,,,} - z_{,,,}\mathrm{d}x_{,,,}}{\mathrm{d}t} = c \sin\theta \cos\varphi + c' (\sin\psi \sin\varphi + \cos\theta \cos\psi \cos\varphi) + c''(\cos\psi \sin\varphi - \cos\theta \sin\psi \cos\varphi)$$

$$\Sigma m \frac{y_{,,,}dz_{,,,}-z_{,,,}dy_{,,,}}{dt} = -c \sin\theta \sin\varphi + c'(\sin\psi \cos\varphi - \cos\theta \cos\psi \sin\varphi) + c''(\cos\psi \cos\varphi + \cos\theta \sin\psi \sin\varphi).$$

"If we determine ψ and θ in such a manner, that $\sin\theta$ $\sin\psi = \frac{c''}{\surd(cc+c'c'+c''c'')}$, and $\sin\theta\cos\psi = \frac{-c'}{\surd(c^2+c'^2+c''^2)}$, whence $\cos\theta = \frac{c}{\surd(c^2+c'^2+c''^2)}$; we shall have

$$\Sigma m \frac{x_{,,,}dy_{,,,}-y_{,,,}dx_{,,,}}{dt} = \surd(c^2+c'^2+c''^2);$$

$$\Sigma m \frac{x_{,,,}dz_{,,,}-z_{,,,}dx_{,,,}}{dt} = 0; \quad \Sigma m \frac{y_{,,,}dz_{,,,}-z_{,,,}dy_{,,,}}{dt} = 0;$$

consequently the values of c' and c'' will vanish when the plane of $x_{,,,}$ and $y_{,,,}$ is thus determined. And there is only one plane which possesses this property: for if there were any other, and x and y were the coordinates, and θ and φ the angles belonging to it, we should have

$$\Sigma m \frac{x_{,,,}dz_{,,,}-z_{,,,}dx_{,,,}}{dt} = c \sin\theta \cos\varphi, \text{ and}$$

$$\Sigma m \frac{y_{,,,}dz_{,,,}-z_{,,,}dy_{,,,}}{dt} = -c \sin\theta \sin\varphi;$$

but since c' and $c''=0$, by the supposition, for the supposed plane; and since these quantities have been shown to be $=0$, for the planes of $x_{,,,}$, $z_{,,,}$, and $y_{,,,}$, $z_{,,,}$, we have $\sin\theta=0$, and the two planes must coincide. The value of $\Sigma m \frac{x_{,,,}dy_{,,,}-y_{,,,}dx_{,,,}}{dt}$ being equal to $\surd(c^2+c'^2+c''^2)$ whatever be the plane of x and y from which it is derived, it follows that this quantity may be deduced equally well from any other system of coordinates, and that the plane of $x_{,,,}$ and $y_{,,,}$, determined by it, will always be that which makes this elementary area a maximum; and since the angle φ remains undetermined, it follows that, whatever may be its value, the projections of the areas on the planes perpendicular to this plane will vanish. Hence we may at any time find the situation of this plane, in the same wav as the centre of gravity may at any time be found, notwithstanding any mutual actions of the system, and for this reason, it is as natural to suppose the coordinates x and y to be situated in this plane, as to make them begin at the centre of gravity."

[This proposition may be much more simply and intelligibly demonstrated by means of Lemma D; for if $c_{,,,}$, $c'_{,,,}$, and $c''_{,,,}$ be the sums of the products of the masses by the projections of the areas described by the revolving radii of the different bodies of the system on the three planes belonging to the svstem of ordinates $x_{,,,}$, $y_{,,,}$, and $z_{,,,}$, the sum of the squares of $c_{,,,}$, $c'_{,,,}$, and $c''_{,,,}$ will be equal to the sum of the squares of c, c′, and c″: and since this sum is a constant quantity for all systems of planes, it is obvious that when the portion belonging to any one plane is equal to the whole, there can be none left for any plane or planes perpendicular to it. We have, therefore, only to determine the inclination of the section of the parallelepiped to either of its faces, and we shall have the angle θ, the cosine of which will be expressed by $\frac{c}{\sqrt{(c^2+c'^2+c''^2)}}$ for the plane of x and y (329): and since ψ is the distance of x from the intersection of the planes (324), that is, the angle of the face c adjoining to x, its tangent will be $\frac{y}{x}=\frac{c''}{c'}$, since the areas of the faces c' and c″ are to each other as the sides x and y to which they are adjacent, the side z being common to both triangles.]

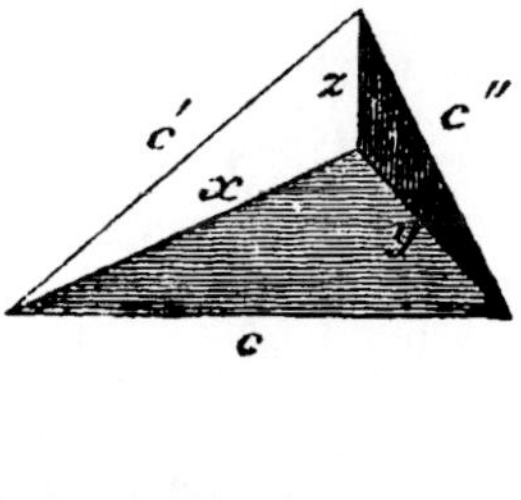

§ 22. *The principles of the preservation of impetus and of areas will still hold good, if the origin of the coordinates be supposed to have a uniform rectilinear motion. In this case, the plane passing through this point, on which the sum of the areas described is a maximum, remains always parallel to itself. These properties may be referred to the relations of the coordinates of the mutual distances of the bodies of the system. The planes which pass through the centre of gravity of each body of the system, parallel to the general mean plane of revolution, are possessed of similar properties.* P. 61.

331. THEOREM. The constancy of the impetus and of the areas described is observed by a system of bodies, referred to a common origin, which moves uniformly in a right line.

If we call the coordinates of the moveable origin of the ordinates of the system X, Y, and Z, and suppose

$x = X + x_{,}$; $y = Y + y_{,}$; $z = Z + z_{,}$.

$x' = X + x'_{,}$; $y' = Y + y'_{,}$; $z' = Z + z'_{,}$; the coordinates of $m, m', \ldots,$ referred to the moveable origin, will be $x_{,}, y_{,}, z_{,}, x'_{,}, \ldots,$ and since by the properties of the centre of gravity (322), we have for any detached system of bodies $0 = \Sigma m\left(\frac{\mathrm{dd}x}{\mathrm{d}t^2} - P\right), \ldots,$ we obtain, by substitution

$0 = \Sigma m\,(\mathrm{d}^2 X + \mathrm{d}^2 x_{,}) - \Sigma m P \mathrm{d}t^2$;

$0 = \Sigma m\,(\mathrm{d}^2 Y + \mathrm{d}^2 y_{,}) - \Sigma m Q \mathrm{d}t^2$;

$0 = \Sigma m\,(\mathrm{d}^2 Z + \mathrm{d}^2 z_{,}) - \Sigma m R \mathrm{d}t^2$; consequently $\Sigma m \mathrm{d}^2 x_{,} - \Sigma m P \mathrm{d}t^2 = 0$, since $\mathrm{d}^2 X = 0$ by the supposition, and in the same manner, $\Sigma m \mathrm{d}^2 y_{,} - \Sigma m Q \mathrm{d}t^2 = 0$, and $\Sigma m \mathrm{d}^2 z_{,} - \Sigma m R \mathrm{d}t^2 = 0$. Then in the equation (P), $0 = \Sigma m \delta x\left(\frac{\mathrm{dd}x}{\mathrm{d}t} \quad P\right) \ldots,$

substituting $\delta X + \delta x_{,}$ for δx, we have [$0 = \Sigma m\,(\delta X + \delta x_{,})\left(\frac{\mathrm{dd}x}{\mathrm{d}t^2} - P\right) + \ldots$ but $\Sigma m \delta X\left(\frac{\mathrm{dd}x}{\mathrm{d}t^2} - P\right) = 0$, because $\Sigma m\left(\frac{\mathrm{dd}x}{\mathrm{d}t^2} - P\right) = 0$, δX being the same for all the quantities of which the sum is denoted by Σ; consequently $0 = \Sigma m\,\delta x_{,}\left(\frac{\mathrm{dd}x}{\mathrm{d}t^2} - P\right) \ldots$, or, d^2x being equal to $\mathrm{d}^2x_{,}$] $0 = \Sigma m$

$$\delta x_{,}\left(\frac{\mathrm{dd}x_{,}}{\mathrm{d}t^2} - P\right) + \Sigma m \delta y_{,}\left(\frac{\mathrm{dd}y_{,}}{\mathrm{d}t^2} - Q\right) + \Sigma m \delta z\left(\frac{\mathrm{dd}z_{,}}{\mathrm{d}t^2} - R\right);$$

an equation which is exactly of the same form with the equation (P), supposing the forces P, Q, R, to depend only on the coordinates $x_{,}$, $y_{,}$, $z_{,}$, $x'_{,}$, ... If, therefore, we apply to it the same reasoning, as was grounded on that equation, we may derive from it the same conclusions, with respect to the preservation of the impetus, aud the description of areas, relative to the moveable origin of the coordinates.

If the system is not subjected to the action of any extraneous force, its centre of gravity will have a rectilinear and uniform motion (322): so that if we suppose the ordinates x, y, and z to begin at the centre of gravity, the laws in question will always we observed: and X, Y, and Z being the coordinates of the centre of gravity, we shall have, by the nature of this point, $0 = \Sigma m x_{,}$, $0 = \Sigma m y_{,}$, and $0 = \Sigma m z_{,}$; whence we have $\Sigma m \frac{x\mathrm{d}y - y\mathrm{d}x}{\mathrm{d}t} = \frac{X\mathrm{d}Y - Y\mathrm{d}X}{\mathrm{d}t} \cdot \Sigma m + \Sigma m \frac{x_{,}\mathrm{d}y_{,} - y_{,}\mathrm{d}x_{,}}{\mathrm{d}t}$; and $\Sigma m\,\frac{\mathrm{d}x^2 + \mathrm{d}y^2 + \mathrm{d}z^2}{\mathrm{d}t^2} = \frac{\mathrm{d}X^2 + \mathrm{d}Y^2 + \mathrm{d}Z^2}{\mathrm{d}t^2}\Sigma m + \Sigma m \frac{\mathrm{d}x_{,}^2 + \mathrm{d}y_{,}^2 + \mathrm{d}z_{,}^2}{\mathrm{d}t^2}$ [, for $\mathrm{d}X$ and $\mathrm{d}Y$ being common to all the system, we have $\Sigma m \mathrm{d}X = \mathrm{d}X\,\Sigma m$, $\Sigma m \mathrm{d}Y = \mathrm{d}Y\,\Sigma m$; and the sums of the squares of

$(dX+dx_{,})\ (dX'+dx'_{,})\ldots$ must be equal to the sums of the squares of dX, dX', ..., $dx_{,}$, $dx'_{,}$, ..., since $(dX+dx_{,})^2=dX^2+dx_{,}^2+2dXdx_{,}$, and $\Sigma m dXdx_{,}=dX\Sigma m dx_{,}=0$, since $\Sigma mx_{,}=0$, and $d(\Sigma mx_{,})=0$]. It appears, therefore, that the quantities, concerned in the impetus and the areas, are composed, first, of the quantities which would have existed, if all the bodies of the system had been united in their common centre of gravity; and secondly, of the similar quantities derived from the centre of gravity, considered as immoveable: and the first system of quantities being constant, it is easily understood that the second must be also constant. It follows, therefore, that if we fix the origin of the coordinates x, y, z, x', ... of the equations (*Z*) (323) at the centre of gravity, the conclusions derived from them will still hold good, and the angles of the planes concerned will remain unaltered; whence it follows that the mean plane of revolution, which affords the maximum of projected areas, must pass through the centre of gravity of the system, and remain always parallel to itself during the motion of the system; and that the sum of the areas, computed for any plane perpendicular to this plane, must always vanish.

[SCHOLIUM. The whole of this elaborate demonstration is rendered perfectly superfluous, if we exclude the distinction of absolute and relative motion from the definitions relating to it (218, 224): but it is satisfactory to find that a complicated analysis is still true when applied to the test of demonstrating by it a very simple proposition.]

332. THEOREM. The sum of the projections of the areas, described by the radii joining each body of a system with each of the

other bodies, considered as at rest, on any given plane, multiplied by the products of the two masses respectively, is equal to the product of the mass of the whole system, into the sum of the projections of the areas described by the separate revolving radii round the common centre of gravity on the same plane, multiplied by the separate masses respectively.

Since $c = \Sigma m \dfrac{x\mathrm{d}y - y\mathrm{d}x}{\mathrm{d}t}$, we obtain, by proper substitutions $c\Sigma m = \Sigma mm'. \dfrac{(x'-x)\ (\mathrm{d}y'-\mathrm{d}y) - (y'-y)\ (\mathrm{d}x'-\mathrm{d}x)}{\mathrm{d}t}$.

[For $c\Sigma m = (m+m'+m''\ldots)\ (m \dfrac{x\mathrm{d}y - y\mathrm{d}x}{\mathrm{d}t} + m' \dfrac{x'\mathrm{d}y' - y'\mathrm{d}x'}{\mathrm{d}t} + \ldots)$ in which all the binary combinations of the different values of m occur once with each of the two corresponding fractions; and taking the first two for example, we have $(m+m').\left(m. \dfrac{x\mathrm{d}y - y\mathrm{d}x}{\mathrm{d}t} + m'. \dfrac{x'\mathrm{d}y' - y'\mathrm{d}x'}{\mathrm{d}t}\right) = (mm+mm')x\mathrm{d}y - (mm+mm')y\mathrm{d}x + (mm'+m'm')x'\mathrm{d}y' - (mm'+m'm')\,y'\mathrm{d}x'$, neglecting the divisor: but $mm'\ (x'-x)\ (\mathrm{d}y'-\mathrm{d}y) - (y'-y)\ (\mathrm{d}x'-\mathrm{d}x) = mm'\ (x'\mathrm{d}y' - x'\mathrm{d}y - x\mathrm{d}y' + x\mathrm{d}y - y'\mathrm{d}x' + y'\mathrm{d}x + y\mathrm{d}x' - y\mathrm{d}x)$, the difference of the two expressions being $mmx\mathrm{d}y - mmy\mathrm{d}x + m'm'x'\mathrm{d}y' - m'm'y'\mathrm{d}x' + mm'x'\mathrm{d}y - mm'y'\mathrm{d}x + mm'x\mathrm{d}y' - mm'y\mathrm{d}x'$, or $m(mx+m'x')\mathrm{d}y - m(my+m'y')\,\mathrm{d}x + m'\,(m'x'+mx)\,\mathrm{d}y' - m'(\,m'y'+my)\mathrm{d}x'$: and if we added together any others of the bodies in pairs, it is obvious that the coefficients of the fluxions in this difference would make the series $mx+m'x'+m''x''+\ldots = \Sigma mx = 0$, by the property of the centre of gravity; consequently the difference would at last vanish, and the two expressions would be equal, as was to be proved, though the transformation is by no means obvious without a demonstration.]

For similar reasons we have

$$c'\Sigma m = \Sigma mm'.\ \frac{(x'-x)(dz'-dz)-(z'-z)(dx'-dx)}{dt}, \text{ and}$$

$$c''\Sigma m = \Sigma mm'.\ \frac{(y'-y)(dz'-dz)-(z'-z)(dy'-dy)}{dt}.$$

It is obvious that the sum, thus ascertained, will be liable to the same conditions of becoming a maximum and vanishing, which have been demonstrated respecting the radii drawn to the common centre of gravity: and that the same mean plane of revolution, determined from it, will always remain parallel to itself.

333. THEOREM. The sum of the squares of the velocities of any system of bodies, taken in pairs, of which the one is considered as moving round the other at rest, and multiplied by the products of the masses of the respective pairs, is expressed by a constant quantity lessened by twice the product of the sum of all the masses into the sum of the reciprocal forces between each pair, combined with the spaces through which they act, and multiplied by the products of the respective masses; or $\Sigma mm'\ \frac{(dx'-dx)^2+(dy'-dy)^2+(dz'-dz)^2}{dt^2}$

$$= C' - 2\Sigma m \Sigma \int mm' F df$$

For since $\Sigma m \frac{dx^2+dy^2+dz^2}{dt^2} = C - 2\Sigma\int mm'Fdf$, (319) [$= m \frac{dx^2+dy^2+dz^2}{dt^2} + m' \frac{dx'^2+dy'^2+dz'^2}{dt^2} + \ldots$; multiplying by $\Sigma m = m+m'+m''+\ldots$, we have $(m+m'+\ldots)(m \frac{dx^2+dy^2+dz^2}{dt^2} + m' \frac{dx'^2+dy'^2+dz'^2}{dt^2} + \ldots) = C' - 2\Sigma m \Sigma\int mm'$ Fdf; but the first member of the equation of the proposition will be found to be equal, when expanded, to the first member of this last, the difference of each part becoming $=0$: thus, taking m and m' for an example, the difference will be $mm'(dx'-dx)^2-(m+m')(mdx^2+m'dx'^2) = mm'dx'^2 + mm\,dx^2 - 2mm'dx'dx - mmdx^2 - mm'dx'^2 - mm'dx^2 - m'm'dx'^2 = -2mm'dx'dx - mmdx^2 - m'm'dx'^2 = -(mdx+m'dx')^2$; and by the successive addition of the different pairs, this difference will become $(\Sigma mdx)^2 = 0$, since $\Sigma mx = 0$ and $\Sigma mdx = 0$, by the properties of the centre of gravity, consequently] the two expressions are equal for the whole system.

§ 23. *Principle of the least action. Combined with that of the preservation of impetus, it gives the general equation of motion.* P. 63.

334. Theorem. The momenta of a system of bodies being multiplied by the fluxions of the spaces respectively described, the sum of the fluents, taken, for the whole system, between any given points of space, is always a minimum.

The equation (R) (319), $\Sigma mv^2 = c + 2\varphi = 2\Sigma m\int(Pdx + Qdy + Rdz)$, affords us the variation $\Sigma mv\delta v = \Sigma m(P\delta x + Q\delta y + R\delta z)$, and combining this with the equation (P) (317)

$0 = \Sigma m\delta x \left(\frac{\mathrm{dd}x}{\mathrm{d}t^2} - P\right) + \ldots$, we obtain $0 = \Sigma m \left(\delta x \, \mathrm{d} \frac{\mathrm{d}x}{\mathrm{d}t} + \delta y \, \mathrm{d} \frac{\mathrm{d}y}{\mathrm{d}t} + \delta z \, \mathrm{d} \frac{\mathrm{d}z}{\mathrm{d}t}\right) - \Sigma m \mathrm{d}tv\delta v$. Now $\mathrm{d}s$ being the fluxion of the path of m, let $\mathrm{d}s'$ be that of the path of $m', \ldots$, and $v\mathrm{d}t = \mathrm{d}s$, $v'\mathrm{d}t = \mathrm{d}s', \ldots$; $\mathrm{d}s$ being $\sqrt{(\mathrm{d}x^2 + \mathrm{d}y^2 + \mathrm{d}z^2)}$: and as it has already been shown (266) with respect to any particular body, that $\delta(v\mathrm{d}s) = \mathrm{d} \frac{\mathrm{d}x\delta x + \mathrm{d}y\delta y + \mathrm{d}z\delta z}{\mathrm{d}t}$, we obtain by adding the results for the different bodies, $\Sigma m\delta\,(v\mathrm{d}s) = \Sigma m\mathrm{d} \frac{\mathrm{d}x\delta x + \mathrm{d}y\delta y + \mathrm{d}z\delta z}{\mathrm{d}t}$: and the fluent, which is taken independently both of the variation, and of the integration expressed by Σ, gives us $\Sigma\delta\int mv\mathrm{d}s = C + \Sigma m \frac{\mathrm{d}x\delta x + \mathrm{d}y\delta y + \mathrm{d}z\delta z}{\mathrm{d}t}$; the variations of the ordinates being those which belong to the extreme points of the curves to be compared. Hence it appears, that when these points are supposed invariable, the equation becomes $\Sigma\delta\int mv\mathrm{d}s = 0$, consequently the quantity $\Sigma\int mv\mathrm{d}s$ is a minimum. And this is the law of the least action, as applied to the motion of a system of bodies, a law which is evidently derivable, by mathematical considerations, from the fundamental principles of equilibrium and of motion.

SCHOLIUM. It is also apparent that this law, combined with that of the preservation of impetus, would afford the equation (P) (317), which includes all that is necessary to the determination of the motions of the system; and it appears from the preceding propositions, that the same principles are applicable to the case of a moveable system of bodies, provided that the motion of its centre of gravity be uniform and rectilinear, and the system be detached from the operation of all foreign forces.

CHAPTER VI.

OF THE LAWS OF THE MOTION OF A SYSTEM OF BODIES, ACCORDING TO ANY RELATION MATHEMATICALLY POSSIBLE BETWEEN FORCE AND VELOCITY.

---

§ 24. *New principles corresponding, on this more enlarged hypothesis, to those of the preservation of impetus, the constancy of areas, the motion of the centre of gravity, and the least action. Forces reduced to a given direction: referred to an axis; and combined with the elements of the spaces.*

335. THEOREM. The sum of the products of the masses of any system of bodies, into the fluent of the product of the velocity into any function of the velocity, which may be supposed to represent the force, is constant with regard to the intervals between any two places of the system.

There are many conceivable relations between force and velocity, which imply no mathematical contradiction, although the simplest is their being directly proportional to each other, as we find that they actually are in nature.

The preceding equations of the motion of a system of bodies have been derived from this law: but the same mode of investigation may be easily extended to all other relations between force and velocity which are mathematically possible, and the principles of motion may thus be exhibited in a new point of view.

For this purpose we may suppose F the force and v the velocity, putting $F=\varphi(v)$, and let $\varphi'(v)=\frac{\mathrm{d}F}{\mathrm{d}v}$, [or, more simply, let φ denote $\varphi(v)$ or F, and $\varphi'\mathrm{d}v$, $\mathrm{d}\varphi$]. If this force be reduced to the direction of x, it will become $\varphi.\frac{\mathrm{d}x}{\mathrm{d}s}$; and in the next instant it will be $\varphi.\frac{\mathrm{d}x}{\mathrm{d}s}+\Delta(\varphi.\frac{\mathrm{d}x}{\mathrm{d}s})$ or $\varphi.\frac{\mathrm{d}x}{\mathrm{d}s}+\Delta\left(\frac{\varphi}{v}.\frac{\mathrm{d}x}{\mathrm{d}t}\right)$, since $\mathrm{d}s=v\mathrm{d}t$.

Now if P, Q, and R be the forces, acting on the body m, in directions parallel to the respective coordinates, the system would remain in equilibrium in virtue of the action of all such forces combined with the elementary differences $\Delta\left(\frac{\mathrm{d}x}{\mathrm{d}t}.\frac{\varphi}{v}\right)$ considered as negative, since these differences are the effects of the results of the forces, and the fluxions are their measures: we shall therefore have, instead of the equation (P) (317),

$$0=\Sigma m\left\{\delta x\left(\mathrm{d}\left(\frac{\mathrm{d}x}{\mathrm{d}t}.\frac{\varphi}{v}\right)-P\mathrm{d}t\right)+\delta y\left(\mathrm{d}\left(\frac{\mathrm{d}y}{\mathrm{d}t}.\frac{\varphi}{v}\right)-Q\mathrm{d}t\right)+\delta z\left(\mathrm{d}\left(\frac{\mathrm{d}z}{\mathrm{d}t}.\frac{\varphi}{v}\right)-R\mathrm{d}t\right)\right\}; \qquad (S)$$

which only differs from it by the substitution of $\frac{\varphi}{v}$ for $\frac{v}{v}$ or unity. This alteration would render its general application to mechanical problems very difficult: we may however derive from it some principles, analogous to those of the preservation of impetus, the constancy of areas, and the motion of the centre of gravity. By substituting dx, dy, and dz, for δx, δy, and δz, we obtain

$$\Sigma m \left\{ \mathrm{d}x\mathrm{d}\left(\frac{\mathrm{d}x}{\mathrm{d}t}.\frac{\varphi}{v}\right) + \mathrm{d}y\mathrm{d}\left(\frac{\mathrm{d}y}{\mathrm{d}t}.\frac{\varphi}{v}\right) + \mathrm{d}z\mathrm{d}\left(\frac{\mathrm{d}z}{\mathrm{d}t}.\frac{\varphi}{v}\right) \right\} = \Sigma mv\mathrm{d}v$$

$\mathrm{d}t\varphi'$[$=\Sigma mv\mathrm{d}\varphi\mathrm{d}t$: for, $\mathrm{d}x\mathrm{d}\left(\frac{\mathrm{d}x}{\mathrm{d}t}.\frac{\varphi}{v}\right) - \frac{\mathrm{d}x^2}{\mathrm{d}t}\mathrm{d}\frac{\varphi}{v} + \frac{\varphi}{v}\mathrm{d}x\mathrm{d}\frac{\mathrm{d}x}{\mathrm{d}t}$, and the three parts together make $\frac{\mathrm{d}s^2}{\mathrm{d}t}\mathrm{d}\frac{\varphi}{v} + \frac{\varphi}{v}\mathrm{d}\frac{\mathrm{d}s^2}{2\mathrm{d}t}$ $= \mathrm{d}\left(\frac{\varphi}{v}.\frac{\mathrm{d}s^2}{\mathrm{d}t}\right) - \frac{\varphi}{v}v\mathrm{d}v\mathrm{d}t = \mathrm{d}\left(\frac{\varphi}{v}v^2\mathrm{d}t\right) - \varphi\mathrm{d}v\mathrm{d}t = \mathrm{d}(\varphi v\mathrm{d}t) - \varphi\mathrm{d}v\mathrm{d}t = v\mathrm{d}\varphi\mathrm{d}t$, and the integral is truly expressed by $\Sigma mv\mathrm{d}\varphi\mathrm{d}t$.] Hence, dividing by d$t$, and taking the fluents, $\Sigma\int mv\mathrm{d}\varphi = c + \Sigma\int m\,(P\mathrm{d}x + Q\mathrm{d}y + R\mathrm{d}z)$; or, supposing the latter member an exact fluxion, and equal to dλ, we have the equation

$$\Sigma\int mv\mathrm{d}\varphi = c + \lambda \qquad (T)$$

an equation resembling (R) (319) and which is converted into it by making $\varphi = v$; consequently the principle of living force is maintained in this hypothesis, if we understand by living force the product of the mass into "twice" the fluent of the velocity multiplied by the fluxion of that function of the velocity, which expresses the force.

336. THEOREM The sum of the finite forces of a system, reduced to any given direction, is constant, and vanishes in the case of equilibrium.

If we substitute, in the equation (S), $\delta x + \delta x'$, for $\delta x'$, $\delta y + \delta y'$, for $\delta y'$, $\delta z + \delta z'$, for δz; $\delta x + \mathrm{d}x''$, for $\delta x''$, ...; we shall obtain, by making the three variations vanish separately, $0 = \Sigma m \left\{ d\left(\frac{dx}{dt}.\frac{\varphi}{v}\right) - Pdt \right\}$; $0 = \Sigma m \left\{ d\left(\frac{dy}{dt}.\frac{\varphi}{v}\right) - Qdt \right\}$, and $0 = \Sigma m \left\{ d\left(\frac{dz}{dt}.\frac{\varphi}{v}\right) - Rdt \right\}$, exactly in the same manner as similar equations have been deduced from (P) in article 322; and as it was inferred in that case, that the motion of the centre of gravity must be uniform, so now, if the system be only subject to the mutual attraction of the bodies comprehended in it, since ΣmP, ΣmQ, and ΣmR are evanescent, on account of the reciprocality of action and reaction, we have $c = \Sigma m \frac{dx}{dt}.\frac{\varphi}{v}$; $c' = \Sigma m \frac{dy}{dt}.\frac{\varphi}{v}$ and $c'' = \Sigma m \frac{dz}{dt}.\frac{\varphi}{v}$; but $m \frac{dx}{dt}.\frac{\varphi}{v} = m\varphi \frac{dx}{ds}$, which is the finite force of the body m, reduced to the direction of x: consequently the sum of the finite forces of the system, reduced to the direction of any given axis, is constant, whatever may be the relation of the force to the velocity; and the state of rest is distinguished by the disappearance of that sum. This result is common to every hypothesis respecting the relation of force to motion, but it is only in the natural state of this relation that the motion of the centre of gravity becomes uniform and rectilinear.

337. THEOREM. The sum of the finite forces, tending to turn the system round any given axis, is constant, and vanishes in the case of equilibrium.

We may make again, in the equation (S),

$$\delta x' = \frac{y'\delta x}{y} + \delta x'_{,}, \ \delta x'' = \frac{y''\delta x}{y} + \delta x''_{,}, \ldots$$

$$\delta y = \frac{-x\delta x}{y} + \delta y_{,}, \quad \delta y' = \frac{-x'\delta x}{y} + \delta y'_{,},$$

so that δx may be made to disappear from the variations of the mutual distances, $f, f', \ldots,$ as in article 323, and from the values of the forces depending on these distances. We shall then have, if the system is free from foreign interference, by making $\delta x = 0$, $0 = \Sigma m. \left\{ x\mathrm{d}\left(\frac{\mathrm{d}y}{\mathrm{d}t} \cdot \frac{\varphi}{v}\right) - y\mathrm{d}\left(\frac{\mathrm{d}x}{\mathrm{d}t} \cdot \frac{\varphi}{v}\right) \right\} + \Sigma m(Py - Qx)\,\mathrm{d}t$; and by taking the fluent,

$c = \Sigma m \dfrac{x\mathrm{d}y - y\mathrm{d}x}{\mathrm{d}t} \cdot \dfrac{\varphi}{v} + \Sigma\!\int\! m\,(Py - Qx)\,\mathrm{d}t$; and in the same manner, taking c' and c'' two other constant quantities,

$$c' = \Sigma m \frac{x\mathrm{d}z - z\mathrm{d}x}{\mathrm{d}t} \cdot \frac{\varphi}{v} + \Sigma\!\int\! m\,(Pz - Rx)\,\mathrm{d}t, \text{ and}$$

$$c'' = \Sigma m \frac{y\mathrm{d}z - z\mathrm{d}y}{\mathrm{d}t} \cdot \frac{\varphi}{v} + \Sigma\!\int\! m(Qz - Ry)\,\mathrm{d}t.$$

If the system is only subject to the mutual actions of its parts, we have $\Sigma m\,(Py - Qx) = 0$; $\Sigma m\,(Pz - Rx) = 0$, and $\Sigma m\,(Qz - Ry) = 0$, as has been shown in article 323: and $m\left(x\dfrac{\mathrm{d}y}{\mathrm{d}t} - y\dfrac{\mathrm{d}x}{\mathrm{d}t}\right)\dfrac{\varphi}{v}$ is the rotatory power of the finite force of the body m, reduced to the plane of x and y, and tending to turn the system round the axis parallel to z; consequently the integral $\Sigma m\left(\dfrac{x\mathrm{d}y - y\mathrm{d}x}{\mathrm{d}t}\right)\dfrac{\varphi}{v}$ is the sum of the rotatory powers of all the finite forces of the bodies of the system, with respect to the same axis; and this sum is shown to be constant: and in the state of equilibrium it vanishes: so that there is here the same difference, between the conditions of motion and of rest, as with respect to the forces parallel to any given axis. In the case of the natural relation of the forces, this property implies the constancy of the areas described by the revolving radii in given times; but this constancy is not observed in the case of any other supposed relation.

[SCHOLIUM. The definition of rotatory power ought properly to have been premised to this demonstration, but its nature is so purely speculative, that it was thought unnecessary to anticipate any part of the important investigation of rotation on this occasion; it is already as intelligible as there is any reason to desire: especially considering the unavoidable confusion attending the idea of a finite force as possessed by a moving body, which is almost incompatible with the true conception of force, as a *cause of a change of motion.*]

338. THEOREM. The sum of the fluents of the finite forces of a system, multiplied by the fluxions of the paths described, is always a minimum, and vanishes in the case of equilibrium.

By taking the variation of the function $\Sigma\int m\varphi ds$, which is here considered, we have $\delta\Sigma\int m\varphi ds = \Sigma\int m\varphi\delta ds + \Sigma\int m ds\delta\varphi$: now $\delta ds = \frac{dx\delta dx + dy\delta dy + dz\delta dz}{ds} = \frac{1}{v}\left(\frac{dx}{dt}d\delta x + \frac{dy}{dt}d\delta y + \frac{dz}{dt}d\delta z\right)$ (265); consequently $\delta\Sigma\int m\varphi ds = \Sigma\frac{m\varphi}{v}\left(\frac{dx}{dt}\delta x + \frac{dy}{dt}\delta y + \frac{dz}{dt}\delta z\right) - \Sigma\int m\left\{\delta x\, d\left(\frac{dx}{dt}.\frac{\varphi}{v}\right) + \delta y\, d\left(\frac{dy}{dt}.\frac{\varphi}{v}\right) + \delta z\, d\left(\frac{dz}{dt}.\frac{\varphi}{v}\right)\right\} + \Sigma\int m\delta\varphi ds$, since $d\left\{\frac{m\varphi}{v}\left(\frac{dx}{dt}\delta x + \frac{dy}{dt}\delta y + \frac{dz}{dt}\delta z\right)\right\} = \frac{m\varphi}{v}.\frac{dx}{dt}d\delta x + \delta x\, d\left(\frac{m\varphi}{v}.\frac{dx}{dt}\right)\ldots$ Now the terminations of the curves, described by the different bodies of the system, being fixed, the part of the variation, not included under the sign $\int$, must vanish for the whole paths, so that we have from the equation (S), $\delta\Sigma\int m\varphi \mathrm{d}s = \Sigma\int m \mathrm{d}s \delta\varphi - \Sigma\int m \mathrm{d}t\ (P\delta x + Q\delta y + R\delta z)$; but the variation of the equation (T), multiplied by dt, affords us $\delta\Sigma\int mv \mathrm{d}t \mathrm{d}\varphi = \Sigma\int m \mathrm{d}t\ (P\delta x + Q\delta y + R\delta z)$; or, $\Sigma\int mv \mathrm{d}t\delta\varphi = \Sigma\int m\delta\varphi \mathrm{d}s = \Sigma\int m \mathrm{d}t\ (P\delta x + Q\delta y + R\delta z)$ [since the variation of any quantity is always the same as its fluxion, with the substitution of the character of a variation for that of a fluxion: the steps, by which a variation and a fluxion are obtained, being always identical and undistinguishable; consequently $\delta\Sigma\int m\varphi \mathrm{d}s = 0$. This equation corresponds to the law of least action, in the natural relation of the force to the velocity, since $m\varphi$ is the total force of the body m; and the principle implies that the sum of the fluents of the finite forces of all the bodies of the system, combined with the elements of the spaces described, is a minimum; and in this form the principle is applicable to any relation between the force and the velocity, that can be supposed to be mathematically possible. In the state of equilibrium, the sum of the forces, multiplied by the elements of their lines of direction, disappears, in consequence of the principle of virtual velocities; so that, in all cases, the same differential function, which disappears in the state of equilibrium, becomes, after taking the fluent, a minimum in the state of motion.

CHAPTER VII.

OF THE MOTIONS OF A SOLID BODY OF ANY GIVEN DIMENSIONS.

---

§ 25, 26. [*Introduction.*] *Equations which determine the progressive and rotatory motion of the body in question.*

[339. THEOREM. "349." When a system of bodies has a rotatory motion round any centre, the effect of each body, in turning the system round a given point, must be estimated by the product of its momentum into the distance of the body from that point; and the power of each body, with respect to the original centre of rotation, will be expressed by the product of the mass into the square of the distance.

Suppose the bodies A and *B*, fixed to the ends of two equal levers, to meet each other, and simply to communicate their motion, and let B be twice A, and moving with half its velocity, then the motion of A will exactly destroy the motion of B, and this effect is therefore the measure of the motion of A: but if the bodies A and B be connected with the arms of an inflexible line, and move with equal velocities in the same direction, they will obviously be totally stopped by the application of a fulcrum at the centre of gravity ; for the propositions respecting equilibrium are as well deducible from the composition of motion as from that of force, and the motion of A is here equivalent to the motion of *B*, which now moves with equal velocity at half the distance from the fulcrum, being still twice as large as A : but it was before shown to be equal to the motion of *B*, when it moved with half the velocity at a distance equal to its own : consequently these two motions of *B* are equivalent, with respect to effect in producing rotatory motion : and the same may be shown when the bodies and their motions are in any other proportions. It is also obvious, that since the velocity is as the distance from the centre of rotation, the power, with respect to that centre, will be as the square of that distance, or as the square of the velocity.

SCHOLIUM. It is therefore of importance to bear in mind, that although the equilibrium of a system of bodies is determined by the equality of the product of their weight into their effective distances on each side of the centre, yet the estimation of the mechanical power of each body, when once in motion, requires the mass to be multiplied by the square of the distance, or of the velocity. For this reason, and for some others, the square of the velocity has been considered by many persons as affording the true measure of force ; but the properties of motion, concerned in the determination of rotatory power, are in reality no more than necessary consequences of the simpler laws, on which the whole theory of mechanics is founded. It is only within about half a century, that the mechanical philosophers of Great Britain have begun to entertain correct notions on this subject; they had been perhaps in some degree misled by an accidental error committed by Newton in computing the precession of the equinoxes: the experiments of Smeaton served to set the question in a clearer point of view, and Dr. Wollaston has more lately removed every remaining obscurity from the subject, in one of his Bakerian Lectures, published in the Philosophical Transactions. Mr. Smeaton's apparatus consisted of a vertical axis, turned by a thread, passing over a pulley, and supporting a scale with weights; the thread was applied to different parts of the axis, having different diameters, and the axis supported two arms, on which two leaden weights were fixed, their distances being variable at pleasure. The experiment being thus arranged, the same force produces, in the same time, but half the velocity, in the same situation of the weights, when the thread is applied to a part of the axis of half the diameter: and if the weights are removed to a double distance from the axis, a quadruple force will be required in order to produce an equal angular velocity in a given time.

340. Definition. "350" The centre of gyration is a point, into which if all the particles of a revolving body were condensed, with its actual velocity, the body would retain the same quantity of rotatory power; and the radius of gyration is the distance of this point from the axis of motion.

341. Definition. The rotatory inertia of a body with respect to any given axis, is the sum of all the products of the elementary

particles, multiplied by the squares of their distances from that axis.

SCHOLIUM 1. Consequently the rotatory inertia is equal to the mass multiplied by the square of the radius of gyration. This product is generally called on the continent the "momentum of inertia," but there is no reason for abandoning the Newtonian acceptation of the word momentum.

SCHOLIUM 2. The elements and the squares of the distances being always positive, the products must be always positive, and any addition to the bulk of a body, wherever applied, will always increase the rotatory inertia.

SCHOLIUM 3. The rotatory inertia will generally be different with respect to different axes, but the various cases are often easily deduced from each other, especially when the axes are parallel.]

342. THEOREM. If x, y, and z be the coordinates of the centre of gravity of a body, of which the particles are subjected to the forces P, Q, and R, acting in the respective directions, the sum of the quantities relating to all the particles being denoted by the characteristic S, m being the mass, and Dm the particle, we shall have the equations $m\frac{ddx}{dt^2}=$ SPDm, $m\frac{ddy}{dt^2}=$SQDm, and $m\frac{ddz}{dt^2}=$SRDm.(A)

The fluxional equations of the progressive and rotatory motions of a solid body may easily be deduced from those which have been demonstrated in the fifth chapter; but their importance in the system of the world makes it convenient to develope them somewhat more in detail.

If the coordinates of the particle Dm, referred to the centre of gravity, be x', y', z', so that its whole motion is determined by the sums $x+x'$, $y+y'$ and $z+z'$; "the forces destroyed at each instant in the particle Dm, in the respective directions, considering the element of the time as constant, will be

$$\frac{\mathrm{dd}x+\mathrm{dd}x'}{\mathrm{d}t}\mathrm{D}m+P\mathrm{d}t\mathrm{D}m\,;$$

$$-\frac{\mathrm{dd}y+\mathrm{dd}y'}{\mathrm{d}t}\mathrm{D}m+Q\mathrm{d}t\mathrm{D}m\,;\text{ and}$$

$$\frac{\mathrm{dd}z+\mathrm{dd}z'}{\mathrm{d}t}\mathrm{D}m+R\mathrm{d}t\mathrm{D}m.$$

It is therefore necessary that all the forces thus destroyed should be in equilibrium with each other" [that is, as causes and effects]: and that the sum of all the forces parallel to any given axis, should vanish (307): hence we have the three following equations

$$\mathrm{S}\,\frac{\mathrm{dd}x+\mathrm{dd}x'}{\mathrm{d}t}\mathrm{D}m=\mathrm{S}P\mathrm{D}m,\ \mathrm{S}\,\frac{\mathrm{dd}y+\mathrm{dd}y'}{\mathrm{d}t}\mathrm{D}m=\mathrm{S}\,Q\mathrm{D}m;$$

and $\mathrm{S}\,\frac{\mathrm{dd}z+\mathrm{dd}z'}{\mathrm{d}t}\mathrm{D}m=\mathrm{S}R\mathrm{D}m.$ Now since x, y, and z are the same for all the particles, they may be excluded from the quantity under the sign S; so that we have $\mathrm{S}\,\frac{\mathrm{dd}x}{\mathrm{d}t^2}\mathrm{D}m=m\,\frac{\mathrm{dd}x}{\mathrm{d}t^2},\ldots$; we have also, by the nature of the centre of gravity $\mathrm{S}x'\mathrm{D}m=0,\ldots$; consequently $\mathrm{S}\,\frac{\mathrm{dd}x'}{\mathrm{d}t^2}\mathrm{D}m=0$, $\mathrm{S}\,\frac{\mathrm{dd}y'}{\mathrm{d}t^2}\mathrm{D}m=0$, and $\mathrm{S}\frac{\mathrm{dd}z'}{\mathrm{d}t^2}\mathrm{D}m=0$: and lastly

$$m\,\frac{\mathrm{dd}x}{\mathrm{d}t^2}-\mathrm{S}P\mathrm{D}m,\ m\frac{\mathrm{dd}y}{\mathrm{d}t^2}=\mathrm{S}Q\,\mathrm{D}m,\text{ and } m\,\frac{\mathrm{dd}z}{\mathrm{d}t^2}=\mathrm{S}\ R\,\mathrm{D}\,m.$$

These three equations determine the motion of the centre

of gravity of the body, and being analogous to the equations of article 323 relating to a system of bodies.

[SCHOLIUM. There can be no objection, in the strictest geometrical sense, to the employment of the character D to denote the element of a material body, as we have no evidence to make it necessary to suppose that the particles of matter are infinitely small, or that one material body is ever incommensurable to another: but then the particular character S must always be applied to the corresponding integral, which is here an actual sum.]

343. THEOREM. Retaining the same notation, (342) and making x', y', and z' the ordinates of the particles with respect to the centre of gravity, we have also

$$S\,\frac{x'dy'-y'dx'}{dt}\,Dm=S\!\int(Qx'-Py')\,dt\;Dm=N;$$

$$S\,\frac{x'dz'-z'dx'}{dt}\,Dm=S\!\int(Rx'-Pz')\,dt\;Dm=N';$$

$$S\,\frac{y'dz'-z'dy'}{dt}\,Dm=S\!\int(Ry'-Qz')\,dt\;Dm=N''. \qquad (B)$$

Since it is necessary for the equilibrium of a solid body, that the sum of the forces parallel to x multiplied by the distances of their lines of direction from the axis parallel to z, diminished by the sum of the forces parallel to y multiplied by their distances from the same axis, should vanish: we shall have

$$S\left\{(x+x')\frac{ddy+ddy'}{dt^2}-(y+y')\frac{ddx+ddx'}{dt^2}\right\}Dm=$$

$S\left\{(x+x')\,Q-(y+y')\,P\right\}Dm$ "(1)" : and since

$S\,(xddy-yddx)\,Dm=m\,(xddy-yddx)$, "[2]"; and

$S\,(Qx-Py)\,Dm=xSQDm-ySPDm$, "[3]": and lastly

S $(x'ddy+xddy'-y'ddx-yddx')$ Dm=ddy Sx'Dm— ddx Sy'D$m+x$ S ddy'D$m-y$ Sddx' Dm ; each of the terms of the second member of this equation being equal to nothing, by the properties of the centre of gravity, the equation "(1)" will become S$\frac{x'ddy'-y'ddx'}{dt^2}Dm$=S$(Qx'$ — $Py')$(Dm), since the parts [2] and [3] destroy each other in consequence of the equation (A) of the preceding proposition; and the fluent of this expression, considered with respect to the time t, gives us S$\frac{x'dy'-y'dx'}{dt}$ Dm=S$\int$ $(Qx'$ — $Py')$ dtDm.

SCHOLIUM 1. These three equations include the principle of the constancy of the areas described; they are sufficient to determine the rotatory motion of the body, round its centre of gravity, and in combination with the three equations of the preceding proposition, they afford us the complete determination of the progressive and rotatory motion of the body.

SCHOLIUM 2. If the body is attached to a fixed point with liberty to move round it, the motions may be determined by means of this proposition, as is obvious from article 308 ; but in that case the coordinates x', y', and z' must be supposed to originate at that point.

344. DEFINITION. The three principal axes of rotation of any body are those, with respect to which the three sums S$x''y''$Dm, S$x''z''$Dm, and S$y''z''$Dm vanish, x'', y'', and z'' being axes moveable with the body.

[SCHOLIUM 1. If the body revolve about x'', the sum S$x''y''$Dm will be the effect of the centrifugal force of all the particles Dm, tending to turn the body about z'', since this force is simply as the distance from x'' (261) or as $\surd(y''^2+z''^2)$, but when reduced to the direction of y'', as y'' only, acting on the lever x''; and the same sum will obviously be the rotatory pressure with regard to the same axis, if y'' instead of x'' be the axis of rotation.

SCHOLIUM 2. The rotatory inertia, with respect to these three axes, is $\mathrm{S}\,(x''^2+y''^2)\,\mathrm{D}m=C$, $\mathrm{S}\,(x''^2+z''^2)\,\mathrm{D}m=B$, and $\mathrm{S}(y''^2+z''^2)\mathrm{D}m=A$ respectively.

SCHOLIUM 3. The evanescence of $\mathrm{S}x''y''\mathrm{D}m$ and $\mathrm{S}x''z''\mathrm{D}m$ determines only the position of the axis x''; but when that of $\mathrm{S}y''z''\mathrm{D}m$ is added, it obviously gives us the two necessary conditions with respect both to y'' and to z'', since we have for the former $\mathrm{S}y''x''\mathrm{D}m=0$ and $\mathrm{S}y''z''\mathrm{D}m=0$, and for the latter $\mathrm{S}z''x''\mathrm{D}m=0$, and $\mathrm{S}z''y''\mathrm{D}m=0$.]

345. THEOREM. If x'', y'', and z'', parallel to the principal axes of rotation of a solid, be the coordinates of the particle Dm, A, B, and C, the rotatory inertia with respect to these axes, θ the angle made by the plane of x'' and y'' with that of x and y, φ the distance of x'' from the intersection, and ψ [that of x, which is also] the complement of the angle made with x by the projection of z'' on the same plane; putting $d\varphi-d\psi\cos\theta=p\,dt$, $d\psi\sin\theta\sin\varphi-d\theta\cos\varphi=q\,dt$, and $d\psi\sin\theta\cos\varphi+d\theta\sin\varphi=r\,dt$, we shall have

$Aq\sin\theta\sin\varphi+Br\sin\theta\cos\varphi-Cp\cos\theta=-N$

$(Aq\cos\theta\sin\varphi+Br\cos\theta\cos\varphi+Cp\sin\theta)\cos\psi$

$+(Br \sin \varphi - Aq \cos \varphi) \sin \psi - -N'$

$(Br \sin \varphi - Aq \cos \varphi) \cos \psi - (Aq \cos \theta \sin \varphi + Br \cos \theta \cos \varphi + Cp \sin \theta) \sin \psi = -N''$;

N being $= S\int(Qx' - Py')\mathrm{d}t\,\mathrm{D}m$, $N' = S\int(Rx' - Pz')\mathrm{d}t\,\mathrm{D}m$, and $N'' = S\int(Ry' - Qz')\mathrm{d}t\,\mathrm{D}m$, (343); and x', y', and z' being the coordinates, referred to the centre of gravity, and parallel to x, y, and z. (C)

We have first, for x', y' and z', which are the x, y, and z of article 324,

$x' = x''(\cos \theta \sin \psi \sin \varphi + \cos \psi \cos \varphi) + y''(\cos \theta \sin \psi \cos \varphi - \cos \psi \sin \varphi) + z'' \sin \theta \sin \psi$.

$y' = x''(\cos \theta \cos \psi \sin \varphi - -\sin \psi \cos \varphi) + y'' \cos \theta \cos \psi \cos \varphi + \sin \psi \sin \varphi) + z'' \sin \theta \cos \psi$.

$z' = z'' \cos \theta - y'' \sin \theta \cos \varphi - x'' \sin \theta \sin \varphi$: [and if we substitute for these equations, in order to shorten a very tedious reduction, $x' = \alpha x'' + \beta y'' + \gamma z''$, $y' = \delta x'' + \varepsilon y'' + \zeta z''$; and then, in order to obtain the value of $x'\mathrm{d}y' - y'\mathrm{d}x'$ (343), make $\mathrm{d}x' = \alpha' x'' + \beta' y'' + \gamma' z''$, and $\mathrm{d}y' = \delta' x'' + \varepsilon' y'' + \zeta' z''$, we may omit in the products all the terms containing $x''y''$, $x''z''$, or $y''z''$, since their sum vanishes for the whole body, and we shall obtain a result in the form $A'x''^2 + B'y''^2 + C'z''^2$, which may be transformed into $\frac{1}{2}(B' + C' - A)A + \frac{1}{2}(A' + C' - B')B + \frac{1}{2}(A' + B' - C')C$, for the whole body, since $A = S(y''^2 + z''^2)\mathrm{D}m$, $B = S(x''^2 + z''^2)\mathrm{D}m$, and $C = S(x''^2 + y''^2)\mathrm{D}m$. Now for $x'\mathrm{d}y' - y'\mathrm{d}x'$, we have $\alpha\delta' x''^2 + \beta\varepsilon' y''^2 + \gamma\zeta' z''^2 - \alpha'\delta x''^2 - \beta'\varepsilon y''^2 - \gamma'\zeta z''^2$, and $A' = \alpha\delta' - \alpha'\delta$, $B' = \beta\varepsilon' - \beta'\varepsilon$, and $C' = \gamma\zeta' - \gamma'\zeta$.

Again,

$\alpha = \cos \theta \sin \psi \sin \varphi + \cos \psi \cos \varphi$

$\alpha' = -d\theta.\ \sin\theta \sin\psi \sin\varphi + d\psi.(\cos\theta \cos\psi \sin\varphi - \sin\psi$
$\cos\varphi)$
$+ d\varphi.(\cos\theta \sin\psi \cos\varphi - \cos\psi \sin\varphi)$

$\delta = \cos\theta \cos\psi \sin\varphi - \sin\psi \cos\varphi$

$\delta' = -d\theta.\ \sin\theta \cos\psi \sin\varphi - d\psi.(\cos\theta \sin\psi \sin\varphi + \cos\psi$
$\cos\varphi)$
$+ d\varphi.(\cos\theta \cos\psi \cos\varphi - \sin\psi \sin\varphi)$

$\beta = \cos\theta \sin\psi \cos\varphi - \cos\psi \sin\varphi$

$\beta' = -d\theta.\ \sin\theta \sin\psi \cos\varphi + d\psi.(\cos\theta \cos\psi \cos\varphi + \sin\psi$
$\sin\varphi)$
$-d\varphi.\ (\cos\theta \sin\psi \sin\varphi + \cos\psi \cos\varphi)$

$\varepsilon = \cos\theta \cos\psi \cos\varphi + \sin\psi \sin\varphi$

$\varepsilon' = -d\theta.\ \sin\theta \cos\psi \cos\varphi - d\psi.\ (\cos\theta \sin\psi \cos\varphi - \cos\psi$
$\sin\varphi)$
$-d\varphi.\ (\cos\theta \cos\psi \sin\varphi - \sin\psi \cos\varphi)$

$\gamma = \sin\theta \sin\psi$

$\gamma' = d\theta.\ \cos\theta \sin\psi + d\psi.\ \sin\theta \cos\psi$

$\zeta = \sin\theta \cos\psi$

$\zeta' = d\theta.\ \cos\theta \cos\psi - d\psi.\ \sin\theta \sin\psi.$ Hence

$\alpha\delta' = -d\theta.\ (\sin\cos\theta \sin\cos\psi \sin^2\varphi + \sin\theta \cos^2\psi \sin\cos\varphi)$
$-d\psi.\ (\cos^2\theta \sin^2\psi \sin^2\varphi + \cos\theta \sin\cos\psi \sin\cos\varphi +$
$\cos\theta \sin\cos\psi \sin\cos\varphi + \cos^2\psi \cos^2\varphi)$
$+d\varphi.\ (\cos^2\theta \sin\cos\psi \sin\cos\varphi + \cos\theta \sin^2\psi \sin^2\varphi +$
$\cos\theta \cos^2\psi \cos^2\varphi + \sin\cos\psi \sin\cos\varphi)$

$\alpha'\delta = -d\theta.\ (\sin\cos\theta \sin\cos\psi \sin^2\varphi - \sin\theta \sin^2\psi \sin\cos\varphi)$
$+d\psi.\ (\cos^2\theta \cos^2\psi \sin^2\varphi - \cos\theta \sin\cos\psi \sin\cos\varphi -$
$\cos\theta \sin\cos\psi \sin\cos\varphi + \sin^2\psi \cos^2\varphi)$
$+d\varphi.\ (\cos^2\theta \sin\cos\psi \sin\cos\varphi - \cos\theta \cos^2\psi \sin^2\varphi -$
$\cos\theta \sin^2\psi \cos^2\varphi + \sin\cos\psi \sin\cos\varphi)$

$$\left.\begin{array}{l} \alpha\delta'\ \ \alpha'\delta = -d\theta.\ \sin\theta \sin\cos\varphi \\ \quad -d\psi.\ (\cos^2\theta \sin^2\varphi + \cos^2\varphi) \\ \quad +d\varphi.\ (\cos\theta \sin^2\varphi + \cos\theta \cos^2\varphi) \text{ or } ..+ \\ \qquad d\varphi.\ \cos\theta \end{array}\right\} = A'$$

$$\beta\varepsilon' = -d\theta.\ (\sin\cos\theta\ \sin\cos\psi\ \cos^2\varphi - \sin\theta\ \cos^2\psi\ \sin\cos\varphi)$$
$$-d\psi.\ (\cos^2\theta\ \sin^2\psi\ \cos^2\varphi - \cos\theta\ \sin\cos\psi\ \sin\cos\varphi$$
$$-\cos\theta\ \sin\cos\psi\ \sin\cos\varphi + \cos^2\psi\ \sin^2\varphi)$$
$$-d\varphi.\ (\cos^2\theta\ \sin\cos\psi\ \sin\cos\varphi - \cos\theta\ \sin^2\psi\ \cos^2\varphi$$
$$-\cos\theta\ \cos^2\psi\ \sin^2\varphi + \sin\cos\psi\ \sin\cos\varphi)$$
$$\beta'\varepsilon = -d\theta.\ (\sin\cos\theta\ \sin\cos\psi\ \cos^2\varphi + \sin\theta\ \sin^2\psi\ \sin\cos\varphi)$$
$$+d\psi.\ (\cos^2\theta\ \cos^2\psi\ \cos^2\varphi + \cos\theta\ \sin\cos\psi\ \sin\cos\varphi$$
$$+\cos\theta\ \sin\cos\psi\ \sin\cos\varphi + \sin^2\psi\ \sin^2\varphi)$$
$$-d\varphi.\ (\cos^2\theta\ \sin\cos\psi\ \sin\cos\varphi + \cos\theta\ \cos^2\psi\ \cos^2\varphi$$
$$+\cos\theta\ \sin^2\psi\ \sin^2\varphi + \sin\cos\psi\ \sin\cos\varphi)$$
$$\left.\begin{array}{l}\beta\varepsilon' - \beta'\varepsilon = d\theta.\ \sin\theta\ \sin\cos\varphi \\ \quad -d\psi.\ (\cos^2\theta\ \cos^2\varphi + \sin^2\varphi) \\ \quad +d\varphi.\ \cos\theta\ (\cos^2\varphi + \sin^2\varphi) \text{ or } ..+d\varphi.\ \cos\theta\end{array}\right\} = B'$$
$$\gamma\zeta' = d\theta.\ \sin\cos\theta\ \sin\cos\psi$$
$$-d\psi.\ \sin^2\theta\ \sin^2\psi$$
$$\gamma'\zeta = d\theta.\ \sin\cos\theta\ \sin\cos\psi$$
$$+d\psi.\ \sin^2\theta\ \cos^2\psi$$
$$\gamma\zeta' - \gamma'\zeta = -d\psi\ \sin^2\theta = C'.$$

Combining these results, we have $B' + C' - A' = 2d\theta.\ \sin\theta\ \sin\cos\varphi - d\psi\ (\cos^2\theta\ \cos^2\varphi + \sin^2\varphi + \sin^2\theta - \cos^2\theta\ \sin^2\varphi - \cos^2\varphi)$ or $..d\psi\ (\cos^2\theta\ (1 - \sin^2\varphi - \sin^2\varphi) + \sin^2\varphi - 1 + \sin^2\varphi + \sin^2\theta) = d\psi(\cos^2\theta - 2\cos^2\theta\ \sin^2\varphi + 2\sin^2\varphi - 1 + \sin^2\theta) = d\psi \left\{ 2\sin^2\varphi\ (1 - \cos^2\theta) \right\} = 2\ d\psi.\ \sin^2\varphi\ \sin^2\theta$, and $\frac{1}{2}(B' + C' - A')\ A = (d\theta.\ \sin\theta\ \sin\cos\varphi - d\psi.\ \sin^2\theta\ \sin^2\varphi)\ A$.

If we subtract this from C', the remainder will be $\frac{1}{2}(A' + C' - B) = -d\theta.\ \sin\theta\ \sin\cos\varphi + d\psi.\ (\sin^2\theta\ \sin^2\varphi - \sin^2\theta) = -d\theta.\ \sin\theta\ \sin\cos\varphi - d\psi.\ \sin^2\theta\ \cos^2\varphi$, the coefficient of B.

Subtracting the same quantity from B', we obtain $\frac{1}{2}(A' + B' - C') = d\psi.\ (\sin^2\theta\ \sin^2\varphi - \cos^2\theta\ \cos^2\varphi - \sin^2\varphi) + d\varphi.\ \cos\theta = d\psi. \left\{ -\cos^2\theta\ (\sin^2\varphi + \cos^2\varphi) \right\} + d\varphi.\ \cos\theta = -$

$d\psi . \cos^2\theta + d\varphi \cos\theta$; which is the coefficient of C in the value of $x'dy' - y'dx'$.

We have next to perform a similar computation for the areas $x'dz' - z'dx'$, and $y'dz' - z'dy'$: and the same characters may again be employed in each of these cases with their appropriate significations: half of them retaining the same values.

$$\alpha = \cos\theta \sin\psi \sin\varphi + \cos\psi \cos\varphi$$

$$\alpha' = -d\theta . \sin\theta \sin\psi \sin\varphi + d\psi (\cos\theta \cos\psi \sin\varphi - \sin\psi \cos\varphi) + d\varphi . (\cos\theta \sin\psi \cos\varphi - \cos\psi \sin\varphi)$$

$$\delta = -\sin\theta \sin\varphi$$

$$\delta' = d\theta . \cos\theta \sin\varphi - d\varphi . \sin\theta \cos\varphi$$

$$\beta = \cos\theta \sin\psi \cos\varphi - \cos\psi \sin\varphi$$

$$\beta' = -d\theta . \sin\theta \sin\psi \cos\varphi + d\psi (\cos\theta \cos\psi \cos\varphi + \sin\psi \sin\varphi) - d\varphi . (\cos\theta \sin\psi \sin\varphi + \cos\psi \cos\varphi)$$

$$\varepsilon = -\sin\theta \cos\varphi$$

$$\varepsilon' = -d\theta . \cos\theta \cos\varphi + d\varphi . \sin\theta \sin\varphi$$

$$\gamma = \sin\theta \sin\psi$$

$$\gamma' = d\theta . \cos\theta \sin\psi + d\psi . \sin\theta \cos\psi$$

$$\zeta = \cos\theta$$

$$\zeta' = -d\theta . \sin\theta$$

$$\alpha\delta' = -d\theta . (\cos^2\theta \sin\psi \sin^2\varphi + \cos\theta \cos\psi \sin\cos\varphi) . - d\varphi . (\sin\cos\theta \sin\psi \sin\cos\varphi + \sin\theta \cos\psi \cos^2\varphi)$$

$$\alpha'\delta = d\theta . \sin^2\theta \sin\psi \sin^2\varphi - d\psi . (\sin\cos\theta \cos\psi \sin^2\varphi \sin\theta \sin\psi \sin\cos\varphi) - d\varphi (\sin\cos\theta \sin\psi \sin\cos\varphi - \sin\theta \cos\psi \sin^2\varphi)$$

$$\left. \begin{aligned} \alpha\delta' - \alpha'\delta = & -d\theta . (\sin\psi \sin^2\varphi + \cos\theta \cos\psi \sin\cos\varphi) \\ & + d\psi (\sin\cos\theta \cos\psi \sin^2\varphi - \sin\theta \sin\psi \sin\cos\varphi) \\ & - d\varphi . \sin\theta \cos\psi \end{aligned} \right\} = A'$$

$\beta\varepsilon' = -d\theta.(\cos^2\theta \sin\psi \cos^2\varphi - \cos\theta \cos\psi \sin\cos\varphi)$
$+ d\varphi.(\sin\cos\theta \sin\psi \sin\cos\varphi - \sin\theta \cos\psi \sin^2\varphi)$.

$\beta'\varepsilon = \quad d\theta. \sin^2\theta \sin\psi \cos^2\varphi - d\psi.(\sin\cos\theta \cos\psi \cos^2\varphi + \sin\theta \sin\psi \sin\cos\varphi)$
$+ d\varphi.(\sin\cos\theta \sin\psi \sin\cos\varphi + \sin\theta \cos\psi \cos^2\varphi)$

$$\left.\begin{aligned}\beta\varepsilon' - \beta'\varepsilon = &- d\theta.(\sin\psi \cos^2\varphi - \cos\theta \cos\psi \sin\cos\varphi)\\ &+ d\psi.(\sin\cos\theta \cos\psi \cos^2\varphi + \sin\theta \sin\psi \sin\cos\varphi)\\ &- d\varphi. \sin\theta \cos\psi\end{aligned}\right\} = B'$$

$\gamma\zeta' = -d\theta. \sin^2\theta \sin\psi$

$\gamma'\zeta - \quad d\theta. \cos^2\theta \sin\psi + d\psi. \sin\cos\theta \cos\psi$

$\gamma\zeta' - \gamma'\zeta = -d\theta. \sin\psi - d\psi \sin\cos\theta \cos\psi \qquad = C'$

Hence $B' + C' \quad A' = --d\theta.(\sin\psi(\cos^2\varphi - \sin^2\varphi) - 2\cos\theta \cos\psi \sin\cos\varphi + \sin\psi) + d\psi.(\sin\cos\theta \cos\psi(\cos^2\varphi - \sin^2\varphi) + 2\sin\theta \sin\psi \sin\cos\varphi - \sin\cos\theta \cos\psi) - d\theta.(2\sin\psi \cos^2\varphi - 2\cos\theta \cos\psi \sin\cos\varphi) - d\psi(2\sin\cos\theta \cos\psi \sin^2\varphi - 2\sin\theta \sin\psi \sin\cos\varphi)$, half of which is in the coefficient of A in the value of $-N'$.

For that of B, we subtract this half from C', and obtain $-d\theta(\sin\psi(1 - \cos^2\varphi) + \cos\theta \cos\psi \sin\cos\varphi) + d\psi(\sin\cos\theta \cos\psi(\sin^2\varphi - 1) - \sin\theta \sin\psi \sin\cos\varphi) - -d\theta(\sin\psi \sin^2\varphi + \cos\theta \cos\psi \sin\cos\varphi) - d\psi(\sin\cos\theta \cos\psi \cos^2\varphi + \sin\theta \sin\psi \sin\cos\varphi)$: and subtracting it from B', we have $d\psi. \sin\cos\theta \cos\psi - d\varphi. \sin\theta \cos\psi$.

It is easy to perceive that these are the coefficients already assigned for the value of N', divided by dt: for qdt being $- -d\theta. \cos\varphi + d\psi. \sin\theta \sin\varphi$, we have for $q \cos\theta \sin\varphi \cos\psi - q \cos\varphi \sin\psi$, $-d\theta.(\cos\theta \cos\psi \sin\cos\varphi - \sin\psi \cos^2\varphi) + d\psi.(\sin\cos\theta \cos\psi \sin^2\varphi - \sin\theta \sin\psi \sin\cos\varphi)$; and, since $rdt = d\theta. \sin\varphi + d\psi. \sin\theta \cos\varphi$, for $r \cos\theta \cos\varphi \cos\psi + r \sin\varphi \sin\psi$, $d\theta.(\cos\theta \cos\psi \sin\cos\varphi + \sin\psi \sin^2\varphi) + d\psi.(\sin\cos\theta \cos\psi \cos^2\varphi + \sin\theta \sin\psi \sin$

$\cos\varphi$); and pdt being $= -d\psi \cos\theta + d\varphi$, we obtain for $p \sin\theta \cos\psi$, $-d\psi . \sin\cos\theta\cos\psi + d\varphi . \sin\theta\cos\psi$. In the last place, for N'',

$\alpha = \cos\theta\cos\psi\sin\varphi - \sin\psi\cos\varphi$

$\alpha' - -d\theta . \sin\theta\cos\psi\sin\varphi - d\psi . (\cos\theta\sin\psi\sin\varphi + \cos\psi \cos\varphi)$

$+ d\varphi . (\cos\theta\cos\psi\cos\varphi + \sin\psi\sin\varphi)$

$\delta = -\sin\theta\sin\varphi$

$\delta' = -d\theta . \cos\theta\sin\varphi - d\varphi . \sin\theta\cos\varphi$

$\beta = \cos\theta\cos\psi\cos\varphi + \sin\psi\sin\varphi$

$\beta' = -d\theta . \sin\theta\cos\psi\cos\varphi - d\psi . (\cos\theta\sin\psi\cos\varphi - \cos\psi\sin\varphi)$

$-d\varphi . (\cos\theta\cos\psi\sin\varphi - \sin\psi\cos\varphi)$

$\varepsilon = -\sin\theta\cos\varphi$

$\varepsilon' = -d\theta . \cos\theta\cos\psi + d\varphi . \sin\theta\sin\varphi$

$\gamma = \sin\theta\cos\psi$

$\gamma' = d\theta . \cos\theta\cos\psi - d\psi . \sin\theta\sin\psi$

$\zeta = \cos\theta$

$\zeta' = -d\theta . \sin\theta$

$\alpha\delta' - -d\theta . (\cos^2\theta\cos\psi\sin^2\varphi - \cos\theta\sin\psi\sin\cos\varphi)$

$-d\varphi . (\sin\cos\theta\cos\psi\sin\cos\varphi - \sin\theta\sin\psi\cos^2\varphi)$

$\alpha'\delta = d\theta . \sin^2\theta\cos\psi\sin^2\varphi + d\psi (\sin\cos\theta\sin\psi\sin^2\varphi + \sin\theta \cos\psi\sin\cos\varphi)$

$-d\varphi . (\sin\cos\theta\cos\psi\sin\cos\varphi + \sin\theta\sin\psi\sin^2\varphi)$

$$\left.\begin{array}{l}\alpha\delta' - \alpha'\delta = -d\theta . (\cos\psi\sin^2\varphi - \cos\theta\sin\psi\sin\cos\varphi) \\ \quad -d\psi . (\sin\cos\theta\sin\psi\sin^2\varphi + \sin\theta\cos\psi \sin\cos\varphi) \\ \quad + d\varphi . \sin\theta\sin\psi\end{array}\right\} = A'$$

$\beta\varepsilon' - -d\theta . (\cos^2\theta\cos\psi\cos^2\varphi + \cos\theta\sin\psi\sin\cos\varphi)$

$+ d\varphi . (\sin\cos\theta\cos\psi\sin\cos\varphi + \sin\theta\sin\psi\sin^2\varphi)$

$\beta'\varepsilon - d\theta . \sin^2\theta\cos\psi\cos^2\varphi + d\psi (\sin\cos\theta\sin\psi\cos^2\varphi - \sin\theta\cos\psi\sin\cos\varphi)$

$+ d\varphi . (\sin\cos\theta\cos\psi\sin\cos\varphi - \sin\theta\sin\psi\cos^2\varphi)$

$$\left.\begin{array}{l}\beta\varepsilon'-\beta'\varepsilon=-d\theta.(\cos\psi\cos^2\varphi+\cos\theta\sin\psi\sin\cos\varphi \\ \quad -d\psi.(\sin\cos\theta\sin\psi\cos^2\varphi-\sin\theta\cos\psi\sin \\ \quad\quad \cos\varphi) \\ \quad +d\varphi.\sin\theta\sin\psi\end{array}\right\}=B'$$

$$\gamma\zeta'=-d\theta.\sin^2\theta\cos\psi$$

$$\gamma'\zeta=\ d\theta.\cos^2\theta\cos\psi-d\psi.\sin\cos\theta\sin\psi$$

$$\gamma\zeta'-\gamma'\zeta=-d\theta.\cos\psi+d\psi\sin\cos\theta\sin\psi \qquad =C'$$

We have here $B'+C'-A'-\ -d\theta\,(\cos\psi\,(\cos^2\varphi-\sin^2\varphi+1)+2\cos\theta\sin\psi\sin\cos\varphi)-d\psi\,(\sin\cos\theta\sin\psi\,(\cos^2\varphi-\sin^2\varphi-1)-2\sin\theta\cos\psi\sin\cos\varphi)=-d\theta\,(2\cos\psi\cos^2\varphi+2\cos\theta\sin\psi\sin\cos\varphi)+d\psi.\ (2\sin\cos\theta\sin\psi\sin^2\varphi\ 2\sin\theta\cos\psi\sin\cos\varphi)$ twice the coefficient of A; whence we obtain, as before, for the other coefficients, $-d\theta.\ (\cos\psi\sin^2\varphi-\cos\theta\sin\psi\sin\cos\varphi)+d\psi.\ (\sin\cos\theta\sin\psi\cos^2\varphi-\sin\theta\cos\psi\sin\cos\varphi)$, and $-d\psi.\sin\cos\theta\sin\psi+d\varphi.\sin\theta\sin\psi$; which we must compare with $-q\cos\varphi\cos\psi-q\cos\theta\sin\varphi\sin\psi$, with $r\sin\varphi\cos\psi-r\cos\theta\cos\varphi\sin\psi$, and with $-p\sin\theta\sin\psi$ respectively: of these the first becomes $+d\theta.\ (\cos^2\varphi\cos\psi+\cos\theta\sin\cos\varphi\sin\psi)-d\psi.\ (\sin\theta\sin\cos\varphi\cos\psi+\sin\cos\theta\sin^2\varphi\sin\psi)$, the second $d\theta.\ (\sin^2\varphi\cos\psi-\cos\theta\sin\cos\varphi\sin\psi)+d\psi.\ (\sin\theta\sin\cos\varphi\cos\psi-\sin\cos\theta\cos^2\varphi\sin\psi)$ and the third $d\psi.\sin\cos\theta\sin\psi-d\varphi.\sin\theta\sin\psi$; agreeing in each instance with the reduction here detailed, which is inserted more for the sake of preserving the uniformity of system, and of leaving nothing undemonstrated, than for its immediate importance to a student.]

346. THEOREM. Retaining the notation of the last propositions, and making ψ infinitely small, or x infinitely near to the plane of x'' and y'', putting also $Cp=p'$, $Aq=q'$, and $Br=r'$, we obtain the equations

$$dp' + \frac{B-A}{AB} \cdot q'r'dt = dN. \cos\theta - dN'. \sin\theta;$$

$$dq' + \frac{C-B}{CB} \cdot r'p'dt = -(dN. \sin\theta + dN' \cos\theta) \sin\phi + dN''. \cos\phi;$$

$$dr' + \frac{A-C}{AC} p'q'dt = -(dN. \sin\theta + dN'. \cos\theta) \cos\phi \quad dN''. \sin\phi. \qquad (D)$$

The equations for N afford us, by taking their fluxions, when $\sin\psi = 0$, and $\cos\psi = 1$,

$d\theta. \cos\theta\,(Br \cos\phi + Aq \sin\phi) + \sin\theta. d\,(Br \cos\phi + Aq \sin\phi) - d\,(Cp \cos\theta) = -dN;$

$d\psi. (Br \sin\phi - Aq \cos\phi) - d\theta. \sin\theta\,(Br \cos\phi + Aq \sin\phi) + \cos\theta\, d\,(Br \cos\phi + Aq \sin\phi) + d\,(Cp \sin\theta) = -dN';$

$d\,(Br. \sin\phi - Aq \cos\phi) - d\psi. \cos\theta\,(Br \cos\phi + Aq \sin\phi) - Cpd\psi \sin\theta = -dN''.$

Hence we have [$dN. \cos\theta = -d\theta. \cos^2\theta\,(Br \cos\phi + Aq \sin\phi) - \sin \cos\theta. d\,(Br \cos\phi + Aq \sin\phi) + \cos\theta. d\,(Cp \cos\theta)$, and $-dN'. \sin\theta = -d\theta. \sin^2\theta\,(Br \cos\phi + Aq \sin\phi) + \sin \cos\theta. d\,(Br \cos\phi + Aq \sin\phi) + \sin\theta. d\,(Cp \sin\theta) + d\psi. \sin\theta\,(Br \sin\phi - Aq \cos\phi)$, and adding these together, the sum will be $-d\theta\,(Br \cos\phi + Aq \sin\phi + d\,(Cp) + d\psi. \sin\theta\,(Br \sin\phi - Aq \cos\phi) + d\,(Cp) = Br\,(\sin\theta \sin\phi. d\psi - \cos\phi. d\theta) - Aq\,(\sin\phi. d\theta + \sin\theta \cos\phi. d\psi) = Brqdt - Aqrdt + dp'$, and $dp' + (B-A)\,qrdt =$] $dN. \cos\theta - dN'. \sin\theta - dp' + \frac{(B-A)}{AB} . q'r'dt$. In the next place [$-dN. \sin\theta \sin\phi = d\theta. \sin \cos\theta \sin\phi\,(Br \cos\phi + Aq \sin\phi) + \sin^2\theta \sin\phi. d\,(Br \cos\phi + Aq \sin\phi) - \sin\theta \sin\phi. d\,(Cp \cos\theta)$; $-dN'. \cos\theta \sin\phi = -d\theta. \sin \cos\theta \sin\phi\,(Br \cos\phi + Aq \sin\phi) + d\psi. \cos\theta \sin\phi\,(Br \sin\phi - Aq \cos\phi) + \cos^2\theta \sin\phi. d\,(Br \cos\phi + Aq \sin\phi) + \cos\theta \sin\phi. d\,(Cp \sin\theta)$; $+dN'' \cos\phi = d\psi. \cos\theta \cos\phi\,(Br \cos\phi + Aq \sin\phi) - \cos\phi. d\,(Br \sin\phi$

$+Aq\cos\varphi)+Cp\mathrm{d}\psi\sin\theta\cos\varphi$; the sum of which is $\sin\varphi.\mathrm{d}\,(Br\cos\varphi+Aq\sin\varphi)$—$\cos\varphi.\mathrm{d}\,(Br\sin\varphi-Aq\cos\varphi)+\mathrm{d}\theta.(\sin^2\theta+\cos^2\theta).\sin\varphi\,Cp+\mathrm{d}\psi\left\{\cos\theta\sin\varphi\,(Br\sin\varphi-Aq\cos\varphi)+\cos\theta\cos\varphi\,(Br\cos\varphi+Aq\sin\varphi)\right\}+Cp\mathrm{d}\psi\sin\theta\cos\varphi$—$-Br\mathrm{d}\varphi+\mathrm{d}\,(Aq)+\mathrm{d}\theta\,.\sin\varphi\,Cp+\mathrm{d}\psi\,(\cos\theta\,Br+Cp\sin\theta\cos\varphi)=\mathrm{d}q'+Cp\,(\mathrm{d}\theta.\sin\varphi+\mathrm{d}\psi.\sin\theta\cos\varphi)-Br(\mathrm{d}\varphi-\mathrm{d}\psi.\cos\theta)=\mathrm{d}q'+(Cpr-Brp)\,\mathrm{d}t=]\,\mathrm{d}q'+\frac{C-B}{CB}r'p'\mathrm{d}t=-(\mathrm{d}N\sin\theta+\mathrm{d}N'\cos\theta)\sin\varphi+\mathrm{d}N''\cos\varphi$. Lastly [($-\mathrm{d}N.\sin\theta\cos\varphi=\mathrm{d}\theta.\sin\cos\theta\cos\varphi\,(Br\cos\varphi+Aq\sin\varphi)+\sin^2\theta\cos\varphi.\mathrm{d}\,(Br\cos\varphi+Aq\sin\varphi)-\sin\theta\cos\varphi\,\mathrm{d}\,(Cp\cos\theta)$;$-\mathrm{d}N'.\cos\theta\cos\varphi=-\mathrm{d}\theta.\sin\cos\theta\cos\varphi\,(Br\cos\varphi+Aq\sin\varphi)+\mathrm{d}\psi\cos\theta\cos\varphi\,(Br\sin\varphi-Aq\cos\varphi)+\cos^2\theta\cos\varphi\mathrm{d}\,(Br\cos\varphi+Aq\sin\varphi+\cos\theta\cos\varphi.\mathrm{d}\,(Cp\sin\theta)$; and$-\mathrm{d}N''\sin\varphi=-\mathrm{d}\psi\cos\theta\sin\varphi\,(Br\cos\varphi+Aq\sin\varphi)+\sin\varphi\mathrm{d}\,(Br\sin\varphi-Aq\cos\varphi)-Cp\mathrm{d}\psi\sin\theta\sin\varphi$; and adding these together we have $\cos\varphi.\mathrm{d}\,(Br\cos\varphi+Aq\sin\varphi)-\cos\varphi.\mathrm{d}\theta.Cp+\mathrm{d}\psi\cos\theta\left\{\cos\varphi\,(Br\sin\varphi-Aq\cos\varphi)-\sin\varphi\,(Br\cos\varphi+Aq\sin\varphi)\right\}+\sin\varphi\mathrm{d}\,(Br\sin\varphi-Aq\cos\varphi)-Cp.\mathrm{d}\psi\sin\theta\sin\varphi=\mathrm{d}\,(Br)+Aq.\mathrm{d}\varphi+\cos\varphi.\mathrm{d}\theta.\,Cp-\mathrm{d}\psi.\cos\theta\,Aq+Cp\,\mathrm{d}\psi\sin\theta\sin\varphi-\mathrm{d}r'+Aq\,(\mathrm{d}\varphi-\mathrm{d}\psi\cos\theta)-Cp\,(\mathrm{d}\psi\sin\theta\sin\varphi-\mathrm{d}\theta\cos\varphi)=\mathrm{d}r'+Aqp\mathrm{d}t-Cp\mathrm{d}t=]\,\mathrm{d}r'+\frac{A-C}{AC}p'q'\mathrm{d}t=-(\mathrm{d}N\sin\theta+\mathrm{d}N'\cos\theta)\cos\varphi-\mathrm{d}N''\sin\varphi$.

SCHOLIUM. These three equations are very convenient for the computation of the rotatory motion of a body, when it turns very nearly round one of its principal axes, which is the case with the rotations of the heavenly bodies.

§ 27. *Of the principal axes of rotation. In general a body has only one system of principal axes. Of [rotatory] inertia. The greatest and least inertia belong to the principal axes, and the least of all belongs to one of the three principal axes passing through the centre of gravity. Case in which there is an infinity of principal axes.* P. 73.

347. THEOREM. Every material body has at least three principal axes of rotation, at right angles to each other.

In order to determine the situation of the axes x'' y'', and z'', in such a manner as to agree with the conditions of the definition (344) we have (324 Cor.)

$x''=x'(\cos\theta\sin\psi\sin\phi+\cos\psi\cos\phi)$
$\quad+y'(\cos\theta\cos\psi\sin\phi-\sin\psi\sin\phi)-z'\sin\theta\sin\phi$;
$y''=z'(\cos\theta\sin\psi\cos\phi-\cos\psi\sin\phi)$
$\quad+y'(\cos\theta\cos\psi\cos\phi+\sin\psi\sin\phi)-z'\sin\theta\cos\phi$;
$z''=x'\sin\theta\sin\psi+y'\sin\theta\cos\psi+z'\cos\theta$. Hence we readily obtain

$x''\cos\phi-y''\sin\phi=x'\cos\psi-y'\sin\psi$
$x''\sin\phi+y''\cos\phi=x'\cos\theta\sin\psi+y'\cos\theta\cos\psi-z'\sin\theta$.

If we now put $Sx'^2Dm=a^2$, $Sy'^2Dm=b^2$, $Sz'^2Dm=c^2$; $Sx'y'Dm=f$, $Sx'z'Dm=g$, $Sy'z'Dm=h$, we shall have $Sx''z''Dm.\cos\phi-Sy''z''Dm\sin\phi=[S(x'^2\sin\theta\sin\cos\psi+x'y'\sin\theta\cos^2\psi+x'z'\cos\theta\cos\psi-x'y'\sin\theta\sin^2\psi-y'^2\sin\theta\sin\cos\psi-y'z'\cos\theta\sin\psi)Dm=](a^2-b^2)\sin\theta\sin\cos\psi+f\sin\theta(\cos^2\psi-\sin^2\psi)+\cos\theta(g\cos\psi-h\sin\psi)$; $Sx''z''Dm.\sin\phi+Sy''z''Dm.\cos\phi[=S(x'^2\sin\cos\theta\sin^2\psi+x'y'\sin\cos\theta\sin\cos\psi+x'z'\cos^2\theta\sin\psi+x'y'\sin\cos\theta\sin\cos\psi+y'^2\sin\cos\theta\cos^2\psi+y'z'\cos^2\theta\cos\psi-x'z'\sin^2\theta\sin\psi-y'z'\sin^2\theta\cos\psi-z'^2\sin\cos\theta)Dm]=\sin\cos\theta(a^2\sin^2\psi+$

$b^2 \cos^2\psi - c^2 + 2f \sin\cos\psi) + (\cos^2\theta - \sin^2\theta).(g \sin\psi + h \cos\psi)$.

Now since the first members of each of these equations must vanish, the second will vanish also, and we have $\sin\theta \left\{(a^2 - b^2 \sin\psi \cos\psi + f(\cos^2\psi - \sin^2\psi) = \cos\theta\,(h \sin\psi - g \cos\psi)\right.$, consequently $\frac{\sin\theta}{\cos\theta} =$

$\frac{h \sin\psi - g \cos\psi}{(a^2 - b^2)\sin\cos\psi + f(\cos^2\psi - \sin^2\psi)}$; and [since $\sin\cos\theta\,(a^2 \sin^2\psi + b^2 \cos^2\psi - c^2 + 2f \sin\cos\psi) = \sin^2\theta - \cos^2\theta\,(g \sin\psi + h \cos\psi)$, whence

$$\frac{a^2 \sin^2\psi + b^2 \cos^2\psi - c^2 + 2f \sin\psi \cos\psi}{g \sin\psi + h \cos\psi} = \frac{\sin\theta}{\cos\theta} - \frac{\cos\theta}{\sin\theta},$$ or

substituting u for tang $\psi = \frac{\sin\psi}{\cos\psi}$, having divided by $\frac{1}{\cos\psi} = \frac{\cos\psi}{\cos^2\psi}$, $\frac{a^2 \sin^2\psi + b^2 \cos^2\psi - c^2 + 2f \sin\psi \cos\psi}{gu+h} =$

$$\frac{hu-g}{(a^2-b^2)u+f(1-u^2)} - \frac{\left\{(a^2-b^2)u+f(1-u^2)\right\}\cos^2\psi}{hu-g},$$ and

$(a^2 \sin^2\psi + b^2 \cos^2\psi - c^2 + 2f \sin\cos\psi)\left\{(a^2-b^2)u + f(1-u^2)\right\}(hu-g) = (hu-g)^2(gu+h) - \left\{(a^2-b^2)u+f(1-u^2)\right\}^2 \cos^2\psi\,(gu+h)$; or $0 = (gu+h).(hu-g)^2 - \left\{(a^2-b^2)u+f(1-u^2)\right\}\left(\left\{(a^2-b^2)u+f(1-u^2)\right\}\cos^2\psi\,(gu+h) + (a^2 \sin^2\psi + b^2 \cos^2\psi - c^2 + 2f \sin\cos\psi)\left\{hu-g\right\}\right)$, in which the latter part becomes $-c^2(hu-g) + a^2(u \cos^2\psi\,(gu+h) + \sin^2\psi\,(hu-g) - b^2(u \cos^2\psi\,(gu+h) - \cos^2\psi\,(hu-g) + f\left\{(1-u^2)\cos^2\psi(gu+h) + 2 \sin\cos\psi(hu-g)\right\} = -c^2(hu-g)$

$+ a^2 (gu^2 \cos^2\psi + hu \cos^2\psi + hu \sin^2\psi - g \sin^2\psi) - b^2 (gu^2 \cos^2\psi + hu \cos^2\psi - hu \cos^2\psi + g \cos^2\psi) + f (gu \cos^2\psi - gu^3 \cos^2\psi + h \cos^2\psi - hu^2 \cos^2\psi + 2 hu \sin \cos\psi - 2g \sin \cos\psi) = -c^2 (hu - g) + a^2 (hu) - b^2 (g) + f(-gu + h)$, whence the whole equation will be] $0 = (gu + h).(hu - g)^2 + \left\{(a^2 - b^2).u + f(1 - u^2)\right\}\left\{(hc^2 - ha^2 + fg)u + gb^2 - gc^2 - hf\right\}$.

By solving this cubic equation, we may always find a value of u, such that both $\cos\phi\, Sx''z''Dm - \sin\phi\, Sy''z''Dm$ and $\sin\phi\, Sx''z''Dm + \cos\phi\, Sy''z''Dm$ may vanish; cousequently their squares and the sum of their squares $(Sx''z''Dm)^2 + (Sy''z''Dm)^2$ will vanish, and each of these integrals must vanish separately.

Having found the angles ψ and θ from this computation, we may determine ϕ by means of the value of $Sx''y''Dm$, which may be obtained in terms of the angles θ and ψ, and of a^2, b^2, c^2, f, g and h, and making this expression vanish, we shall have the value of $\frac{\sin \cos \phi}{\cos^2\phi - \sin^2\phi} = \frac{1}{2} \text{tang}\, 2\phi$ [, since $2 \sin \cos \phi = \sin 2\phi$, and $(1 - 4\sin^2 \cos^2\phi) = \cos^2 2\phi = (1 - 4 \sin^2\phi (1 - \sin^2\phi) = 1 - 4 \sin^2\phi + 4 \sin^4\phi = (1 - 2 \sin^2\phi)^2 - (\cos^2\phi - \sin^2\phi)^2$].

By these means we may find the angles θ, ψ, and ϕ, such as to make $Sx''y''Dm = 0$, $Sx''z''Dm = 0$, and $Sy''z''Dm = 0$. It might indeed be expected that the equation of the third degree would afford three values of u, and three systems of axes; [since in general the equation $(x - a).(x - b).(x - c) = 0$ must vanish when x is equal to a, to b or c;] but we must observe that u is the tangent of the angle ψ formed by x with the intersection of the two first planes, and as there is no condition to distinguish the plane of x'' and y'' from the planes of x'' and z'' and of y'' and z'', the solution must be equally applicable to the three intersections formed by the plane of x and y with the three principal planes of the body respectively: hence it follows that all the three roots of the equation are possible, and that they determine in all cases the same system of three axes in the body: [although, as Euler observes, it would be difficult to demonstrate, from a direct consideration of the equation, that all its roots must necessarily be possible.]

348. COROLLARY. If C' be the rotatory inertia with respect to the axis z', we shall find $C' = A \sin^2\theta \sin^2\varphi + B \sin^2\theta \cos^2\varphi + C \cos^2\theta$.

This may be shown by substituting the values of x' and y' in the expression $C' = S(x'^2 + y'^2)Dm$ [; or rather in $C' = S \left\{ (x''^2 + y''^2 + z''^2) - z'^2 \right\} Dm$, which is equal to it, z'^2 being $= z''^2 \cos^2\theta + y''^2 \sin^2\theta \cos^2\varphi + x''^2 \sin^2\theta \sin^2\varphi$; since all the products of the cross multiplications vanish in the integral, and $C' = S \left\{ x''^2 (1 - \sin^2\theta \sin^2\varphi) + y''^2 (1 - \sin^2\theta \cos^2\varphi) + z''^2 (1 - \cos^2\theta) \right\} Dm$, whence, by adding together two of the coefficients, and subtracting the third, as in article 345, we have $\sin^2\theta(\sin^2\varphi - \cos^2\varphi) + \sin^2\theta = \sin^2\theta (2 \sin^2\varphi - 1 + 1)$ and half this, or $\sin^2\varphi \sin^2\theta$, is the coefficient of A; hence we have, secondly, $\sin^2\theta - \sin^2\varphi \sin^2\theta = \sin^2\theta \cos^2\varphi$; and thirdly, $1 - \sin^2\theta \cos^2\varphi - \sin^2\theta \sin^2\varphi = 1 - \sin^2\theta = \cos^2\theta$ for the coefficient of C.].

SCHOLIUM. The quantities $\sin^2\theta \sin^2\varphi$, $\sin^2\theta \cos^2\varphi$, and $\cos^2\theta$, are the squares of the cosines of the angles made by z' with x'', y'', and z'': [for with respect to z'', it is ob-

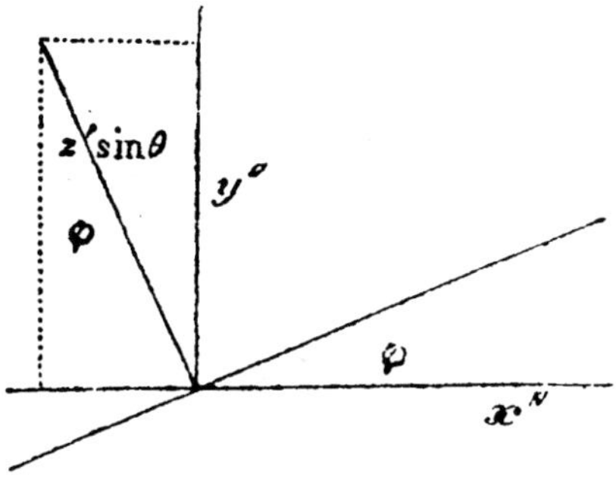

vions that the angle θ is equal to the angle formed by the two axes perpendicular to the planes of which it is the inclination; and the projection of z' on the equatorial plane of the body, or on that of x'' and y'', being z' sin θ, and the perpendiculars falling from its extremity, on x'' and on y'', z' sin θ sin φ and z' sin θ cos φ respectively, these perpendiculars will be the cosines of the angles formed with z', when z' is the radius.]

SCHOLIUM 2. Hence it follows in general, that if we multiply the rotatory inertia belonging to each principal axis by the square of the cosine of the angle which it makes with any other axis, the sum of the three products will be the rotatory inertia with respect to this axis.

349. COROLLARY 2. The greatest and the least rotatory inertia belong to two of the principal axes of rotation.

For the quantity C' is always less than the greatest of the three quantities A, B, and C, [, because their joint coefficients are always equal to unity]; it is also greater than the smallest, for a similar reason.

350. COROLLARY 3. The minimum of rotatory inertia belongs to one of the principal axes passing through the centre of gravity.

Let the coordinates of the centre of gravity, reckoned from the common origin at the centre of motion be X, Y, and Z, then the coordinates of the particle Dm as referred to the centre of gravity, will be $x'-X$, $y'-Y$, and $z'-Z$; consequently the rotatory inertia with respect to an axis parallel to z', and passing through the centre of gravity, will be $S\left\{(x'-X)^2+(y'-Y)^2\right\}.\mathrm{D}m$; now, by the properties of the centre of gravity we have $Sx'\mathrm{D}m=mX$, and $S\,y'\mathrm{D}m=mY$; consequently $S(x'^2-2x'X+X^2+y'^2-2y'Y-Y^2).\mathrm{D}m=-m(X^2+Y^2)+S(x'^2+y'^2).\mathrm{D}m$[, X and Y being invariable in the integration]. We may thus obtain the rotatory inertia of the solid with regard to an axis passing through any point whatever, when it is known with regard to the axes passing through the centre of gravity: and it is obvious that [when X and Y vanish, and the centres of motion and of gravity coincide, the rotatory inertia with respect to the centre of motion is only equal to that which belongs to the centre of gravity, exceeding it in other cases by $m\,(X^2+Y^2)$, so that] the minimum of the rotatory inertia takes place with respect to one of the principal axes passiug through the centre of gravity.

351. THEOREM. If the rotatory inertia with respect to two of the principal axes of a solid is of equal magnitude, it will also be the same for any other axis situated in the same plane with them.

If $A=B$, we have $C'=A\sin^2\theta\sin^2\varphi+B\sin^2\theta\cos^2\varphi+C\cos^2\varphi=A\sin^2\theta+C\cos^2\theta$; and whenever the new axis z' is in the plane of x'' and y'', it forms a right angle with z'', and $C'=A$.

It is easy to understand that in this case, for the axis z'' and for any two x', and y', that are perpendicular to it, we have $S\,x'y'\mathrm{D}m=0$, $S\,x'z''\mathrm{D}m=0$, and $S\,y'z''\mathrm{D}m=0$, for taking x'' and y'' for the two given principal axes, we have by the supposition $S(x''^2+z''^2)Dm=S(y''^2+z''^2)Dm$, whence $Sx''^2Dm=Sy''^2Dm$: now if ε be the angle made by x' and x'', we shall have, as in article 324, $x'=x''\cos\varepsilon+y''\sin\varepsilon$, $y'=y''\cos\varepsilon-x''\sin\varepsilon$, whence $Sx'y'Dm=Sx''y''Dm(\cos^2\varepsilon-\sin^2\varepsilon)+S(y''^2-x''^2)Dm\sin\cos\varepsilon=0$. And in the same manner it may be shown that $Sx'z''Dm=0$, and $Sy'z''Dm=0$; so that all the axes perpendicular to z'' will be principal axes; and their number will be unlimited.

352. COROLLARY If $A=B=C$, we have in general $C'=A$, and the rotatory inertia is equal for every axis.

We have here $S\,x'y'Dm=0$, $S\,x'z'Dm=0$, $S\,y'z'Dm=0$ whatever may be the position of the axes x' and y', so that all the lines passing through the centre of gravity are principal axes. This is the case with the sphere, and we shall hereafter find that the property belongs to an infinite number of other solids, of which the general equation will be demonstrated.

§ 28. *Investigation of the momentary axis of rotation of a body: the quantities, which determine its position with respect to the principal axes, give at the same time the velocity of rotation.* P. 79.

353. THEOREM. There is always one axis at rest, in every body of which any point is at rest, although the same axis may only be at rest for a moment.

[We may readily conceive the nature of a momentary axis, by considering that a rolling cylinder revolves round every line of its surface in succession as an axis; but in more complicated cases it is not so obvious, without a demonstration, that the whole of some one line must ncces-sarily be at rest at each instant. Now] the quantities p, q, r, which have been introduced into the equations (C') (345), are remarkable for affording the situation of the true momentary axis of rotation with regard to the principal axes. For if we take the fluxions of the values of x', y', and z' (345 or 324) and make them vanish, and afterwards take also $\psi=0$, which is always allowable, since the position of the fixed ordinates is wholly arbitrary, we shall have, [retaining the notation of article 345, $dx'=\alpha' x''+\beta' y'' +\gamma' z''$, and $dy'=\delta' x''+\varepsilon' y''+\zeta' z''$, or $dx'=x''\left\{-d\theta.\sin\theta\sin\psi\sin\varphi+d\psi.(\cos\theta\cos\psi\sin\varphi-\sin\psi\cos\varphi)+d\varphi.(\cos\theta\sin\psi\cos\varphi-\cos\psi\sin\varphi)\right\}+y''\left\{-d\theta.\sin\theta\sin\psi\cos\varphi+d\psi.(\cos\theta\cos\psi\cos\varphi+\sin\psi\sin\varphi)-d\varphi.(\cos\theta\sin\psi\sin\varphi+\cos\psi\cos\varphi)\right\}+z''(d\theta.\cos\theta\sin\psi+d\psi.\sin\theta\cos\psi)$; or putting $\psi=0$,] $dx'=x''(d\psi.\cos\theta\sin\varphi-d\varphi.\sin\varphi)+y''(d\psi.\cos\theta\cos\varphi-d\varphi.\cos\varphi,+z''(d\psi.\sin\theta)=0$. In the same manner we obtain $dy'=$

$$x''(-d\theta.\sin\theta\sin\varphi-d\psi.\cos\varphi+d\varphi.\cos\theta\cos\varphi)+y''(-d\theta.\sin\theta\cos\varphi+d\psi.\sin\varphi-d\varphi.\cos\theta\sin\varphi)+z''d\theta.\cos\theta=0;$$ and $dz'=$

$$x''(-d\theta.\cos\theta\sin\varphi-d\varphi.\sin\theta\cos\varphi)+y''(-d\theta.\cos\theta\cos\varphi+d\varphi.\sin\theta\sin\varphi)+z''(-d\theta.\sin\theta)=0.$$

If we multiply the first of these equations by $-\sin\varphi$, the second by $\cos\theta\cos\varphi$, and the third by $-\sin\theta\cos\varphi$, and then add them together, we obtain

$$[x''\left\{-d\theta.(\sin\cos\theta\sin\cos\varphi-\sin\cos\theta\sin\cos\varphi)\right.$$

$-d\psi.(\cos\theta\sin^2\varphi+\cos\theta\cos^2\varphi)$

$+d\varphi.(\sin^2\varphi+\cos^2\theta\cos^2\varphi+\sin^2\theta\cos^2\varphi)\Big\}+$

$y''\Big\{-d\theta.(\sin\cos\theta\cos^2\varphi-\sin\cos\theta\cos^2\varphi)$

$-d\psi.(\cos\theta\sin\cos\varphi-\cos\theta\sin\cos\varphi)$

$+d\varphi.(\sin\cos\varphi-\cos^2\theta\sin\cos\varphi-\sin^2\theta\sin\cos\varphi)\Big\}+$

$z''\Big\{d\theta.(\cos^2\theta\cos\varphi+\sin^2\theta\cos\varphi)$

$-d\psi.\sin\theta\sin\varphi\Big\}=$

$x''(-d\psi.\cos\theta+d\varphi)+z''(d\theta.\cos\varphi-d\psi\sin\theta\sin\varphi)=dt.](px''-qz'')=0.$

Secondly, multiplying the first equation by cos φ, the second by cos θ sin φ, and the third by —sin θ sin φ, we have

$[x''\Big\{d\theta.(-\sin\cos\theta\sin^2\varphi+\sin\cos\theta\sin^2\varphi)$

$+d\psi.(\cos\theta\sin\cos\varphi-\cos\theta\sin\cos\varphi)$

$+d\varphi.(-\sin\cos\varphi+\cos^2\theta\sin\cos\varphi+\sin^2\theta\sin\cos\varphi)\Big\}+$

$y''\Big\{d\theta.(-\sin\cos\theta\sin\cos\varphi+\sin\cos\theta\sin\cos\varphi)$

$+d\psi(\cos\theta\cos^2\varphi+\cos\theta\sin^2\varphi)$

$+d\varphi(-\cos^2\varphi-\cos^2\theta\sin^2\varphi-\sin^2\theta\sin^2\varphi)\Big\}+$

$z''\Big\{d\theta.(\cos^2\theta\sin\varphi+\sin^2\theta\sin\varphi)$

$+d\psi.\sin\theta\cos\varphi\Big\}=$

$y''(d\psi.\cos\theta-d\varphi)+z''(d\theta.\sin\varphi\quad d\psi\sin\theta\cos\varphi)=(-py''+rz'')dt=0$; and] $py''-rz''=0$.

Lastly, if we multiply the second equation by sin θ, and the third by cos θ, and add them together, or more simply, [if we multiply $px''-qz''$ by r, and $py''-rz''$ by q, we have $prx''-qrz''=0$ and $pqy''-qrz''=0$; whence, by subtraction,] $qy''-rx''=0$. Thus the evanescence of the three fluxions is reduced to the two conditions $px''=qz''$, and $py''=rz''$, which belong to a right line, forming angles with x'', y'', and z'' of which the cosines are $\frac{q}{\sqrt{(p^2+q^2+r^2)}}$, $\frac{r}{\sqrt{(p^2+q^2+r^2)}}$, and $\frac{p}{\sqrt{(p^2+q^2+r^2)}}$; consequently this line is at rest, and forms the true momentary axis of rotation, since the equations hold good equally with respect to all its points, whatever may be the actual magnitude of their coordinates x', y', and z'.

354. THEOREM. Retaining the notation of article 345, the angular velocity of rotation is $\sqrt{(p^2+q^2+r^2)}$.

We may consider the motion of a point so situated, that z'' may be 1, $x''=0$, and $y''=0$; we shall then have the velocity of this point, in the directions of x', y', and z', by dividing the respective fluxions by dt, and we shall thus obtain $\frac{d\psi}{dt}\sin\theta$, $\frac{d\theta}{dt}\cos\theta$, and $\frac{-d\theta}{dt}\sin\theta$ respectively; consequently the whole velocity of the point in question will be $\frac{\sqrt{(d\theta^2+d\psi^2.\sin^2\theta)}}{dt}=\sqrt{(q^2+r^2)}$[, since $qdt=d\psi\sin\theta\sin\varphi-d\theta\cos\varphi$, and $rdt=d\psi\sin\theta\cos\varphi+d\theta\sin\varphi$]. Now dividing this velocity by the distance of the point in question from the momentary axis of rotation, which is evidently the sine of the angle made by that axis with z'', of which $\frac{p}{\sqrt{(p^2+q^2+r^2)}}$ is the cosine, that is, by $\sqrt{\frac{q^2+r^2}{p^2+q^2+r^2}}$, we shall have $\surd(p^2+q^2+r^2)$ for the angular velocity of rotation.

SCHOLIUM. It is obvious that the quantities p, q, and r, of which the determination is extremely important in all inquiries respecting rotatory motion, are independent of the situation of the plane of x' and y', and that they are sufficient to express the relation of the momentary axis of rotation to the principal axes of the body, being however themselves susceptible of perpetual variation at successive times.

§ 29. *Equations for determining the position of the momentary axis, and of the principal axes, in terms of the time. Case of rotation derived from an impulse not passing through the centre of gravity. Formula for determining the direction of the primitive impulse. Example of the rotation of the planets and of the earth in particular.* P. 80.

355. THEOREM. When the body revolves freely, without any foreign disturbance, we have, with respect to the plane of greatest rotatory power, $\cos\theta = \frac{p'}{k}$, ta $\varphi = \frac{q'}{r'}$ and $d\psi = \frac{-kdt(H^2-ABp'^2)}{ABC(k^2-p'^2)}$; k^2 being $=p'^2+q'^2+r'^2$, and H being a constant quantity; dt being also in general $=$

$$\frac{ABCdp'}{\surd\left\{(ACk^2-H^2+(AB-AC)p'^2).(H^2-BCk^2-(AB-BC)p'^2)\right\}}$$

The equations (D) ,(346), afford us, by making the fluxions dN, dN', and dN'', which depend on the forces, vanish, and multiplying them respectively by p', q', and r', $p'\mathrm{d}p'=\frac{B-A}{AB}\cdot p'q'r'\mathrm{d}t$; $q'\mathrm{d}q'=\frac{C-B}{BC}q'r'p'\mathrm{d}t$, and $r'\mathrm{d}r'$ — $\frac{A-C}{AC}\cdot r'p'q'\mathrm{d}t$; but $BC-AC+AC-AB+AB-BC=0$, and the sum of the three equations becomes $0=p'\mathrm{d}p'+q'\mathrm{d}q'+r'\mathrm{d}r'$; or taking the fluent, $p'^2+q'^2+r'^2=k^2$, k being a constant quantity, to be determined by the conditions of the motion.

Again, if we multiply the three equations by ABp', BCq', and ACr', and add them together, we obtain, by taking the fluent, $ABp'^2+BCq'^2+ACr'^2=H^2$, an equation which includes the condition of the preservation of the impetus of the system, [being equivalent to $ABC(p^2+q^2+r^2)=H^2$, which implies that the square of the angu lar velocity of rotation is constant; and $H^2=ABCk'^2$, if k' be the angular velocity.]

Now since $AC(p'^2+q'^2+r'^2)=ACk^2$, we have $ACk^2-H^2=AC(p'^2+q'^2)-ABp'^2-BCq'^2$, and q'^2 — $\frac{ACk^2-H^2+(AB-AC)p'^2}{AC-BC}$; and in the same manner we find $r'^2=\frac{H^2-BCk^2+(BC-AB)p'^2}{AC-BC}$: whence we may find q' and r' from p' if H and k are known. Now the first of the equations (D) gives in this case $\mathrm{d}t=\frac{AB\mathrm{d}p'}{(A-B)q'r'}$, and by substituting for q' and r' we obtain the equation of the theorem, which, however, can only be integrated when to of the three quantities, A, B, and C, are equal.

The determination of p', q', and r' from t includes therefore that of three independent quantities H, k, and the constant quantity to be introduced in the fluent of t. But this determination relates only to the situation of the momentary axis of rotation with regard to the principal axes, and to the angular velocity of rotation. In order to ascertain the true motion of the body with respect to a quiescent space, the position of the principal axes, with regard to that space must be known; and for this purpose three new independent quantities are required, and three more integrations, which united to the former, afford the complete solution of the problem. The equations (*C*), of article **345**, include three independent quantities, N, N', and N'', but they are not altogether distinct from H and k, for if we add together the squares of the first members of the equations (*C*), we have $p'^2+q'^2+r'^2, = N^2+N'^2+N''^2 = k^2$. [For these equations are $q' \sin\theta \sin\varphi + r' \sin\theta \cos\varphi - p' \cos\theta = -N$; $(q' \cos\theta \sin\varphi + r' \cos\theta \cos\varphi + p' \sin\theta) \cos\psi + (r' \sin\varphi - q' \cos\varphi) \sin\psi = -N'$, and $-(q' \cos\theta \sin\varphi + r' \cos\theta \cos\varphi + p' \sin\theta) \sin\psi +) r' \sin\varphi - q' \cos\varphi) \cos\psi = -N''$, which may be called $(\alpha+\beta-\gamma)$, $(\delta+\varepsilon+\zeta) \cos\psi + \eta \sin\psi$, and $-(\delta+\varepsilon+\zeta) \sin\psi + \eta \cos\psi$; now the sum of the squares of the two latter quantities is $(\delta+\varepsilon+\zeta)^2 + \eta^2$, and the whole becomes $(\alpha+\beta-\gamma)^2 + (\delta+\varepsilon+\zeta)^2 + \eta^2$, which, since here $(\alpha+\beta) \cdot \gamma = (\delta+\varepsilon)\zeta$, is equal to $(\alpha+\beta)^2+\gamma^2+(\delta+\varepsilon)^2+\zeta^2+\eta^2$: now $(\alpha+\beta)^2 = q'^2 \sin^2\theta \sin^2\varphi + r'^2 \sin^2\theta \cos^2\varphi + 2q'r' \sin^2\theta \sin\cos\varphi$, and $(\delta+\varepsilon)^2 = q'^2 \cos^2\theta \sin^2\varphi + r'^2 \cos^2\theta \cos^2\varphi + 2q'r' \cos^2\theta \sin\cos\varphi'$, their sum being $q'^2 \sin^2\varphi + r'^2 \cos^2\varphi + 2q'r' \sin\cos\varphi$, to which adding $\gamma^2+\zeta^2+\eta^2$ or $p'^2 \cos^2\theta + + p'^2 \sin^2\theta + r'^2 \sin^2\varphi + q'^2 \cos^2\varphi - 2q'r' \sin\cos\varphi$, we have finally $p'^2+q'^2+r'^2 = N^2+N'^2+N''^2$.]

The constant quantities, N, N', and N'', correspond to c, c', and c'' of article **323** [c being there $\Sigma m \dfrac{x\mathrm{d}y - y\mathrm{d}x}{\mathrm{d}t}$, and N here $\mathrm{S}\dfrac{x'\mathrm{d}y' - y'\mathrm{d}x'}{\mathrm{d}t}\mathrm{DM}$]; and the quantity $\frac{1}{2}t \sqrt{(p'^2+}$

$q'^2+r'^2$) expresses the sum of the areas described in the time t by the projections of the revolving radii of all the molecules on the plane with regard to which this sum is a maximum, and with respect to which N' and N'' vanish. For this plane we obtain, by making N' and $N''=0$, $0-Br\sin\varphi-Aq\cos\varphi$, [or $r'\sin\varphi=q'\cos\varphi$, and $\frac{\sin\varphi}{\cos\varphi}=\text{ta}\varphi=\frac{q'}{r'}$], and $Aq\cos\theta\sin\varphi-Br\cos\theta\cos\varphi+Cp\sin\theta=0$, [or $-p'\sin\theta=q'\cos\theta\sin\varphi+r'\cos\theta\cos\varphi$, whence $-\frac{\sin\theta}{\cos\theta}=-\text{ta}\,\theta=\frac{q'}{p'}\sin\varphi+\frac{r'}{p'}\cos\varphi$: now $\sin\varphi=\frac{\text{ta}\varphi}{\text{sec}\varphi}=\frac{q'}{r}\cdot\sqrt{\frac{r^2}{q'^2+r^2}}=\frac{q}{\sqrt{(q'^2+r'^2)}}$, and $\cos\varphi=\frac{r}{\sqrt{(q'^2+r'^2)}}$; consequently $-\text{ta}\,\theta=\frac{q'^2+r'^2}{p'\sqrt{(q'^2+r'^2)}}=\frac{\sqrt{(q'^2+r'^2)}}{p'}$,] whence $\cos\theta=\frac{1}{\sec\theta}=\frac{p'}{\sqrt{(p'^2+q'^2+r'^2)}}=\frac{p'}{k}$, $\sin\theta\sin\varphi=-\frac{\sqrt{(q'+^2r'^2)}}{k}\cdot\frac{q'}{\sqrt{(q'^2+r'^2)}}=\frac{-q'}{k}$, and $\sin\theta\cos\varphi=\frac{\sqrt{(q'^2+r'^2)}}{k}\cdot\frac{r'}{\sqrt{(q'^2+r'^2)}}=-\frac{r'}{k}$.

By means of these equations we obtain the values of θ and φ for any given time with regard to the plane of greatest rotatory power. We have only further to determine the angle ψ, made by x' with the common intersection of the fixed plane and that of the two principal axes x'' and y'', which requires a distinct integration. Now since $q\mathrm{d}t=\mathrm{d}\psi.\sin\theta\sin\varphi-\mathrm{d}\theta.\cos\varphi$, and $r\mathrm{d}t=\mathrm{d}\psi.\sin\theta\cos\varphi.+\mathrm{d}\theta.\sin\varphi$, we have $q\mathrm{d}t.\sin\theta.\sin\varphi+r\mathrm{d}t.\sin\theta\cos\varphi=\mathrm{d}\psi.\sin^2\theta$, and $\mathrm{d}\psi-\,-q\mathrm{d}t.\frac{q'}{q'^2+r'^2}-r\mathrm{d}t.$

$\frac{r'}{q'^2+r'^2} = \frac{-dt}{q'^2+r'^2} \cdot \left(\frac{q'^2}{A}+\frac{r'^2}{B}\right)$: but since $q'^2+r'^2=k^2-p'^2$, and $Bq'^2 + Ar'^2 = \frac{H^2-ABp'^2}{C}$, we have $d\psi =$

$$\frac{-kdt.(H^2-ABp'^2)}{+ABC(k^2-p'^2)}.$$

If we substitute in this equation the value of dt already found, we shall be able to find ψ in terms of p': and we shall thus obtain the three angles θ, ϕ, and ψ in terms of p', q', and r', which will also be derived from the time t. Having therefore computed in this manner the values of these angles, with regard to the plane of x' and y' which has been considered, it will be easy to deduce from them, by spherical trigonometry, the similar quantities which belong to any other plane, and of which the determination will introduce two new independent quantities, which, with the three already mentioned, and that which belongs to the fluent of ψ, will constitute the six independent quantities required in the complete solution of the problem: but the investigation is obviously simplified by referring it to the fixed plane of greatest rotatory power.

SCHOLIUM. The position of the three principal axes with respect to the body being supposed to be known, if we are acquainted with that of the momentary axis of rotation for any instant, and with the angular velocity of rotation, we shall have the values of p, q, and r, for the given time, since their values are equal to the products of the angular velocity into the cosines of the angles formed by the momentary axis with the principal axis: hence we shall obtain the values of p', q', and r', which are proportional to the sines of the angles formed by the principal axes with the plane of greatest rotatory power, which is supposed in this proposition to be that of x' and y', and with respect to which the sum of the projections of the areas described by the revolving radii, multiplied by the masses of the respective particles, is a maximum. We may therefore determine at every instant the intersection of the surface of the body with this plane, and may consequently find its situation by the actual conditions of the motion of the body.

[356. LEMMA. The square of the radius of gyration of a sphere is $\frac{2}{5}$ of the square of the semidiameter.

The fluxion of the surface of a sphere is as $dx\frac{r}{y}.y=rdx$, that of a great circle being $dx\frac{r}{y}$, where the sine is x, and the cosine y: and at last, when $x=r$, the surface of the hemisphere becomes equal to that of the corresponding semicylinder (183): the fluxion of the rotatory inertia of the surface will be represented by $rdx.y^2=(r^2-x^2)\,rdx=r^3dx-rx^2dx$, and the fluent by $r^3x-\frac{1}{3}rx^3$ or, for the hemisphere, by $\frac{2}{3}r^2$ which, divided by r^2, gives the square of the radius of gyration $\frac{2}{3}r^2$, and the rotatory inertia $\frac{2}{3}r^2M$, M being the content or mass of the surface of which the radius is r.

If the sphere be now supposed to increase by concentric surfaces, the fluxion of the mass will be as $r^2dr.\varrho$, if ϱ be the density, and that of the rotatory inertia as $\frac{2}{3}r^4dr.\varrho$, and the square of the radius of gyration will be $\frac{2}{3}\frac{\int\varrho r^4dr}{\int\varrho r^2dr}$, which, when $\varrho=1$, becomes $\frac{2}{3}.\frac{\frac{1}{5}r^5}{\frac{1}{3}r^3}=\frac{2}{5}r^2$, and the rotatory inertia of the homogeneous sphere will be $\frac{2}{5}r^2m$.]

357. THEOREM. In a homogeneous sphere, the distance f, at which an impulse must have been given, in order to cause a revolution and a rotation at once, must be $\frac{2}{5} \cdot \frac{R^2}{r} \cdot \frac{p}{U}$, R being the radius of the sphere, r its distance from the centre of revolution, p the angular velocity of rotation, and U that of revolution.

An impulse acting on any part of the body will produce the same progressive motion as if it were immediately applied to the centre of gravity itself (322, 331) and the same rotatory motion as if the centre of gravity were fixed. [Thus if we imagined the force to be communicated by a particle moving with a given velocity, and attaching itself to the substance, it is evident, from the properties of the centre of gravity, that the velocity of this point will be the same, whatever be the part of the body to which the particle attaches itself; and, with respect to the velocity of rotation round the centre, it is obvious that this velocity would not be affected by the subsequent application of any force to the centre of gravity capable of destroying the progressive motion, neither will it be affected by the interference of the obstacle, either immediately after, or at, the very beginning of the motion.] The sum of the areas described round the centre of gravity, by the projections of the revolving radii of the different particles on a fixed plane, multiplied by their masses, will always be proportional to the rotatory power of the primitive force, projected on the same plane; and the plane, with respect to which the projection of the momentum is greatest, must obviously be the plane in which the force itself acts, and which passes through the centre of gravity: this plane is therefore the invariable plane of rotation. Now calling the distance of the direction of the primitive impulse from the centre of gravity f, and v the velocity communicated to the centre of gravity, m being the mass of the body, the rotatory power of the impulse must have been mfv; and multiplying this by $\frac{1}{2}t$, the product will be equal to the sum of the areas described during the time t, which has already been found equal to $\frac{1}{2}t\sqrt{(p'^2+q'^2+r'^2)}$ (355); consequently $\sqrt{(p'^2+q'^2+r'^2)}=mfv$. Hence if we know the origin of the motion, and the position of the principal axes of the body with regard to the invariable plane, as determining the angles θ and ϕ, we shall have the values of p', q', and r' in the first instance, and consequently those of p, q, and r, whence the values of the same quantities may be found for any other time.

Now if we imagine any one of the planets to be a homogeneous sphere, deriving its rotation and its annual motion round the sun from a single impulse, the radius being R, and the angular velocity of revolution U; r being the distance from the sun, we shall have $v=rU$: and if f be the distance of the direction of the impulse from the centre, it is plain that the planet will acquire a rotatory motion round an axis perpendicular to the invariable plane. If therefore we consider this axis as the third principal axis z'', we shall have $\theta=0$, and consequently $q'=0$ and $r'=0$, and $p'=mfv$, or $Cp=mfrU$. Now, in the sphere, $C=\frac{2}{5}mR^2$ (356), consequently $f=\frac{2}{5}\frac{R^2}{r}\cdot\frac{p}{U}$, whence we have f, the distance of the direction of the impulse from the centre of the planet, which corresponds to the proportion between the two velocities. With regard to the earth, $\frac{p}{U}$ being $=366.25638$, and $\frac{R}{r}$, the sun's parallax, $=.000042665$, f is found very nearly $\frac{1}{160}R$, [or about 25 miles].

SCHOLIUM. The planets not being homogeneous, they may here be considered as formed of concentric spherical strata of different densities, and in this case we have $C=\frac{2}{3}m\frac{\int\rho R^4 dR}{\int\rho R^2 dR}$, (355) whence $f=\frac{2}{3}\frac{p}{rU}\cdot\frac{\int\rho R^4 dR}{\int\rho R^2 dR}\cdot$ and if, as it is natural to suppose, the strata nearest the centre are the densest, the quantity $\frac{\int\rho R^4 dR}{\int\rho R^2 dR}$ will be less than $\frac{3}{5}R^2$, and the value of f will be less than for a homogeneous body.

§ 30. *Of the oscillations of a body which turns very nearly round one of the principal axes. Stability of the motion round the principal axes of which the rotatory inertia is the greatest and the least: instability with respect to the third axis.* P. 85.

[358. LEMMA The cosine of an imaginary arc may be expressed by a real exponential quantity: thus we have $\sin\sqrt{-1}\,vt-\frac{e^{-vt}-e^{vt}}{2\sqrt{-1}}$ and $\cos\sqrt{-1}\,vt=\frac{e^{-vt}+e^{vt}}{2}$.

If $\Gamma=\cos\gamma t+\sqrt{-1}\sin\gamma t$, $d\Gamma=-\sin\gamma t\,.\,\gamma dt+\sqrt{-1}\cos\gamma t\,.\,\gamma dt=\Gamma\sqrt{-1}\,\gamma dt$, and $\frac{d\Gamma}{\Gamma}=dhl\Gamma=\sqrt{-1}\gamma dt$, whence $\Gamma=e^{\sqrt{-1}\,\gamma t}$, if $hle=1$. Again, if $\Gamma'=\cos\gamma t-$

$\sqrt{-1}\sin\gamma dt$, $d\Gamma' = (-\sin\gamma t - \sqrt{-1}\cos\gamma dt)\gamma dt = -\Gamma'\sqrt{-1}$ γdt and $\Gamma' = e^{-\sqrt{-1}\gamma t}$ consequently $\Gamma - \Gamma' = 2\sqrt{-1}$ $\sin\gamma t = e^{\sqrt{-1}\gamma t} - e^{-\sqrt{-1}\gamma t}$, and $\Gamma + \Gamma' = 2\cos\gamma t = e^{\sqrt{-1}\gamma t} + e^{-\sqrt{-1}\gamma t}$ And if we substitute $\sqrt{-1}\nu$ for the indeterminate quantity γ, we have $2\sqrt{-1}\sin\sqrt{-1}\nu t$ $= e^{-\nu t}\ e^{\nu t}$, and $2\cos\sqrt{-1}\nu t = e^{-\nu t} + e^{\nu t}$.]

359. THEOREM The permanency of rotation round two of the principal axes of every irregular body is stable, and round the third unstable.

We might deduce the laws of the oscillations of a body turning round an axis very near to the third principal axis from the fluents found in the preceding propositions; but it is more simple to derive them at once from the differential equations (*D*) of article 346. The forces acting on the body being supposed to vanish, we have $dp' = \frac{B-A}{AB}$ $q'r'dt = 0$, $dq' + \frac{C-B}{CB}r'p'dt = 0$, and $dr' + \frac{A-C}{AC}p'q'dt =$ 0; and substituting Cp, Aq, and Br, for p', q', and r', $dp + \frac{B-A}{C}qrdt = 0$, $dq + \frac{C-B}{A}rpdt = 0$, and $dr + \frac{A-C}{B}pq$ $dt = 0$.

Now supposing the solid to perform its rotation very nearly round the third principal axis, so that q and r may be very small, their squares and their products may obviously be neglected in comparison with the other quantities concerned; we shall therefore have $dp = 0$, and if we substitute in the other equations the indeterminate values $q = \mu\sin(nt + \gamma)$, and $r = \mu'\cos(nt + \gamma)$ [in order to obtain a particular solution of the problem], we shall have $n = p$

$\sqrt{\frac{(C-A).(C-B)}{AB}}$, and $\mu' = -\mu\sqrt{\frac{AC-AA}{BC-BB}}$, μ and γ being two constant independent quantities; and the angular velocity of rotation, which is $\sqrt{(p^2+q^2+r^2)}$, will be reduced simply to p, by neglecting the squares of q and r, so that this velocity may be considered as constant, and the sine of the minute angle formed by the momentary axis of rotation with the third principal axis will be $\frac{\sqrt{(q^2+r^2)}}{p}$. [For the value of d$q$, being, according to the substitution, μ cos $(nt+\gamma)n$dt, and that of d$r = -\mu'$ sin $(nt+\gamma)n$dt, we have μ cos $(nt+\gamma)n + p\mu'$ cos $(nt+\gamma)\frac{C-B}{A} = 0$, or $\mu n + p\mu'\frac{C-B}{A} = 0$ and $-\mu'$ sin $(nt+\gamma)n + p\mu$ sin $(nt+\gamma)\frac{A-C}{B} = 0$, or $-\mu' n + p\mu\frac{A-C}{B} = 0$; whence $\mu' = -\mu n\frac{A}{p(C-B)} = p\mu\frac{A-C}{nB}$, and $n\frac{A}{p(C-B)} = p\frac{C-A}{nB}$, consequently $n^2AB = p^2(C-A).(C-B)$; and $\mu' = -\frac{\mu}{p}.p\sqrt{\frac{(C-A).(C-B)}{AB}}.\frac{A}{C-B} = -\mu\sqrt{\frac{A(C-A)}{B(C-B)}}$.] Now if, at the beginning of the motion, $q = 0$, and $r = 0$, that is, if the momentary axis of rotation coincides with the principal axis, we shall have $\mu = 0$, $\mu' = 0$, and q and r will always remain $= 0$, the axis of rotation always coinciding with the third principal axis; whence it follows that if the body begins to turn round one of the principal axes, it will continue to turn uniformly round the same axis. This remarkable property, belonging to the principal axes, has caused them to be denominated axes of permanent rotation, and it belongs to them exclusively; for if the momentary axis of rotation be supposed invariable with respect to the body, we must have $dp=0$, $dq=0$, and $dr=0$, whence. from the equations (*D*) we have $\frac{B-A}{C}rq=0$, $\frac{C-B}{A}rp=0$, and $\frac{A-C}{B}pq=0$: and, in the general extent of the theorem, *A*, *B*, and *C* being all unequal, it follows that two of the three quantities *p*, *q*, and *r* must vanish, which supposes the momentary axis of rotation to coincide with one of the principal axes.

If two of the three quantities, *A*, *B* and *C*, are equal, for example if $A=B$, these three equations only give us $rp=0$ and $pq=0$, which will be true if *p* only be supposed to vanish, so that the axis of rotation may be perpendicular to the third principal axis, and it has been already shown that, in this case (351), all the axes so situated are principal axes. And again, if *A*, *B*, and *C* are all equal, the three equations will be true, whatever may be the values of *p*, *q*, and *r*; but in this case all the axes are principal axes (352).

Hence it follows that the principal axes only can be permanent axes of rotation: but they do not possess this property in the same manner: the rotation round that axis, with regard to which the rotatory inertia is intermediate between the two others, may be disturbed in a sensible degree by the slightest cause, so that such a motion is possessed of no stability.

Stability consists in such a state of a system, that when it is very slightly deranged, the derangement can only remain extremely slight, and the system will oscillate about the state of stability. Thus if we imagine the momentary axis of rotation to be infinitely little removed from the third principal axis, in this case the values of *q* and *r* will always remain infinitely small, and the momentary axis will only make excursions of the same order about the third principal axis. But if the value of n^2 became negative, and n were consequently imaginary, the values of sin $(nt+\gamma)$ and cos $(nt+\gamma)$ would be changed into exponential or logarithmic quantities (358), and the expressions for q and r might then increase indefinitely, and these quantities would no longer be infinitely small, so that the motion would have no stability. Now the value of n is real if C is the greatest or the smallest of the three quantities A, B, and C, for then the product $(C-A) \,.\, (C-B)$ is positive, but this product is negative when C is of intermediate magnitude, and n then becomes imaginary.

360. COROLLARY. Retaining the same notation, if θ be very small, we shall have

$$\sin\theta \sin\phi = \epsilon \sin(pt+\lambda) - \frac{A}{Cp}\mu \sin(nt+\gamma), \text{ and}$$

$$\sin\theta \cos\phi = \epsilon \cos(pt+\lambda) - \frac{B}{Cp}\mu' \cos(nt+\gamma);\ \epsilon$$

and λ being two new constant quantities.

In order to determine the position of the axes with regard to a quiescent space, we may suppose the third principal axis very nearly perpendicular to the plane of x' and y', so that we may be able to neglect the square of θ, and to make $\cos\theta = 1$, we shall then find for the value of pdt, instead of $d\phi - d\psi . \cos\theta$, $d\phi - d\psi$, whence $\psi = \phi - pt - \varepsilon$, ε being a constant quantity. We have then, since $qdt = d\psi . \sin\theta \sin\phi - d\theta . \cos\phi$, and $rdt = d\psi . \sin\theta \cos\phi + d\theta . \sin\phi$, putting $\sin\theta \sin\phi = s$, and $\sin\theta \cos\phi = u$, $ds = \cos\theta \sin\phi\, d\theta + \sin\theta \cos\phi d\phi = \sin\phi d\theta + \sin\theta \cos\phi d\phi$, $pudt = \sin\theta \cos\phi\, (d\phi - d\psi)$, $ds - pudt = \sin\phi d\theta + \sin\theta \cos\phi d\psi = rdt$; and $du = \cos\phi d\theta - \sin\theta \sin\phi d\phi$, $psdt = \sin\theta \sin\phi\, (d\phi - d\psi)$ and $du + psdt = \cos\phi d\theta - \sin\theta \sin\phi d\psi = -qdt$. Now the conditions of this fluxional equation are fulfilled by putting $s = \mathrm{G} \sin (pt+\lambda) - \frac{A}{Cp} \mu \sin (nt+\gamma)$ or $= \mathrm{G} \sin (pt+\lambda) - \frac{A}{Cp} q$, and $u = \mathrm{G} \cos (pt+\lambda) - \frac{B}{Cp} \mu' \cos (nt+\gamma)$, or $= \mathrm{G} \cos (pt+\lambda) - \frac{B}{Cp} r$ [, since ds becomes $= \mathrm{G} \cos (pt+\lambda) p\dot{d}t - \frac{A}{Cp} dq$, and $pudt = \mathrm{G} \cos (pt+\lambda) pdt - \frac{B}{C} rdt$: but since $dq - \frac{B-C}{A} rpdt$(360), $\frac{A}{Cp} dq = \frac{B-C}{C} rdt$, and $ds - pudt = \left(\frac{B}{C} - \frac{B-C}{C}\right) rdt = rdt$, and $du = -\mathrm{G} \sin (pt+\lambda) pdt - \frac{B}{Cp} dr$, $psdt = \mathrm{G} \sin (pt+\lambda) pdt - \frac{A}{C} qdt$, and $\frac{B}{Cp} dr = \frac{C-A}{C} qdt$, whence $du + psdt = -qdt$].

In this manner the problem is completely resolved, since the values of s and u afford us θ and φ in terms of the time, and since ψ is deduced from φ and t. If the quantity $\mathrm{G} = 0$, the plane of x' and y' becomes the invariable plane, to which the angles θ, φ, and ψ have been referred in the preceding section (355).

§ 31. *Of the motion of a solid body round a fixed axis. Determination of the simple pendulum oscillating in the same time with the body.* P. 88.

361. THEOREM. The vibrations of a gravitating body, whatever may be its form, are synchronous with those of a simple pendulum of the length $\frac{C}{mh}$, C being the rotatory inertia with respect to the axis of motion, or S $(y''^2 +$

$z''^2)$ Dm, m the mass, and h the distance of the centre of gravity from the axis of motion.

The preceding investigations are sufficient for determining the motion of a solid round its centre of gravity, when it is either at liberty, or fixed to a single point of suspension only: it now remains for us to consider the motion of a solid round a fixed axis.

We may call the axis of motion x', and suppose its direction to be horizontal: the last of the equations (B) (343), will be sufficient to determine the motion; that is $S\,\frac{y'dz'-z'dy'}{dt}\,Dm = S\int(Ry'-Qz')\,dt\,Dm = N''$. We may suppose y' to be also horizontal, and z' vertical, or perpendicular to the horizon, the plane of y' and z' passing through the centre of gravity of the body, and a moveable axis being supposed to pass constantly through this centre and the origin of the coordinates. Now θ being the angle which this new axis makes with z', and y'' and z'' being the coordinates perpendicular and parallel to this new axis in the plane of y' and z', we have $y' = y''\cos\theta + z''\sin\theta$, and $z' = z''\cos\theta - y''\sin\theta$, consequently $[y'dz' - z'dy' = d\theta\left\{(y''\cos\theta + z''\sin\theta).(-z''\sin\theta - y''\cos\theta) - (z''\cos\theta - y''\sin\theta).(-y''\sin\theta + z''\cos\theta)\right\} = d\theta\left\{-(y''\cos\theta + z''\sin\theta)^2 - (z''\cos\theta - y''\sin\theta)^2\right\} = -d\theta\left\{y''^2(\cos^2\theta + \sin^2\theta) + z''^2(\sin^2\theta + \cos^2\theta)\right\}$ and$]\ S\,\frac{y'dz'-z'dy'}{dt}\,Dm = -\frac{d\theta}{dt}\,S$

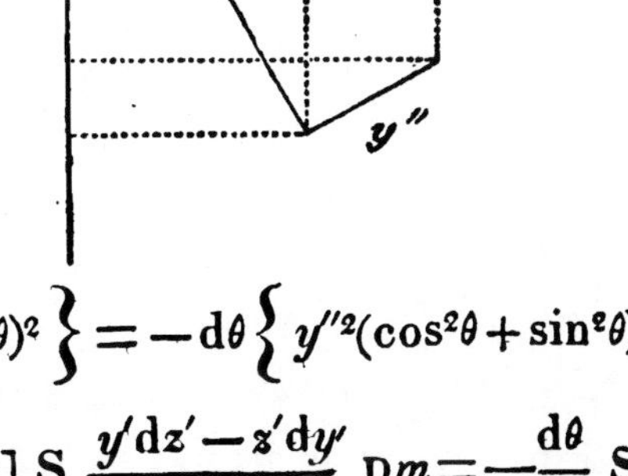

$(y''^2+z''^2)\,\mathrm{D}m - \frac{\mathrm{d}\theta}{\mathrm{d}t}C = N''$; and taking the fluxion, $\frac{\mathrm{d}N''}{\mathrm{d}t} = \frac{-\mathrm{dd}\theta}{\mathrm{d}t^2}C$, $\mathrm{d}t$ being constant: and the body being subject only to the force of gravitation, P and Q will be $=0$, and R will be constant; therefore $\mathrm{d}N'' = \mathrm{d}\mathrm{S}\!\int\! Ry'\mathrm{d}t\,\mathrm{D}m = \mathrm{S}Ry'\mathrm{d}t\,\mathrm{D}m$; $\frac{\mathrm{d}N''}{\mathrm{d}t} = \mathrm{S}Ry'\mathrm{D}m = R\mathrm{S}y'\mathrm{D}m = R\cos\theta\,\mathrm{S}y''\mathrm{D}m + R\sin\theta\,\mathrm{S}z''\mathrm{D}m$: but since z'' passes through the centre of gravity of the body, we have $\mathrm{S}y''\mathrm{D}m=0$; and if h be the distance of the centre of gravity from the axis of motion x', we have $\mathrm{S}z''\mathrm{D}m=mh$, m being the mass of the body, whence $\frac{\mathrm{d}N''}{\mathrm{d}t} = mhR\sin\theta$, and $\frac{\mathrm{dd}\theta}{\mathrm{d}t^2} = \frac{-mhR\sin\theta}{C}$.

If we now consider a second body, of which all the atoms are united in a single point at the distance l from the axis x'', we have in this case $C=m'l^2$, m' being the mass: and $h=l$; consequently $\frac{\mathrm{dd}\theta}{\mathrm{d}t^2} = \frac{-m'hR\sin\theta}{m'l^2} = \frac{R}{l}\sin\theta$. The two bodies will therefore have exactly the same oscillatory motion, if their initial angular velocities, when their centres of gravity are in the vertical line, are equal, and if $l = \frac{C}{mh}$. The second body here taken iuto consideration is the simple pendulum, of which the oscillations have been determined in § 11 (280); and we may always assign, by means of this formula, the length l of the simple pendulum, of which the oscillations are isochronous with that of the solid here considered, which constitutes a compound pendulum. It is thus that the length of the simple pendulum, vibrating in a second, has been determined from observations on the vibrations of compound pendulums.

CHAPTER VIII.

OF THE MOTIONS OF FLUIDS.

---

§ 32. [*Introduction.*] *Equations of the motion of fluids: condition relating to their continuity.*

[INTRODUCTION. The subject of this section being somewhat intricate, and involving a variety of connected quantities, it may probably be of advantage to premise, as a detached illustration of the mode of treating it, the investigation of Poisson, which is nearly similar, but reduced to more elementary principles, and in some instances more clearly expressed. Traité de Mécanique, 1811, Vol. II. P. 472.

"We are now about to consider the motion of fluids in the most general point of view, and to examine the conditions of the motion of the fluid mass, for which we have already investigated the laws of equilibrium. The fluid may be either homogeneous or heterogeneous, either incompressible or elastic; all its particles are supposed to be actuated by given forces, such as their mutual attractions, and other attractive forces directed either to fixed or to moveable centres. But all these forces we suppose to be reduced to three, parallel to three fixed orthogonal axes, and to the coordinates x, y, and z; and we may call these three forces X, Y, and Z. These forces are simply dependent on x, y, and z, when their intensity is invariable in magnitude and direction; but when they are directed to moveable centres of attraction, or are dependent on the mutual actions of the particles, their values will comprehend the time that has elapsed: so that calling the time t, we may consider the forces X, Y, Z, in general as functions of x, y, z, and t.

"Now if we call the velocity of the element, to which the ordinates x, y, and z belong, reduced to the direction of the axes, u, v, and w, these quantities will be unknown functions of x, y, z, and t; they must depend on the ordinates x, y, z, because, at the same instant, or for the same value of t, the velocity may vary between one particle and another in magnitude and in direction: they must also depend on the time t, because in the same place, and for the same original values of x, y, z, the velocity may change, from one instant to another. If we wish to compare the velocities of any one particle in two consecutive instants, we must suppose that the variable quantity t becomes $t+dt$, [or rather $t+\Delta t$]; and in the same time the coordinates of the particles x, y, and z, will become [$x+u\Delta t$, $y+v\Delta t$, and $z+w\Delta t$]; for in virtue of the velocities u, v, w, the same particle which belonged to the coordinates x,y,z, at the end of the time t, will correspond to $x+u\Delta t$, $y+v\Delta t$, and $z+w\Delta t$, at the end of the time $t+\Delta t$. It follows, then, that in order to obtain the variation of the quantities u, v, and w, with regard to the same particle at the different instants, we must take the differences with regard to t, and with regard to x, y, and z, considering $u\Delta t$, $v\Delta t$, and $w\Delta t$ as the elementary variations of these quantities. We have therefore

$$du = \frac{d'u}{dt}dt + \frac{d'u}{dx}udt + \frac{d'u}{dy}vdt + \frac{d'u}{dz}wdt\,;$$

$$dv = \frac{d'v}{dt}dt + \frac{d'u}{dx}udt + \frac{d'v}{dy}vdt + \frac{d'v}{dz}wdt\,; \text{ and}$$

$$dw = \frac{d'w}{dt}dt + \frac{d'w}{dx}udt + \frac{d'w}{dy}vdt + \frac{d'w}{dz}wdt.$$

"The fluid being supposed to be divided into infinitely small rectangular parallelepipeds, of which the sides are parallel to the coordinates, we have, for the volume of the element corresponding to x, y, and z, [$DxDyDz$, using the characteristic D with regard to the variations of space for the same instant of time, while Δ and d are employed for the successive changes only.] The density of the fluid may be considered as constant throughout this space, and may be called ρ, so that the mass will be $\rho DxDyDz$. We may also designate by p the pressure, on each unit of the surface, exerted by the fluid in contact with the different faces of the parallelepiped, and which, according to the fundamental property of fluids, is the same in all directions. The two quantities, ρ and p, as well as the velocities u, v, w, are unknown functions of x, y, z, and t; the five quantities, u, v, w, ρ, and p, are required to be found for the solution of the problem; and when these have been obtained, in terms of x, y, z, and t, the state of the fluid will be known for every instant, the velocity and direction of the motion of each particle being determined, together with the density of the fluid and the pressure exerted, whether at the surface or within the substance of the fluid. We must therefore proceed to seek for the equations expressing these five quantities.

"Now three of these equations are immediately afforded us by the principle of Dalembert. The velocities "lost" during the instant Δt, by the particle subjected to the action of the forces X, Y, and Z, are $X\Delta t - \Delta u$, $Y\Delta t - \Delta v$, and $Z\Delta t - \Delta w$; for Δu, Δv, and Δw, express the augmentations of velocity which really take place in the given instant, and $X\Delta t$, $Y\Delta t$, and $Z\Delta t$, those which would be produced by the forces X, Y, and Z, if the particle were free and insulated. These supposed velocities, divided by Δt, will give the measures of the forces capable of producing them; and calling the quotients X', Y', Z', we have

$$X - \frac{d'u}{dt} - \frac{d'u}{dx}u - \frac{d'u}{dy}v - \frac{d'u}{dz}w = X';$$

$$Y - \frac{d'v}{dt} - \frac{d'v}{dx}u - \frac{d'v}{dy}v - \frac{d'v}{dz}w = Y'; \text{ and}$$

$$Z - \frac{d'w}{dt} - \frac{d'w}{dx}u - \frac{d'w}{dy}v - \frac{d'w}{dz}w = Z'.$$

"Now, according to the principle in question, the fluid mass would be in equilibrium, if all the particles were actuated by forces capable of communicating to them the velocities lost or gained at each instant; [or in other words the unemployed forces of the whole system must hold each in equilibrium:] we may therefore satisfy the general conditions of equilibrium by considering X' Y' and Z' as the forces, parallel to the coordinates, acting on each particle, instead of X, Y, and Z, which represent the whole forces in those directions. Hence we have

$\frac{d'p}{dx} = \rho X'$; $\frac{d'p}{dy} = \rho Y'$; and $\frac{d'p}{dz} = \rho Z'$: or substituting for these quantities, and dividing by ρ;

$$\frac{d'p}{\varrho dx} = X - \frac{d'u}{dt} - \frac{d'u}{dx}u - \frac{d'u}{dy}v - \frac{d'u}{dz}w;$$

$$\frac{d'p}{\varrho dy} = Y - \frac{d'v}{dt} - \frac{d'v}{dx}u - \frac{d'v}{dy}v - \frac{d'v}{dz}w;$$

$$\frac{d'p}{\varrho dz} = Z - \frac{d'w}{dt} - \frac{d'w}{dx}u - \frac{d'w}{dy}v - \frac{d'w}{dz}w.$$

"Each of the elements, into which the fluid is supposed to be divided, will change its form during the instant Δt, and it may also change its volume, if the fluid is compressible: but since the mass must always remain constant, it follows that if we find its volume and its density at the end of the time $t+\Delta t$, their product must be the same as at the end of the time t: and by making the variation of this product vanish, we shall obtain a new equation for the motion.

"In order to form this equation, we may consider the rectangular parallelepiped, of which the volume was expressed by $\mathrm{D}x\mathrm{D}y\mathrm{D}z$ at the end of the time t, and examine the form which it will assume at the end of the time $t+\Delta t$. supposing **M** to be the summit of the parallelepiped which corresponds to the coordinates x, y, z, and **MN, ML, MK**, the three sides or edges which meet in it, and which are parallel to the axes Θz Θy and Θx respectively, so that we have **MN=DZ, ML=DY**, and **MK=DX**: supposing also **E,F,G**, and **H**, to be the four other angles of the parallelepiped; and the points **M,N,L,K,E,F,G,H**, to be removed, during the instant Δt, to **M′,N′,L′,K′, E′,F′, G′, H′**. The new solid will still be a parallelepiped, as may be thus demonstrated.

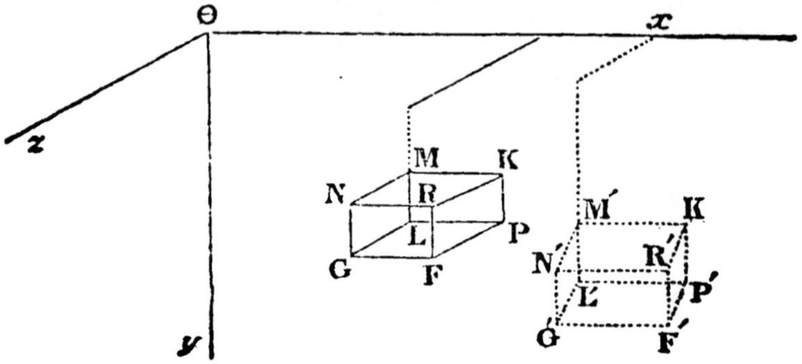

"The coordinates, x, y, z, of the point M, become, at the end of the instant Δt, $x+udt$, $y+v\Delta t$, and $z+w\Delta t$, which are therefore the coordinates of the point M′, and those of any other angular point may be found by substituting the corresponding variations: thus for the point N′, the ordinates are at first x, y, and $z+\mathrm{D}z$, and afterwards, u being changed to $u+\mathrm{D}_z u$ in each instance, we have for the new ordinates $x+u\Delta t+\frac{d'u}{dz}\mathrm{D}z\Delta t$; $y+v\Delta t+\frac{d'v}{dz}\mathrm{D}z\Delta t$, and $z+\mathrm{D}z+w\Delta t+\frac{d'w}{dz}\mathrm{D}z\Delta t$. The differences are $\frac{d'u}{dz}\mathrm{D}z\Delta t$, $\frac{d'v}{dz}\mathrm{D}z\Delta t$, and $\mathrm{D}z+\frac{d'w}{dz}\mathrm{D}z\Delta t$, and the sum of their squares will be the square of M′N′: but the two former being infinitely small in comparison with the latter, their squares may be neglected, and $M'N'=\mathrm{D}z+\frac{d'w}{dz}\mathrm{D}z\mathrm{D}t$.

"The coordinates of the point E′ must be deduced from those of M′, and the coordinates of F′ from those of N′, by substituting $x+\mathrm{D}x$ and $y+\mathrm{D}y$ in the place of x and y: consequently the length of E′F′ may be deduced in the same manner from that of M′N′; hence we have

$$E'F' = Dz + \frac{d'w}{dz} Dz\Delta t + \frac{d'^2w}{dzdx} DxDz\Delta t + \frac{d'^2w}{dzdy} DyDz\Delta t;$$

which only differs from M′N′ by quantities evanescent in comparison with itself; and in the same manner K′H′ and L′G′ may be shown to be ultimately equal to M′N′

Precisely in the same manner, by substituting first y and v, and then x and u, for z and w, we obtain

$M'L' = Dy + \frac{d'v}{dy} Dy\Delta t$, and $M'K' = Dx + \frac{d'u}{dx} Dx\Delta t$; and the opposite sides of the parallelepiped will be found to be respectively equal to them, so that the figure still remains that of a parallelepiped, although its angles are rendered oblique; but the obliquity produced in the instant Δt is infinitely small, so that, without neglecting the cosine of the angles, their sines may still be considered as unity, and the volume of the solid will be expressed by the product of its three sides M′N′.M′L′.M′K′. This product, neglecting the terms involving the higher powers of the differences, which are comparatively evanescent, becomes $DxDyDz(1 + \left(\frac{d'u}{dx} + \frac{d'v}{dy} + \frac{d'w}{dz}\right)\Delta t)$: and this is the volume of the element which, at the end of the time t, was $DxDyDz$. Now the density ϱ being a function of x, y, z, and t, it follows that when t becomes $t + \Delta t$, and x, y, and z are changed to $x + u\Delta t$, $y + v\Delta t$, and $z + w\Delta t$, it becomes $\varrho + \frac{d'\varrho}{dt}\Delta t + \frac{d'\varrho}{dx} u\Delta t + \frac{d'\varrho}{dy} v\Delta t + \frac{d'\varrho}{dz} w\Delta t$: and if we multiply this density by the corresponding volume, the product will express the mass at the end of the time: from which if we subtract $\varrho DxDyDz$, the initial mass, the remainder will be the variation of the mass: and this must vanish. Hence, neglecting the

terms which contain the square of Δt, and dividing by $\mathrm{D}x\mathrm{D}y\mathrm{D}z\Delta t$, we obtain
$\frac{d'\varrho}{dt}+\frac{d'\varrho}{dx}u+\frac{d'\varrho}{dy}v+\frac{d'\varrho}{dz}w+\varrho\left(\frac{d'u}{dx}+\frac{d'v}{dy}+\frac{d'w}{dz}\right)=0$; or, what amounts to the same, $\frac{d'\varrho}{dt}+\frac{d'(\varrho u)}{dx}+\frac{d'(\varrho v)}{dy}+\frac{d'(\varrho w)}{dz}=0$." It is unnecessary to pursue Mr. Poisson's investigation any further, since it is only introduced as an illustration of some of the less perspicuous parts of Laplace's mode of considering the subject, to which we are now to return.]

363. THEOREM The motions of fluids in general may be deduced from the equation

$$\delta V-\frac{\delta p}{\varrho}=\delta x\frac{ddx}{dt^2}+\delta y\frac{ddy}{dt^2}+\delta z\frac{ddz}{dt^2}; \qquad (F)$$

δV being $=P\delta x+Q\delta y+R\delta z$, p the pressure, ϱ the density, and P, Q, R the external forces acting in the directions of the coordinates x, y, and z.

It will be convenient to deduce the laws of the motions of fluids from those of their equilibrium, in the same manner as, in Chapter V, the laws of the motions of a system of solid bodies have been deduced from those of the equilibrium of the system. For this purpose we may resume the equation $\delta p=\varrho\,(P\delta x+Q\delta y+R\delta z)$ from the demonstration of article 316.

Now when the fluid is in motion, the forces unemployed in generating motion are $P-\frac{ddx}{dt^2}$, $Q-\frac{ddy}{dt^2}$, and $R-\frac{ddz}{dt^2}$, which must hold each other in equilibrium: we must therefore substitute these forces for the P, Q, and R of the equation of equilibrium, and it will become $\delta p = \varrho \left\{ \delta x \left(P - \frac{\mathrm{dd}x}{\mathrm{d}t^2} \right) + \delta y \left(Q - \frac{\mathrm{dd}y}{\mathrm{d}t^2} \right) + \delta z \left(R - \frac{\mathrm{dd}z}{\mathrm{d}t^2} \right) \right\}$; or, supposing $P\delta x + Q\delta y + R\delta z$ to be an exact variation, and equal to δV, $\delta V - \frac{\delta p}{\varrho} = \delta x \frac{\mathrm{dd}x}{\mathrm{d}t^2} + \delta y \frac{\mathrm{dd}y}{\mathrm{d}t^2} + \delta z \frac{\mathrm{dd}z}{\mathrm{d}t^2}$.

364. COROLLARY. Since the three variations are independent, their coefficients may be made to vanish separately, and the theorem may be resolved into three distinct equations.

365. THEOREM The condition of the continuity of the fluid is expressed by the equation $\varrho \mathfrak{C} = (\varrho)$; (G)
(ϱ) being the initial value of the density ϱ, and $\mathfrak{C} = \frac{\mathrm{d}'x}{\mathrm{d}a} \cdot \frac{\mathrm{d}'y}{\mathrm{d}b} \cdot \frac{\mathrm{d}'z}{\mathrm{d}c} - \frac{\mathrm{d}'x}{\mathrm{d}a} \cdot \frac{\mathrm{d}'y}{\mathrm{d}c} \cdot \frac{\mathrm{d}'z}{\mathrm{d}b} + \frac{\mathrm{d}'x}{\mathrm{d}b} \cdot \frac{\mathrm{d}'y}{\mathrm{d}c} \cdot \frac{\mathrm{d}'z}{\mathrm{d}a} \cdot - \frac{\mathrm{d}'x}{\mathrm{d}b} \cdot \frac{\mathrm{d}'y}{\mathrm{d}a} \cdot \frac{\mathrm{d}'z}{\mathrm{d}c} + \frac{\mathrm{d}'x}{\mathrm{d}c} \cdot \frac{\mathrm{d}'y}{\mathrm{d}a} \cdot \frac{\mathrm{d}'z}{\mathrm{d}b} - \frac{\mathrm{d}'x}{\mathrm{d}c} \cdot \frac{\mathrm{d}'y}{\mathrm{d}b} \cdot \frac{\mathrm{d}'z}{\mathrm{d}a}$; the initial values of x, y, and z being expressed by a, b, and c, which are variable from particle to particle only.

The coordinates, x, y, and z, are functions of the primitive coordinates a, b, c, and of the time t: [it is evident, for example, in the propagation of a wave, that the motion of any particle, to which the ordinates x, y, and z belong, depends entirely on the initial state of other ordinates of the surface of the fluid, in combination with the time elapsed from the beginning of the motion:] consequently, [if the variations δ be taken with respect to any one instant of time,]

$$\delta x - \frac{d'x}{da}\delta a + \frac{d'x}{db}\delta b + \frac{d'x}{dc}\delta c;$$

$$\delta y = \frac{d'y}{da}\delta a + \frac{d'y}{db}\delta b + \frac{d'y}{dc}\delta c; \text{ and}$$

$$\delta z - \frac{d'z}{da}\delta a + \frac{d'z}{db}\delta b + \frac{d'z}{dc}\delta c.$$

By substituting these values in the equation (F) (363), we may obtain three separate equations of the coefficients of δa, δb, and δc, considered as vanishing separately; these equations expressing the relations of the partial fluxions of the coordinates x, y, and z, the primitive ordinates a, b, and c, and the time t.

We must next investigate the conditions required for the continuity of the fluid. For this purpose we may consider the elementary portion of the fluid, at the beginning of the motion, as a rectangular parallelepiped, of which the sides are Da, Db, and Dc, and the mass (ρ) DaDbDc. We may call this parallelepiped (A): and it is easy to see that after the time t it will be changed into an oblique parallelepiped; for all the molecules at first situated in any face of the parallelepiped (A) will still be in the same plane, at least if we neglect the infinitely small effect of curvature on the infinitely small faces; and all the particles situated in the parallel edges of (A) will be found in elementary right lines equal and parallel to each other. We may call this new parallelepiped (B), and we may conceive two planes, parallel to that of x and y, to pass through the extremities of its edge formed by the particles which in (A) occupied the edge Dc. Then if all the edges of (B) be prolonged, until they meet these two planes, they will form a new parallelepiped (C), equal to (B); for it is clear that as much as one of these planes cuts off from the parallelepiped (B), so much is added to it by the other. The

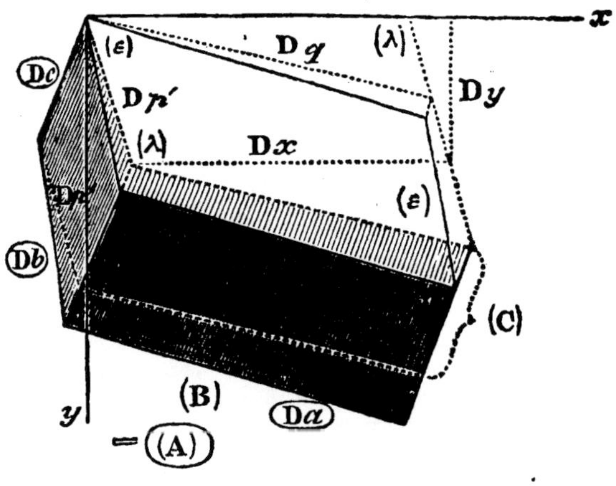

parallelepiped (C) will have its two bases parallel to the plane of x and y: its height between the bases will evidently be equal to the element of z; and since in this point of view x, y, and t may all be considered as constant, and the same values only of a and b enter into the determination, the element will be merely $\frac{d'z}{dc}$ Dc [, which must be equivalent to the D$z+\frac{d'w}{dz}$ D$z\Delta t$ of Poisson]. The base of the parallelepiped (C) will be found by observing that it is equal to the section of (B) by a plane parallel to that of x and y; and we may call this section (ε): with respect to the particles situated in it, the value of z will be the same for all, and we shall have D$z=0-\frac{d'z}{da}$ D$a+\frac{d'z}{db}$ D$b+\frac{d'z}{dc}$ Dc.

Now if Dp' and Dq be two contiguous sides of the section (ε), the first derived from the face answering to DbDc of (A), the second from DaDc: if through the extremities of the side Dp' we imagine two right lines to be drawn parallel to x, and the side opposite to Dp' to be produced so as to meet these lines, they will intercept a new parallelogram (λ) equal to (ε), having its base parallel to x. The side Dp' is formed by some of the particles belonging to the face DbDc, that is, by those particles with regard to which the value of z is invariable, and it is easy to see that the height of the parallelogram (λ) is the element of y, taken on the supposition that b and c only vary, while a, t, and z remain constant. Hence

$$\text{D}y = \frac{d'y}{db}\,\text{D}b + \frac{d'y}{dc}\,\text{D}c;\quad 0 = \frac{d'z}{db}\,\text{D}b + \frac{d'z}{dc}\,\text{D}c.$$

consequently $\text{D}c = -\dfrac{\frac{d'z}{db}}{\frac{d'z}{dc}}\,\text{D}b$, and $\text{D}y = \text{D}b\left(\dfrac{d'y}{db} - \dfrac{\frac{d'y}{dc}\cdot\frac{d'z}{db}}{\frac{d'z}{dc}}\right)$

$$= \text{D}b \cdot \frac{\frac{d'y}{db}\cdot\frac{d'z}{dc} - \frac{d'y}{dc}\cdot\frac{d'z}{db}}{\frac{d'z}{dc}}:$$ which is the height of the parallelogram (λ). Its base is equal to the section of the parallelogram formed by a line parallel to x, belonging to a plane in which those particles of the parallelepiped (A) are found, with respect to which z and y are constant: the length of this section is therefore equal to the element of x, supposing z, y, and t to be constant.

We have therefore, for the element $\text{D}x$, the three equations

$$\text{D}x = \frac{d'x}{da}\,\text{D}a + \frac{d'x}{db}\,\text{D}b + \frac{d'x}{dc}\,\text{D}c$$

$$0 - \frac{d'y}{da}\,\text{D}a + \frac{d'y}{db}\,\text{D}b + \frac{d'y}{dc}\,\text{D}c;\quad 0 - \frac{d'z}{da}\,\text{D}a + \frac{d'z}{db}\,\text{D}b + \frac{d'z}{dc}\,\text{D}c:$$

[and multiplying the second by $\frac{d'z}{dc}$, and the third by $\frac{d'y}{dc}$, we have

$$0 = \frac{d'y}{da}\cdot\frac{d'z}{dc}\,\text{D}a + \frac{d'y}{db}\cdot\frac{d'z}{dc}\,\text{D}b - \frac{d'y}{dc}\cdot\frac{d'z}{da}\,\text{D}a - \frac{d'y}{dc}\cdot\frac{d'z}{db}\,\text{D}b:$$ whence

$$\text{D}b - \frac{\frac{d'y}{da}\cdot\frac{d'z}{dc} - \frac{d'y}{dc}\cdot\frac{d'z}{da}}{\frac{d'y}{dc}\cdot\frac{d'z}{db} - \frac{d'y}{db}\cdot\frac{d'z}{dc}}\,.\text{D}a;$$ and in a similar manner we obtain

$$\text{D}c = \frac{\frac{d'y}{da}\cdot\frac{d'z}{db} - \frac{d'y}{db}\cdot\frac{d'z}{da}}{\frac{d'y}{db}\cdot\frac{d'z}{dc} - \frac{d'y}{dc}\cdot\frac{d'z}{db}}.\text{D}a\text{:]}$$ consequently $\text{D}x = \frac{d'x}{da}.\text{D}a +$

$$\frac{\frac{d'x}{db}\cdot\frac{d'y}{dc}\cdot\frac{d'z}{da} - \frac{d'x}{db}\cdot\frac{d'y}{da}\cdot\frac{d'z}{dc} + \frac{d'x}{dc}\cdot\frac{d'y}{da}\cdot\frac{d'z}{db} - \frac{d'x}{dc}\cdot\frac{d'y}{db}\cdot\frac{d'z}{da}}{\frac{d'y}{db}\cdot\frac{d'z}{dc} - \frac{d'y}{dc}\cdot\frac{d'z}{db}}.\text{D}a;$$ and

$$\text{D}x = \frac{\varsigma\text{D}a}{\frac{d'y}{db}\cdot\frac{d'z}{dc} - \frac{d'y}{dc}\cdot\frac{d'z}{db}};$$ which is the base of the parallelogram (λ); and its height being $\text{D}y$, its area is $= \frac{\varsigma\text{D}a\text{D}b}{\frac{d'z}{dc}}$: which is also the area of the parallelogram (ε), and which, multiplied by $\frac{d'z}{dc}.\text{D}c$, will become $\varsigma\text{D}a\text{D}b\text{D}c$, for the volume of the parallelepipeds (C) and (B): and ρ being the density after the time t, the mass must be $\rho\varsigma\text{D}a\text{D}b\text{D}c$, which being equal to $(\rho)\,\text{D}a\text{D}b\text{D}c$, we shall have $\rho\varsigma = (\rho)$ for the equation implying the continuity of the fluid.

§ 33. *Transformation of these equations: shown to be integrable provided that the density be any function of the pressure, and that the sum of the velocities parallel to three orthogonal coordinates, each being multiplied by the element of its direction, make an exact variation. This condition fulfilled at every instant if it is at a single one.* **P. 94.**

366. THEOREM. If u, v, and w be the velocities of a particle in the directions of x, y, and z, we have $\delta V - \frac{\delta p}{\rho} = \delta x\left(\frac{d'u}{dt} + u\frac{d'u}{dx} + v\frac{d'u}{dy}\right.$

$+w\frac{d'u}{dz}\Big)+\delta y\Big(\frac{d'v}{dt}+u\frac{d'v}{dx}+v\frac{d'v}{dy}+w\frac{d'v}{dz}\Big)+\delta z\Big(\frac{d'w}{dt}+u\frac{d'w}{dx}+v\frac{d'w}{dy}+w\frac{d'w}{dz}\Big).$ (*H*)

For since $\frac{dx}{dt}=u, \frac{dy}{dt}=v$, and $\frac{dz}{dt}=w$, if we take the fluxions of these equations, regarding u, v, and w as functions of the coordinates x, y, and z of the particle, and of the time t, we shall have

$$\frac{ddx}{dt^2}=\frac{d'u}{dt}+u\frac{d'u}{dx}+v\frac{d'u}{dy}+w\frac{d'u}{dz};$$

$$\frac{ddy}{dt^2}=\frac{d'v}{dt}+u\frac{d'v}{dx}+v\frac{d'v}{dy}+w\frac{d'v}{dz}; \text{ and}$$

$$\frac{ddz}{dt^2}=\frac{d'w}{dt}+u\frac{d'w}{dx}+v\frac{d'v}{dy}+w\frac{d'w}{dz}.$$ [For since $du=\frac{d'u}{dt}dt+\frac{d'u}{dx}dx+\frac{d'u}{dy}dy+\frac{d'u}{dz}dz$, and $dx=udt$, $dy=vdt$, and $dz=wdt$, the truth of the equations is manifest; and by substituting these values in the equation (*F*) 363, we obtain the equation (*H*) of this proposition.]

367. THEOREM. For the equation of continuity we have also $0=\frac{d\rho}{dt}+\frac{d'(\rho u)}{dx}+\frac{d'(\rho v)}{dy}+\frac{d'(\rho w)}{dz}$. (*K*)

If we suppose the coordinates, x, y, and z, to be infinitely near to a, b, and c, we may conceive a, b, and c in the value of ϵ, to be equal to x, y, and z, and x, y, and z to become $x+u\Delta t$, $y+v\Delta t$, and $z+w\Delta t$: we shall then have $\epsilon=1+\Delta t\left(\frac{d'u}{dx}+\frac{d'v}{dy}+\frac{d'w}{dz}\right)$; [since $\frac{d'x}{da}$ becomes $\frac{d'x}{da}+\frac{d'u}{da}\Delta t=\frac{d'x}{dx}+\frac{d'u}{dx}\Delta t=1+\Delta t\frac{du}{dx}$, $\frac{d'y}{db}=1+\Delta t\cdot\frac{d'v}{dy}$, and $\frac{d'z}{dc}=$

$1+\Delta t.\frac{d'w}{dz}$, so that the first term of the value of C becomes equal to the product of these three quantities, and the five other terms vanish, since $\frac{d'x}{db}=0, \frac{d'x}{dc}=0, \frac{d'y}{da}=0, \frac{d'y}{dc}=0,$ $\frac{d'z}{da}=0, \frac{d'z}{db}=0.$] The equation ($G$) becomes therefore

$$\varrho\Delta t\left(\frac{d'u}{dx}+\frac{d'v}{dy}+\frac{d'w}{dz}\right)+\varrho-(\varrho)=0$$: and if ϱ be considered as a function of x, y, z, and t, we have

$$(\varrho)=\varrho-\Delta t\,\frac{d'\varrho}{dt}-u\Delta t\,\frac{d'\varrho}{dx}-v\Delta t\,\frac{d'\varrho}{dy}-w dt\frac{d'\varrho}{dz}$$; so that the preceding equation becomes

$$\frac{d'\varrho}{dt}+\frac{d'(\varrho u)}{dx}+\frac{d'(\varrho v)}{dy}+\frac{d'(\varrho w)}{dz}=0 \qquad (K)$$

It is easy to see that this equation is the fluxion of the equation (G) (365) taken with regard to the time t [: for it has been deduced from G by taking the difference of its terms with regard to the evanescent element of time Δt].

368. THEOREM. If $u\delta x+v\delta y+w\delta z=\delta\varphi$, ϱ being any function of the pressure p, we shall have $V-\int\frac{\delta p}{\varrho}=\frac{d\varphi}{dt}+\frac{1}{2}\left\{\left(\frac{d'\varphi}{dx}\right)^2+\left(\frac{d'\varphi}{dy}\right)^2+\left(\frac{d'\varphi}{dz}\right)^2\right\}$, and, for homogeneous fluids, $\frac{dd'\varphi}{dx^2}+\frac{dd'\varphi}{dy^2}+\frac{dd'\varphi}{dz^2}=0.$

When $u\delta x+v\delta y+w\delta z$ is an exact variation of x, y, and z (313) and ϱ is also a function of the pressure, the equation (H) is susceptible of integration, for it becomes $\delta V-\frac{\delta p}{\varrho}-\delta\frac{d\varphi}{dt}+\frac{1}{2}\delta\left\{\left(\frac{d'\varphi}{dx}\right)^2+\left(\frac{d'\varphi}{dy}\right)^2+\left(\frac{d'\varphi}{dz}\right)^2\right\}$ [; since

$\frac{d'\varphi}{dx} - u$, $\delta u - \frac{d'u}{dx}\delta x + \frac{d'u}{dy}\delta y + \frac{d'u}{dz}\delta z$, $= \frac{d'u}{dx}\delta x + \frac{d'v}{dx}\delta y + \frac{d'w}{dx}\delta z$ (314), and $\frac{1}{2}\delta\left(\frac{d'\varphi}{dx}\right)^2 = u\delta x \frac{d'u}{dx} + u\delta y \frac{d'v}{dx} + u\delta z \frac{d'w}{dx}$; and the variations of the other parts of the expressions being transformed in a similar manner, the sum will obviously be equal to the corresponding terms of (H)]. The fluent of this equation is $V - \int \frac{\delta p}{\varrho} = \frac{d'\varphi}{dt} + \frac{1}{2}\left\{\left(\frac{d'\varphi}{dx}\right)^2 + \left(\frac{d'\varphi}{dy}\right)^2 + \left(\frac{d'\varphi}{dz}\right)^2\right\}$. It would be necessary to add an independent constant quantity, expressed in terms of t, to this fluent, but this quantity may be supposed to be included in φ. The velocity of the particles, in the directions of the coordinates, is obtained from the quantity φ; since $u = \frac{d'\varphi}{dx}$, $v = \frac{d'\varphi}{dy}$, and $w = \frac{d'\varphi}{dz}$.

The equation (K), expressing the continuity of the fluid, or $0 = \frac{d'\varrho}{dt} + \frac{d'(\varrho u)}{dx} + \frac{d'(\varrho v)}{dy} + \frac{d'(\varrho w)}{dz}$, becomes $0 = \frac{d'\varrho}{dt} + \frac{d'\varrho}{dx} \cdot \frac{d'\varphi}{dx} + \frac{d'\varrho}{dy} \cdot \frac{d'\varphi}{dy} + \frac{d'\varrho}{dz} \cdot \frac{d'\varphi}{dz} + \varrho\left(\frac{dd'\varphi}{dx^2} + \frac{dd'\varphi}{dy^2} + \frac{dd'\varphi}{dz^2}\right)$; consequently, with regard to homogeneous fluids, since $d\varrho = 0$, we have $0 = \frac{dd'\varphi}{dx^2} + \frac{dd'\varphi}{dy^2} + \frac{dd'\varphi}{dz^2}$ $\left[-\frac{d'u}{dx} + \frac{d'v}{dy} + \frac{d'w}{dz}\right]$.

369. THEOREM. **If the quantity $u\,x + v\delta y + w\delta z$ is an exact variation of x, y, and z, at any one instant, it will always remain so.**

If, for example, this variation be at any one instant equal to $\delta\varphi$: it will be at the next instant equal to $d\varphi + \Delta t \left(\frac{d'u}{dt}\delta x + \frac{d'v}{dt}\delta y + \frac{d'w}{dt}\delta z\right)$ which will still be an exact variation if $\frac{d'u}{dt}\delta x+\frac{d'v}{dt}\delta y+\frac{d'w}{dt}\delta z$ is an exact variation in the first instance: now we have from the equation (*H*) in this case $\frac{d'u}{dt}\delta x+\frac{d'v}{dt}\delta y+\frac{d'w}{dt}\delta z=\delta V-\frac{1}{2}\delta\left\{\left(\frac{d'\varphi}{dx}\right)^2+\left(\frac{d'\varphi}{dy}\right)^2+\left(\frac{d'\varphi}{dz}\right)^2\right\}-\frac{\delta p}{\varrho}$: the first member of the equation is consequently an exact variation of a function of x, y, and z; the function $u\delta x+v\delta y+w\delta z$ is therefore an exact variation in the subsequent instant if it is in the preceding: it is therefore an exact variation at every instant.

370. THEOREM. When the motion of the fluid is infinitely small, we have $V-\int\frac{dp}{\varrho}=\frac{d\varphi}{dt}$.

Neglecting the squares and products of u, v, and w, the partial velocities, the latter part of the equation (*H*) (366, 368) will vanish; and in this case $u\delta x+v\delta y+w\delta z=\delta\varphi$ must be an exact variation whenever p is a function of ϱ: and when the fluid is homogeneous, the equation of continuity remains $0=\frac{dd\varphi}{dx^2}+\frac{dd\varphi}{dy^2}+\frac{dd\varphi}{dz^2}$. These two equations contain the whole theory of infinitely small undulations of homogeneous fluids.

§ 34. *Case of the rotation of a homogeneous fluid mass, with a uniform velocity, round one of the axes of the coordinates.* P. 97.

371. THEOREM. In the case of a homogeneous fluid, revolving round an axis with a uniform velocity, the equation of the pressure

becomes $\frac{\delta p}{\rho} - \delta V + n^2(y\delta y + z\delta z)$; and the quantity $u\delta x + v\delta y + w\delta z$ is not an exact variation.

Supposing x to be the axis of motion, and the angular velocity n, at a distance considered as unity, we shall have $v = -nz$, $w = ny$, and the equation (H)(366) becomes $\frac{\delta p}{\rho} - \delta V - \frac{d'v}{dz} w\delta y - \frac{d'w}{dy} v\delta y = \delta V + n\,.\,ny.\delta y + n.nz.\delta z = \delta V + n^2(y\delta y + z\delta z)$: an equation of which both the members are exact variations, and which is therefore possible. The equation (K) (367), will become $0 = \Delta t \frac{d'\rho}{dt} + u\Delta t \frac{d'\rho}{dx} + v\Delta t \frac{d'\rho}{dy} + w\Delta t \frac{d'\rho}{dz}$; and it is obvious that this equation will be satisfied if the fluid is homogeneous. Both equations therefore being true, the supposed motion is possible, and a fluid may move uniformly round an axis, [without any internal change of the disposition of its particles.]

The centrifugal force, at the distance $\sqrt{(y^2+z^2)}$ from the axis of rotation, is expressed by the square of the velocity $n^2(y^2+z^2)$, divided by the distance; consequently the quantity $n^2(y\delta y + z\delta z)$ is the product of the centrifugal force $n^2\sqrt{(y^2+z^2)}$ into the element of its direction $\frac{y\delta y + z\delta z}{\sqrt{(y^2+z^2)}}$. it is evident, therefore, by comparing this equation with the general equation of the equilibrium of fluids in § 17 (316) that the conditions of the motion are reduced to those of the equilibrium of a fluid actuated by the same forces, and by the centrifugal force in addition to them: which is also sufficiently obvious from the nature of the case.

If the external surface of the mass is at liberty, we have here $\delta p = 0$, and consequently $0 = \delta V + n^2(y\delta y + z\delta z)$; consequently the result of all the forces that act on the external surface must be perpendicular to that surface; it must also be directed towards the interior of the fluid: and when these conditions are fulfilled, a homogeneous fluid mass may be in equilibrium, whatever may be the form of the solid which it covers.

This case is one of those in which the variation $u\delta x + v\delta y + w\delta z$ is not exact; for this variation becomes equal to $-n(z\delta y - y\delta z)$: and $z\delta y - y\delta z$ is not an integrable quantity. Consequently in the theory of the tides we cannot suppose the variation $\delta\varphi$ to be exact, since it is not so in the very simple case of the sea having no other motion than its rotation in common with the earth.

§ 35. *Determination of the very small oscillations of a homogeneous fluid, covering a revolving spheroid.* P. 98.

372. THEOREM. If r be the primitive distance of a particle from the centre, θ the angle formed by r with the axis x, ϖ the angle formed by the plane of x and r with that of x and y; and if, after the time t, r become $r + \alpha s$, θ, $\theta + \alpha u$, and ϖ, $nt + \varpi + \alpha v$, α being very small, we shall have

$$\alpha r^2 \delta\theta . \left(\frac{ddu}{dt^2} - 2n \sin \cos \theta . \frac{dv}{dt}\right) +$$

$$\alpha r^2 \delta\varpi \left(\sin^2\theta \frac{ddv}{dt^2} + 2n \sin \cos \theta \frac{du}{dt} + \frac{2n \sin^2\theta}{r} . \frac{ds}{dt}\right) +$$

$$a\delta i'\left(\frac{dds}{dt^2} - 2nr\sin^2\theta\frac{dv}{dt}\right) - \frac{n^2}{2}\delta\left\{(r+as)\sin \quad +au)\right\}^2$$
$$+\delta V - \frac{\delta p}{\rho}. \qquad (L)$$

It is obvious that the coordinate x will be, at the end of the time t, $(r+as)\cos(\theta+au)$; and the projection of the radius, on the plane perpendicular to x, being $(r+as)\sin(\theta+au)$ we shall have

$y=(r+as)\sin(\theta+au)\cos(nt+\varpi+av)$,

$z=(r+as)\sin(\theta+au)\sin(nt+\varpi+av)$; and substituting these values in the equation (F) (363), that is $\delta V - \frac{\delta p}{\rho} - \delta x\frac{ddx}{dt^2} + \delta y\frac{ddy}{dt^2} + \delta z\frac{ddz}{dt^2}$; neglecting the square of a, [and calling x, $\lambda\cos\mu$; y, $\nu\cos\xi$; and z, $\nu\sin\xi$], we have

[$\delta x=\delta\lambda.\cos\mu - \delta\mu.\lambda\sin\mu$

$d^2x=d^2\lambda.\cos\mu - d^2\mu\lambda\sin\mu$, since $d\lambda\, d\mu=0$, and $d\mu^2=0$, these quantities being multiplied by a^2.

$\delta x d^2x=\delta\lambda(d^2\lambda\cos^2\mu - d^2\mu\lambda\sin\cos\mu) - \delta\mu(d^2\lambda.\lambda\sin\cos\mu - d^2\mu\lambda^2\sin^2\mu)$

$\delta y=\delta\nu.\cos\xi - \delta\xi.\nu\sin\xi$

$d^2y=d^2\nu.\cos\xi - 2d\nu d\xi.\sin\xi - d^2\xi.\nu\sin\xi - d\xi^2.\nu\cos\xi$

$\delta y d^2y=\delta\nu(d^2\nu.\cos^2\xi - 2d\nu d\xi.\sin\cos\xi - d^2\xi.\nu\sin\cos\xi - d\xi^2.\nu\cos^2\xi)$

$-\delta\xi(d^2\nu.\nu\sin\cos\xi - 2d\nu d\xi.\nu\sin^2\xi - d^2\xi.\nu^2\sin^2\xi - d\xi^2\nu^2\sin\cos\xi)$

$\delta z=\delta\nu.\sin\xi + \delta\xi.\nu\cos\xi$

$d^2z=d^2\nu.\sin\xi + 2d\nu d\xi.\cos\xi + d^2\xi.\nu\cos\xi - d\xi \quad \nu\sin\xi$

$\delta z d^2z=\delta\nu(d^2\nu.\sin^2\xi + 2d\nu d\xi.\sin\cos\xi + d^2\xi.\nu\sin\cos\xi - d\xi^2.\nu\sin^2\xi)$

$+\delta\xi(d^2\nu.\nu\sin\cos\xi + 2d\nu d\xi.\nu\cos^2\xi + d^2\xi.\nu^2\cos^2\xi - d\xi^2.\nu^2\sin\cos\xi)$

$\delta y d^2y + \delta z d^2z. = \delta\nu(d^2\nu - d\xi^2\nu) + \delta\xi(2d\nu d\xi.\nu + d^2\xi.\nu^2)$

Now $\delta\lambda = \delta r + \alpha\delta s$

$d\lambda = \alpha ds$

$d^2\lambda = \alpha d^2 s$

$\delta\mu = \delta\theta + \alpha\delta u$

$d\mu = \alpha du$

$d^2\mu = \alpha d^2 u$

$\delta\nu = (\delta r + \alpha\delta s)\sin\theta + (r + \alpha s)\cos\theta\,(\delta\theta + \alpha\delta u)$

$d\nu = r\cos\theta\,\alpha du + \alpha ds\,.\sin\theta$

$d^2\nu = r\cos\theta\,\alpha d^2 u + \alpha d^2 s\,.\sin\theta$

$\delta\xi = \delta\varpi + \alpha dv$

$d\xi = ndt + \alpha\delta v$

$d^2\xi = \alpha d^2 v$

Hence $\delta x\, d^2x = (\delta r + \alpha\delta s).(\alpha d^2 s.\cos^2\theta - \alpha d^2 u.r\sin\cos\theta) - (\delta\theta + \alpha\delta u).(\alpha d^2 s.r\sin\cos\theta - \alpha d^2 u\, r^2\sin^2\theta)$; in which $\alpha\delta s$ and $\alpha\delta u$ may obviously be omitted: again, $\delta y d^2 y + \delta z d^2 z = \left\{(\delta r + \alpha\delta s)\sin\theta + (r + \alpha s)\cos\theta\,(\delta\theta + \alpha\delta u)\right\}.\left\{(r\cos\theta\,\alpha d^2 u + \alpha d^2 s.\sin\theta) - 2\alpha n dt dv.r\sin\theta\right\} - \frac{1}{2}\delta(\nu^2).n^2dt^2 + (\delta\varpi + \alpha\delta v)\left\{2(r\cos\theta\,\alpha du + \alpha ds.\sin\theta)ndt.r\sin\theta + \alpha d^2 v.r^2\sin^2\theta\right\} = (\delta r.\sin\theta + r\cos\theta\,\delta\theta)\,\alpha\left\{(r\cos\theta\, d^2u + d^2s.\sin\theta) - 2ndtdv.r\sin\theta\right\} - \frac{1}{2}\delta(\nu^2).n^2dt^2 + d\varpi.\alpha(2r^2\sin\cos\theta\, du.ndt + r\sin^2\theta\, ndt.ds + d^2v.r^2\sin^2\theta)$; consequently $\delta x d^2x + \delta y d^2 y + \delta z d^2 z =$

$$\delta r.\alpha(d^2s.\cos^2\theta - d^2u.r\sin\cos\theta + d^2u.r\sin\cos\theta + d^2s.\sin^2\theta - 2dt - dv.rn\sin\theta)$$
$$+\delta\theta.\alpha(-d^2s.r\sin\cos\theta + d^2u.r^2\sin^2\theta + d^2u.r^2\cos^2\theta + d^2s.r\sin\cos\theta - 2dtdv.r^2n\sin\cos\theta)$$
$$-\tfrac{1}{2}\delta(\nu^2).n^2dt^2$$
$$+\delta\varpi.\alpha(2dudt.\ r^2n\sin\cos\theta + 2dsdt.rn\sin^2\theta + d^2v.r^2\sin^2\theta);$$

and this, divided by dt^2, becomes equivalent to] the expression contained in the proposition.

373. COROLLARY. At the surface of the sea, we have $r^2\delta\theta\left(\frac{ddu}{dt^2} - 2n \sin \cos \theta \frac{dv}{dt}\right) + r^2\delta\varpi$ $\left(\sin^2\theta \frac{ddv}{dt^2} + 2n \sin \cos \theta \frac{du}{dt} + 2n \sin^2\theta \frac{ds}{dt}\right) = -$ $g\delta y' + \delta V'$: g being the force of gravity, $\alpha\delta y'$ the elevation above the surface of equilibrium, and $\alpha\delta V'$ the part of δV which relates to the disturbing forces only.

At the external surface of the fluid, we have $\delta p = 0$, and in the state of equilibrium

$$0 = \tfrac{1}{2}n^2\delta\left\{(r + \alpha s)\sin(\theta + \alpha u)\right\}^2 + (\delta V),$$ (δV) being the value of δV which belongs to this state: [since the centrifugal force, together with the force contained in V, must in this state balance each other; and the quantities s, u, and v being constant, the first member of the equation (L) must necessarily vanish.] If the fluid in question be the sea, the variation (δV) at its surface will be the force of gravity multiplied by the element of its direction: and calling this force g, and making $\alpha y'$ the elevation of a particle of the surface above the surface of equilibrium, which may in this case be considered as the true level of the sea; it will be evident that the variation (δV) will be increased, in the state of motion, by the quantity $-\alpha g\delta y'$, because the force of gravity acts very nearly in the direction of y', and tends towards its origin [: the y' here intended being however very different from the y of the former part of the proposition, which is an immoveable line, and the force

considered being referred to the particles situated at the surface of equilibrium, and not at the momentary surface, on which the gravitation of the particles below it can have no effect.] Then if we denote by $\alpha\delta V'$ the part of δV which relates to the new forces depending on the state of motion, whether they arise from the changes produced by the motion, or from the attractions of the solid or the fluid, or of any foreign body, we shall have, at the surface [of equilibrium], $\delta V = (\delta V) - \alpha g \delta y' + \alpha \delta V'$.

The variation $\frac{1}{2}n^2\delta\left\{(r+\alpha s)\sin(\theta+\alpha u)\right\}^2$ is increased by the quantity $\alpha n^2\delta y.\, r\sin^2\theta$, in virtue of the elevation of the particle of water above the level of the sea; [since δr becomes $=\alpha\delta y'$, and $\delta(r^2\sin^2\theta.) = 2\delta r.r\sin^2\theta$,]: but this quantitv may be neglected in comparison with $-\alpha g\delta y'$, because even $\frac{n^2 r}{g}$, the value of the centrifugal force, at the equator, where it is greatest, is only a very small fraction, equal to $\frac{1}{289}$. Lastly, the variation of the radius r is so inconsiderable, for the different parts of the surface, in comparison with its whole magnitude, that, for the present purpose, we may make $\delta r = 0$; and dividing the equation (L) thus modified, by the coefficient α, we obtain the equation of the proposition.

374. COROLLARY 2. The equation of continuity will become $0 = r^2\left\{\rho' + (\rho)\cdot\left(\frac{du}{d\theta} + \frac{dv}{d\varpi} + u \cot\theta\right)\right\} + (\rho)\frac{d(r^2 s)}{dr}$; the density, after the time t, being expressed by $(\rho) + \alpha\rho'$.

The initial dimensions of an elementary rectangular parallelepiped being here $\text{D}r$, $r\text{D}\varpi \sin\theta$, and $r\text{D}\theta$, calling the values of r, θ, and ϖ after the time t, r', θ' and ϖ', and following the steps of article 365, we shall find that the volume of the elementary figure will become equal to a rectangular parallelepiped, of which the height is $\frac{dr'}{dr}\text{D}r$; the breadth $r'\sin\theta'\left(\frac{d'\varpi}{d\varpi}\text{D}\varpi+\frac{d\varpi'}{dr}\text{D}r\right)$, from which $\text{D}r$ may be exterminated by the equation $\frac{dr'}{d\varpi}\text{D}\varpi+\frac{dr'}{dr}\text{D}r=0$; and lastly, the length $r'\left(\frac{d\theta'}{dr}.dr+\frac{d\theta'}{d\theta}\text{D}\theta+\frac{d\theta'}{d\varpi}\text{D}\varpi\right)$, provided that we make

$$\frac{dr'}{dr}\text{D}r+\frac{dr'}{d\theta}\text{D}\theta+\frac{dr'}{d\varpi}\text{D}\varpi=0, \text{ and}$$

$$\frac{d\varpi'}{dr}\text{D}r+\frac{d\varpi'}{d\theta}\text{D}\theta+\frac{d\varpi'}{d\varpi}\text{D}\varpi=0;$$ [r, ϖ and θ, and r', ϖ' and θ' being here substituted for a, b, c, x, y and z]: and making $$\mathfrak{C}'=\frac{dr'}{dr}.\frac{d\theta'}{d\theta}.\frac{d\varpi'}{d\varpi}-\frac{dr'}{dr}.\frac{d\theta'}{d\varpi}.\frac{d\varpi'}{d\theta}+\frac{dr'}{d\theta}.\frac{d\theta'}{d\varpi}.\frac{d\varpi'}{dr}-\frac{dr'}{d\theta}.\frac{d\theta'}{dr}.\frac{d\varpi'}{d\varpi}+\frac{dr'}{d\varpi}.\frac{d\theta'}{dr}.\frac{d\varpi'}{d\theta}.-\frac{dr'}{d\varpi}.\frac{d\theta'}{d\theta}.\frac{d\varpi'}{dr},$$ the volume of the element, after the time t, will be $\mathfrak{C}'r'^2\sin\theta\,\text{D}r\,\text{D}\theta\,\text{D}\varpi$; consequently, if we call the primitive density (ρ), and ρ the density corresponding to the time t, we shall have, since the masses must be equal, $\rho\mathfrak{C}'r'^2\sin\theta'=(\rho)r^2\sin\theta$, which is the equation of continuity; and substituting for r', $r+\alpha s$; for θ', $\theta+\alpha u$, and for ϖ', $nt+\varpi+\alpha v$, we shall have, if we neglect the quantities of the order α^2, $\mathfrak{C}'=1+\alpha\frac{ds}{dr}+\alpha\frac{du}{d\theta}+\alpha\frac{dv}{d\varpi}$ [in the same manner as $\mathfrak{C}$ was found equal

to $1+\Delta t\left(\frac{d'u}{dx}+\frac{d'v}{dy}+\frac{d'w}{dz}\right)$ in article 365]. Hence we obtain the equation

$$0=r^2\left\{\varrho'+(\varrho).\left(\frac{du}{d\theta}+\frac{dv}{d\varpi}+\frac{u\cos\theta}{\sin\theta}\right)\right\}+(\varrho)\frac{d(r^2s)}{dr}.$$ [For ϱ being $=(\varrho)+\alpha\varrho'$, $\left\{(\varrho)+\alpha\varrho'\right\}\left(1+\alpha\frac{ds}{dr}+\alpha\frac{du}{d\theta}+\alpha\frac{dv}{d\varpi}\right)(r^2+2\alpha rs)\sin(\theta+\alpha u)=(\varrho)\,r^2\sin\theta$, $(\varrho)\,r^2(\sin\theta+\alpha u\cos\theta)-(\varrho)\,r^2\sin\theta+\alpha\varrho'\,r^2\sin\theta+(\varrho)\,r^2\sin\theta\left(\alpha\frac{ds}{dr}+\alpha\frac{du}{d\theta}+\alpha\frac{dv}{d\varpi}\right)+(\varrho)2\alpha rs\sin\theta=0$ (140), $0=r^2\left\{\varrho'+(\varrho)\left(\frac{du}{d\theta}+\frac{dv}{d\varpi}+\frac{u\cos\theta}{\sin\theta}\right)+(\varrho)\,r^2\frac{ds}{dr}+(\varrho)\frac{2rsdr}{dr}.\right.$]

36. *Case of the motion of the sea, supposing it to be deranged from the state of equilibrium by the action of very small forces.* P. 101.

375. THEOREM. Retaining the notation of the preceding propositions, and supposing the sea of inconsiderable depth, we have, for the surface, $r^2\delta\theta.\left(\frac{ddu}{dt^2}-2n\sin\cos\theta\frac{dv}{dt}\right)+r^2\delta\varpi.\left(\sin^2\theta\frac{ddv}{dt^2}+2n\sin\cos\theta\frac{du}{dt}\right)=-g\delta y'+\delta V'$. (M)

Since the density of the sea is uniform, we have $\varrho'=0$, and consequently $\frac{d(rrs)}{dr}+r^2\left(\frac{du}{d\theta}+\frac{dv}{d\varpi}+\frac{u\cos\theta}{\sin\theta}\right)=0$. Now we may suppose the depth of the sea inconsiderable in comparison with the radius r of the terrestrial spheroid; and calling this depth γ, we may imagine γ to be a very small function of θ and ϖ, determined by the law of the depth. If we consider the nature of the fluent of this equation with regard to the variable quantity r, between the surface of the solid spheroid and that of the sea, it is obvious that the value of s will be a function of θ, ϖ, and t, independent of r, together with a very small function of r, standing in the same relation to u and v, as γ does to r. Now at the surface of the solid covered by the sea, when the angles θ and ϖ are changed into $\theta + \alpha u$ and $nt + \varpi + \alpha v$, it is easy to see that the distance of a particle of water contiguous to that surface, from the centre of gravity of the earth, can only vary by a quantity which is very small with respect to αu and αv, and which is of the same order as the products of these quantities into the eccentricity of the spheroid covered by the sea; consequently the function, independent of r, that enters into the expression of s, must therefore be of the same order, and very minute, so that we may in general neglect s as inconsiderable in comparison with u and v. [Thus if the sea were 4 miles deep, γ would be about $\frac{1}{1000}$ of r, and the ascent and descent of a particle even at the surface of the sea would in general be little more than $\frac{1}{1000}$ of its horizontal motion, supposing the neighbouring particles, for a considerable extent in comparison with the radius, to be moving in the same direction.] We may therefore omit the quantity ds in the equation of article 373, and it affords the equation of this proposition.

376. Theorem. The equation of continuity becomes $y = -\frac{d(\gamma u)}{d\theta} - \frac{d(\gamma v)}{d\varpi} - \frac{\gamma u \cos\theta}{\sin\theta}$, (N) y being the elevation above the surface of equilibrium, and γ the depth of the sea.

The equation (L), article 372, which is applicable to every particle of the fluid, affords us, in the case of equilibrium, $0 = \frac{1}{2}n^2\delta\left\{ r+\alpha s) \sin(\theta+\alpha u)\right\}^2 + (\delta V) - \frac{(\delta p)}{\varrho}$; (δV) and (δp) being the values of δV and δp which belong, in the state of equilibrium, to the quantities $r+\alpha s$, $\theta+\alpha u$, and $\varpi+\alpha v$, and which, in the state of motion, we may suppose to become $\delta V = (\delta V) + \alpha\delta V'$, and $\delta p = (\delta p) + \alpha\delta p'$; and [since the variations and forces in the three different directions afford independent equations,] we have

$$\frac{\mathrm{d}'\left(V' - \frac{p'}{\varrho}\right)}{\mathrm{d}r} = \frac{\mathrm{dd}s}{\mathrm{d}t^2} - 2nr\ \sin^2\theta\frac{\mathrm{d}v}{\mathrm{d}t}:$$

[the other parts of the equation remaining the same as in the case of equilibrium, and therefore balancing each other]. Now it appears from the equation (M) (375), that $n\frac{\mathrm{d}v}{\mathrm{d}t}$ is of the same order with y or with s, and consequently with $\frac{\gamma u}{r}$; the value of the first member of the present equation must therefore be of the same order; and if we multiply this value by $\mathrm{d}r$, and find the fluent for the whole depth of the sea, we shall have for $V' - \frac{p'}{\varrho}$ a very small function, of the order $\frac{\gamma s}{r}$, besides a function of θ, ϖ, and t independent of r, which we may call λ; consequently if in the equation (L) we only consider the two variable quantities θ and ϖ, it will afford us the equation (M), with this difference only, that the second member will become $\delta\lambda$. But since λ is independent of the depth of the particle, this equation becomes equally applicable to the surface and its neighbourhood, and the equations (M) and (L) must in this case coincide with each other: hence we have $\delta\lambda = \delta V' - g\delta y$, and consequently $\delta\left(V' - \frac{p'}{\varrho}\right) = \delta V - g\delta y$; the $\delta V'$ in the second member of the equation relating to the surface of the sea. It will appear, in the theory of the tides, that this value is very nearly the same for all the particles situated in the same radius of the earth, from the bottom of the sea to the surface: we have therefore, for all these particles, $\frac{\delta p'}{\varrho} = g\delta y$, consequently p' must be equal to $\varrho g y$, with the addition of some function independent of θ, ϖ, and r, as a correction of the fluent: now at the surface of equilibrium of the sea, the quantity $\alpha p'$ must be equal to the pressure of the little column of water αy, which is elevated above this surface, and this pressure is expressed by $\alpha \varrho g y$: hence it follows, that throughout the interior of the fluid mass, from the surface of the spheroid covered by the sea, to the surface of the sea itself, $p' = \varrho g y$, or that, in other words, any point of the surface of the solid spheroid is more pressed than in the state of equilibrium, by all the weight of the little column of water, contained between the surface of the sea and the surface of equilibrium; and that this excess of pressure becomes negative at the parts in which the sea is depressed below this surface of equilibrium. [There seems, however, to be wanting in this theory, the consideration of the time required for the transmission of pressure, as well as of the possibility of the divergence of pressure from a direction completely vertical. It cannot be supposed that every ripple, which curls the surface of the ocean, produces an instantaneous diversity of pressure at the depth of several miles; nor is it very probable that each inch of the bottom of the sea at such a depth, is, after any interval of time, affected separately by the transitory inequalities of the surface exactly above it. With respect to the gradual transmission of pressure, it can scarcely be slower in a fluid than it would be in the same substance if congealed into a solid mass: for the effect must depend on the ultimate elasticity of the particles themselves, and not on the rigidity of the aggregate; although Mr. Poisson seems disposed to consider the primary transmission of the pressure as depending on the same conditions as the propagation of a small wave of finite magnitude. With regard to the want of verticallty of the pressure, depending perhaps on a want of perfect fluidity, it seems to be difficult to make any allowance for it in a correct computation: but fortunately, in the great problem of the tides, the depth being inconsiderable in comparison with the extent of a similar and synchronous state of the surface, neither of these sources of inaccuracy can have any material effect.]

It may in general be observed, that having regard to the variations of θ and ϖ only, [and neglecting the slight vertical motion] the equation (L) becomes equivalent to (M) for all the interior parts of the fluid. The values of u and v, relative to all the particles of the sea, situated in the same radius of the earth, are therefore determined by the same differential equations: consequently, if we suppose, as it will be convenient to do in the theory of the tides, that at the origin of the motion, the values of u, $\frac{du}{dt}$, v, and $\frac{dv}{dt}$ were the same for all the particles situated in the same radius, these particles will still remain in the radius, during the oscillations of the fluid: the values of r, u, and v may therefore be supposed very nearly the same throughout the small portion of the radius intervening between the bottom and the surface of the sea: we may therefore consider $r^2_{\prime}s$ as the fluent of $\frac{d'(rrs)}{dr}\,dr$, and calling the value of r^2s at the bottom of the sea (r^2s) we shall have, for the fluent of the equation $0=\frac{d'(rrs)}{dr}+r^2\left(\frac{du}{d\theta}+\frac{dv}{d\varpi}+\frac{u\cos\theta}{\sin\theta}\right)$, taken with respect to r, $0=r^2s-(r^2s)+r^2\gamma\left(\frac{du}{d\theta}+\frac{dv}{d\varpi}+\frac{u\cos\theta}{\sin\theta}\right)$, since γ is the particular value of $\int dr\cdot$ between these limits. The quantity $r^2s-(r^2s)$ is also very nearly equal to $r^2\left\{s-(s)\right\}+2r\gamma(s)$, (s) being the value of s at the bottom of the sea, and considering the minuteness of γ and of s, the latter part of this expression may be neglected in comparison with the former, and we may call $r^2s-(r^2s)=r^2\left\{s-(s)\right\}$. Now the depth of the sea, corresponding to the angles $\theta+\alpha u$, and $nt+\varpi+\alpha v$, is $\gamma+\alpha\left\{s-(s)\right\}$: and if we consider the angles θ and " $nt+$" ϖ as beginning at a fixed point and a fixed meridian on the surface of the earth, which will soon appear to be admissible, this depth will be $\gamma+\alpha u\,\frac{d\gamma}{d\theta}+\alpha v\,\frac{d\gamma}{d\varpi}$, besides the elevation αy of the particle above the surface of equilibrium, [for since γ is, by the supposition, a function of θ and ϖ, it is necessary to comprehend in the equation its variations dependent on those of these angles;] consequently $s-(s)=y+u\,\frac{d\gamma}{d\theta}+v\,\frac{d\gamma}{d\varpi}$. The equation of the continuity of the fluid will therefore become $y=-\frac{d(\gamma u)}{d\theta}-\frac{d(\gamma v)}{d\varpi}-\frac{\gamma u\cos\theta}{\sin\theta}$ $\left[=-\gamma\,\frac{du}{d\theta}-u\,\frac{d\gamma}{d\theta}-\gamma\,\frac{dv}{d\varpi}-v\,\frac{d\gamma}{d\varpi}-\frac{\gamma u\cos\theta}{\sin\theta}\right.$; since $s-(s)=-\gamma\left(\frac{du}{d\theta}+\frac{dv}{d\varpi}+\frac{u\cos\theta}{\sin\theta}\right)=y+u\,\frac{d\gamma}{d\theta}+v\,\frac{d\gamma}{d\varpi}$, which amounts to the same].

It may be observed that, in this equation, the angles θ and " $nt+$"ϖ are reckoned from a fixed point, and a fixed meridian, on the earth, while in the equation (M) the same angles are referred to the axis x, and to a plane passing through that axis, and having a rotatory motion round it expressed by n: now this axis and this plane are not precisely fixed with regard to the surface of the earth, because the attraction and the pressure of the fluid, which covers it, must alter in a slight degree their position on the surface, as well as the rotatory motion of the spheroid. But it is easy to see that these alterations must be to the values of au and av, almost in the proportion of the mass of the sea to that of the solid spheroid: consequently in order to refer the angles θ and " $nt+$" ϖ to an invariable point and an invariable meridian on the surface of the spheroid, in the two equations (M) and (N), it must be sufficient to alter u and v by quantities of the order $\frac{\gamma u}{r}$ and $\frac{\gamma v}{r}$, which we have neglected in this computation: it may therefore be assumed, in these equations, that au and av are the motions of the fluid in latitude and longitude. [It seems more natural to call the angle made by the plane in question with the first meridian ϖ or $\varpi+av$ only, and to express by nt the rotatory motion of the earth only: and perhaps $nt+\varpi$ may have been an error of the pen only for $\varpi+av$.]

It may also be remarked, that the centre of gravity of the spheroid being supposed immoveable, we must transfer to the particles of the fluid in a contrary direction, the effect of the reaction of the sea on that spheroid: but since the place of the common centre of gravity of the solid spheroid and of the sea is not changed in consequence of this reaction, it is evident that the relation of this velocity to that with which the particles are impressed, by the action of the spheroid, is of the same order with the relation of the mass of the fluid to that of the spheroid, or of the order $\frac{\gamma}{r}$, and that it may therefore be neglected in the calculation of $\delta V'$.

§ 37. *Of the earth's atmosphere, considered first in the state of equilibrium. Of the oscillations which it undergoes in the state of motion, having regard only to the regular causes which agitate it; and of the variations which these motions produce in the height of the barometer.* P. 105.

377. THEOREM The oscillations of the atmosphere may be determined by the equations $r^2\delta\theta.\left(\frac{ddu'}{dt^2}-2n\sin\cos\theta\frac{dv'}{dt}\right)+r^2\delta\varpi.\left(\frac{ddv'}{dt^2}.\sin^2\theta\right.$ $\left.+2n\sin\cos\theta\frac{du'}{dt}\right)=\delta V'-g\delta y'-g\delta y$; and $y'=-l\left(\frac{du'}{d\theta}+\frac{dv'}{d\varpi}+\frac{u'\cos\theta}{\sin\theta}\right)$: the quantities u' and v' being analogous to u and v in the case of a homogeneous fluid (372), $\delta V'$ being the portion of δV which belongs to the state of motion only, αy the elevation above the level of the sea, y' the variation of height corresponding to the temporary change of density, and g the force of gravitation.

In examining the motions of the atmosphere, we may omit the consideration of the variation of heat, in different latitudes and at different heights, as well as all the irregular causes of agitation, including in the computation those forces only which act regularly upon it as upon the sea. We may therefore consider the sea as covered by an elastic fluid of a uniform temperature: and we may suppose the density of this fluid proportional to the pressure, as it is found to be by actual experience. This supposition implies that the height of the atmosphere must be infinite, but it is easy to see that, at a very moderate height, the density is so small that it may be regarded as evanescent.

If we now call the quantities s, u, and v, for the particles of the atmosphere, s', u', and v', the equation (L) (372) will become

$$a r^2 \delta\theta . \left(\frac{ddu'}{dt^2} - 2n \sin \cos\theta \frac{dv'}{dt}\right)$$

$$+ a r^2 \delta\varpi . \left(\sin^2\theta \frac{ddv'}{dt^2} + 2n \sin \cos\theta \frac{du'}{dt} + \frac{2n \sin^2\theta}{r} . \frac{ds'}{dt}\right)$$

$$+ a\delta r . \left(\frac{dds'}{dt^2} - 2nr \sin^2\theta \frac{dv'}{dt}\right) = \tfrac{1}{2} n^2 \delta \left\{ (r + as') \sin(\theta + au') \right\}^2$$

$$+ \delta V - \frac{\delta p}{\rho}:$$ which, in the state of equilibrium, affords us, when integrated, $\frac{1}{2} n^2 r^2 \sin^2\theta + V - \int \frac{\delta p}{\rho} = C$, a constant quantity. But since the pressure p is supposed to be proportional to the density ρ, we may call $p = lg\rho$, g being the force of gravity in a determinate place, for instance at the equator, and l a constant quantity, which expresses the height of an atmosphere supposed homogeneous, and of the same density as at the surface of the sea; a height which is very small in comparison with the radius of the earth, being less than $\frac{1}{720}$ of this radius. The fluent $\int \frac{\delta p}{\rho}$ or $\int lg \frac{\delta\rho}{\rho}$ is therefore lg hlρ: and the equation of the equilibrium of the at-

mosphere becomes $lghl\rho = C + V + \frac{1}{2}n^2r^2 \sin^2 \theta$. Now at the surface of the sea, the value of V, expressing the force, must be the same for a particle of air as for the particle of water in contact with it, the same forces acting in both cases: but from the condition of the equilibrium of the sea, we have $V + \frac{1}{2}n^2 r^2 \sin^2\theta$ constant; consequently $lghl\rho$ must be constant, and ρ, the density of the stratum of air, contiguous to the sea, must be every where the same in the state of equilibrium. [It is not intended by this constancy of the force, to imply that gravitation is equal throughout the surface of the sea, but that the pressure on it must be every where equal.]

If we make R equal to the part of the radius r comprehended between the centre of the spheroid and the surface of the sea, and r' the part between that surface and a particle of air elevated above it, we may consider r' as the vertical height of the particle above the surface, which it will be with only an error of the order $\left(\frac{n^2}{g}r'\right)^2 : R$; and quantities of this order may be neglected without iuaccuracy. Then if V', $\frac{dV'}{dr}$, and $\frac{ddV'}{dr^2}$ be the values of these quantities at the surface of the sea, we shall have, for the elevation r', $V = V' + r'\frac{dV'}{dr} + \frac{r'^2}{2} \cdot \frac{ddV'}{dr^2}$, [by Taylor's theorem (247)] and the equation $lghl\rho = C + V + \frac{1}{2} n^2r^2 \sin^2\theta$ will become $lghl\rho = C + V' + r'\frac{dV'}{dr} + \frac{r'^2}{2} \cdot \frac{ddV'}{dr^2} + \frac{1}{2} n^2 R^2 \sin^2\theta + n^2 Rr' \sin^2\theta$: and for the value of V' at the surface of the sea, we have $V + \frac{1}{2}n^2R^2 \sin^2\theta =$ a constant quantity: the effect of gravitation at this surface being $-\frac{dV}{dr} - n^2 R \sin^2\theta$, which we may call g'. The quantity

$\frac{ddV'}{dr'^2}$ being multiplied by the very small square r'^2, we may find its value upon the supposition that the earth is spherical, and we may also neglect the density of the atmosphere in comparison with that of the earth. We may therefore take [, by anticipating the law of gravitation],—

$\frac{dV}{dr} = g = \frac{m}{R^2},$ m being the mass of the earth, consequently $\frac{ddV}{dr^2} = -\frac{2m}{R^3} = -\frac{2g'}{R}$, [since $d\frac{m}{r^2} = -\frac{2mdr}{r^3}$,],

we have therefore $lg\,\mathrm{hl}\rho = C - r'g' - \frac{r'^2}{R}g'$: consequently

$-\frac{r'g'}{lg}\left(1 + \frac{r'}{R}\right) = \mathrm{hl}\rho$, and $\rho = \Pi e^{-\varkappa}$, if $\varkappa = \frac{r'g'}{lg}\left(1 + \frac{r'}{R}\right)$

Π being a constant multiplier representing the density of the air at the surface of the sea, and $\mathrm{hl}e = 1$. If we make h and h' equal to the length of the pendulum vibrating seconds, at the surface of the sea under the equator, and in the latitude of the particle of the atmosphere in question, we shall have $\frac{g'}{g} = \frac{h'}{h}$ and consequently $\varkappa = \frac{r'h'}{lh}\left(1 + \frac{r'}{R}\right)$. Hence it appears that the strata of air of equal density are every where equally elevated above the sea, with the exception of the quantity $\frac{r'(h' - h)}{h}$; but in the exact calculation of the heights of mountains, by observations of the barometer, this quantity must not be neglected.

We may now proceed to determine the oscillations of a stratum which is on a level, or of the same density, in the state of equilibrium. If we make $\alpha\varphi$ the elevation of a particle of air above the level of the surface to which it belongs in the state of equilibrium, it is obvious that in virtue of this elevation the value of δV will be augmented by the differential variation $-\alpha g\delta\varphi$, and that $\delta V=(\delta V)-\alpha g\delta\varphi+\alpha\delta V'$; (δV) being the value of δV which belongs to the stratum in the state of equilibrium, and to the angles $\theta+\alpha u$ and $nt+\varpi+\alpha v$, and $\delta V'$ being the part of δV belonging to the new forces, which act on the atmosphere in the state of motion.

Let ρ be$=(\rho)+\alpha\rho'$, (ρ) being the density of the level stratum in the state of equilibrium. If we make $\frac{l\rho'}{(\rho)}=y'$, we shall have $\frac{\delta p}{\rho}=\frac{lg\delta(\rho)}{(\rho)}+\alpha g\delta y'$; [since $p=lg\rho=lg\left\{(\rho)+\alpha\rho'\right\}$, and $\frac{\delta p}{\rho}=\frac{lg\delta(\rho)}{\rho}+\alpha g\frac{\delta(l\rho')}{\rho}$, or, substituting (ρ) for ρ in the denominator, their difference being inconsiderable in comparison with the whole quantity, $=lg\frac{\delta(\rho)}{\rho}+\alpha g\delta y'$. Now in the state of equilibrium

$$0=\tfrac{1}{2}n^2\delta\left\{r+\alpha s)\sin(\theta+\alpha u)\right\}^2+(\delta V)-\frac{lg\delta(\rho)}{(\rho)}.$$ Consequently the general equation of the motion of the atmosphere will become, in relation to the level strata, with regard to which δr is nearly evanescent,

$$r^2\delta\theta.\left(\frac{ddu'}{dt^2}-2n\sin\cos\theta\,\frac{dv'}{dt}\right)$$

$$+r^2\delta\varpi.\left(\sin^2\theta\frac{ddv'}{dt^2}+2n\sin\cos\theta\,\frac{du'}{dt}+\frac{2n\sin^2\theta}{r}\cdot\frac{ds'}{dt}\right)=\delta V'$$

$-g\delta\varphi-g\delta y'+n^2r\sin^2\theta\,\delta\left\{s'-(s')\right\}$, $\alpha\,(s')$ being the variation of r corresponding, in the state of equilibrium, to the variations $\alpha u'$ and $\alpha v'$ of the angles θ and ϖ. [For $\delta V-\frac{\delta p}{\rho}$ becoming $=(\delta V)-\alpha g\delta\varphi+\alpha\delta V'-\frac{\delta p}{\rho}$, and $(\delta V)-\frac{\delta p}{\rho}$ being $-\tfrac{1}{2}n^2\delta\left\{(r+\alpha s)\sin(\theta+\alpha u)\right\}^2-\alpha g\delta y'=-\alpha n^2r$

$\sin^2\theta\,\delta\,(s')-ag\delta y'$, this part of the second member of the equation derived from (L), combined with the former part, which is here $an^2r\sin^2\theta\delta s'$, affords us the equation here laid down.]

If we suppose that all the particles of air, originally situated on the same radius of the earth, remain constantly on the same radius during their motion, as has been shown to take place with respect to the sea, we may proceed to examine whether this supposition is consistent with the equations of motion and of continuity. For this purpose, it is necessary that the values of u' and v' [representing the motions in latitude and longitude,] should be the same for all these particles: now it will appear hereafter, when we consider the forces concerned, that these forces are very nearly the same for all the particles: the variations $\delta\varphi$ and $\delta y'$ must therefore necessarily be the same for all the particles, and the quantities $2nr\delta\varpi\sin^2\theta$, and $n^2r\sin^2\theta\,\delta\left\{s'-(s')\right\}$ must be so small as to be capable of being neglected in the preceding equation.

At the surface of the sea, we have $\varphi=y$, ay being the elevation of the surface of the sea above the surface of equilibrium. We may therefore inquire whether the suppositions of $\varphi=y$, and of y being constant for all the particles of air situated on the same radius, are consistent with the equation of continuity of the fluid, which, by article 374, is $0=r^2\left\{\rho'+(\rho)\left(\frac{du'}{d\theta}+\frac{dv'}{d\varpi}+\frac{u'\cos\theta}{\sin\theta}\right)+(\rho).\frac{d(r^2s')}{dr}\right.$: hence we have $y'=-l\left(\frac{d(r^2s')}{r^2dr}+\frac{du'}{d\theta}+\frac{dv'}{dw}+\frac{u'\cos\theta}{\sin\theta}\right)$, [since $y'=\frac{l}{(\rho)}\rho'$]. Now, $r+as'$ is equal to the value of r at the surface of equilibrium, corresponding to the angles $\theta + \alpha u$ and $\varpi + \alpha v$, increased by the elevation of the particle of air above this surface; the part of $\alpha s'$, which depends on the variation of the angles θ and ϖ, being of the order $\frac{\alpha n^2 u}{g}$, it may be neglected in the preceding value of y', and we may consequently suppose in this expression $s' = \varphi$; and if we then make $\varphi = y$, we shall have $\frac{d\varphi}{dr} = 0$, since the value of φ is then the same, with regard to all the particles situated in the same radius: besides, y itself is obviously of the same order as l, or as $\frac{nn}{g}$.: we shall therefore have, for the value of y',

$$y' = -l\left(\frac{du'}{d\theta} + \frac{dv'}{d\varpi} + \frac{u' \cos\theta}{\sin\theta}\right)$$

: consequently, since u' and v' are the same, for all particles originally situated in the same radius, y' must be the same for all these particles. It follows also, from these considerations, that the quantities $2nr\delta\varpi . \sin^2\theta \frac{ds'}{dt}$, and $n^2 r \sin^2\theta\, \delta \left\{ s' - (s') \right\}$, may be neglected in the preceding equation of the motion of the atmosphere, which may then be fulfilled by supposing u' and v' the same for all the particles of air originally situated in the same radius; and that the supposition of the continuance of all these particles in the same radius, during the oscillations of the fluid, is consistent with the equations both of motion and of continuity. In this case, the oscillations of the different level strata are the same, and may be determined by means of these equations;

$$r^2\delta\theta . \left(\frac{ddu'}{dt} - 2n \sin\cos\theta \frac{dv'}{dt}\right) + r^2\delta\varpi . \left(\sin^2\theta \frac{ddu'}{dt^2} + 2n\right.$$

$\sin \cos \theta \frac{du'}{dt}\Big) = \delta V' - g\delta y' - g\delta y$; and $y' = -l\left(\frac{du'}{d\theta} + \frac{dv'}{d\varpi}\right.$ $\left. + \frac{u' \cos \theta}{\sin \theta}\right)$.

SCHOLIUM 1. These oscillations of the atmosphere must produce analogous oscillations in the heights of the barometer. In order to determine these from those of the atmosphere, we may consider a barometer fixed at any given height above the surface of the sea. The height of the mercury is proportional to the pressure, to which the surface exposed to the air is subjected; it may therefore be represented by $lg\rho$: but this surface is successively exposed to the pressure of different level strata, which rise and fall like the surface of the sea: consequently the value of ρ at the surface of the mercury varies, first so far as it belongs to a level stratum which in the state of equilibrium was less elevated by a quantity αy, and secondly because, in the state of motion, the density of a given stratum is increased by the quantity $\alpha\rho'$ or $\frac{\alpha(\rho)y'}{l}$. In virtue of the first cause the variation of ρ is $-\alpha y \frac{d\rho}{dr}$, or $\frac{\alpha(\rho)y}{l}$; [since this variation must be to (ρ) the whole density, as the elementary column αy to the height l]; consequently the total variation of the density ρ, at the surface of the mercury, is $\alpha(\rho)\frac{(y+y')}{l}$. Hence, if we call the height of the mercury k in the state of equilibrium, its oscillations in the state of motion will be expressed by the quantity $\frac{\alpha k(y+y')}{l}$; consequently these oscillations are similar at all heights above the sea, and proportional in their extent to the heights of the barometer.

SCHOLIUM 2. It now only remains, for the determination of the oscillations of the sea, and of the atmosphere, to investigate the forces which act on their respective fluids, and to find the fluents of the preceding fluxional equations with regard to those forces; which will be done in a subsequent part of this work.

[SCHOLIUM 3. Instead of attempting to shorten and simplify the steps of this refined investigation, which will hereafter appear to be unnecessarily general, it will be sufficient to insert some collateral considerations on the simplest cases of the transmission of motion through fluids, adapted to a notation resembling that which is employed by the author.

378. THEOREM. "395." When the surface of an incompressible fluid, contained in a narrow prismatic canal, is elevated or depressed a little at any part above the general level; if we suppose a point to move in the surface each way, with a velocity equal to that of a heavy body falling through half M the depth of the fluid, the surface of the fluid, at the part first affected, will always be in a right line between the two moveable points.

The particles constituting any column of the fluid, extending across the canal, are actuated by two forces, derived from the hydrostatic pressures of the columns on each side, these pressures being supposed to extend to the bottom of the canal, with an intensity regulated only by the height of the columns themselves; and this supposition would be either perfectly or very nearly true, if the particles of the fluid were infinitely elastic, that is, absolutely incompressible; and if the fluidity were at the same time so perfect, that no particle of the fluid should be affected by any pressure not tending directly towards it. A distinguished mathematician of the present day appears indeed to have assumed, that the pressure is transmitted downwards with a velocity determined by the depth, and related to the velocity of the horizontal transmission, if not identical with it: but it seems sufficiently obvious, that if the canal be supposed incompressible, the pressure must descend in it, as it confessedly would do in an organ pipe, with a velocity dependent only on the intimate elasticity of the medium, which in this proposition is supposed infinite.

Now the difference of the forces on each side of the thin transverse section of the canal, constituting a partial pressure, is the immediate cause of the horizontal motion; and the vertical motion is the effect of the modification of the horizontal motion: and the difference of the pressures is every where to the weight of the column or section, or of any of its parts, as the difference of the heights to the thickness of the column, or as the fluxion of the height y to that of the horizontal length of the canal x. Hence, if the weight of any particle be called g, the horizontal force acting on it will be $\frac{\mathrm{d}y}{\mathrm{d}x}g$. Such therefore is the force acting horizontally on any elementary column: but the elongation or abbreviation of the column depends on the difference of the velocities, with which its two transverse surfaces are made to advance, and this elevation or depression of the upper surface is therefore to the whole height, as the variation of the fluxion of the length, or thickness, produced by the operation of the force, is to the whole fluxion of the length; that is, δy is to y as $\delta \mathrm{d}x$ to $\mathrm{d}x$, or as $\delta \mathrm{D}x$ to $\mathrm{D}x$. But the force which produces the change being $\mathrm{d}\frac{\mathrm{d}y}{\mathrm{d}x}g=\frac{\mathrm{dd}y}{\mathrm{d}x}g$, making $\mathrm{d}x$ constant, it may be supposed to be increased, with reference to the acceleration of the upper surface of the fluid, in the ratio of the synchronous variations $\delta\mathrm{d}x$ and δy, or that of $\mathrm{d}x$ to y, and it will then become $\frac{y}{\mathrm{d}x}\cdot\frac{\mathrm{dd}y}{\mathrm{d}x}g=\frac{\mathrm{dd}y}{\mathrm{d}x^2}gy$, which will be the measure of the acceleration of the surface, and the surface will ascend or descend precisely as if immediately subjected to the operation of such a force. We may therefore inquire what must be the velocity of a body moving along the curved surface, or what must be the horizontal velocity of a similar surface moving along through the body, in order that the vertical motion should represent the effect of the force $\frac{\mathrm{dd}y}{\mathrm{d}x^2}gy$. Now in the common expression of the magnitude of a force acting in the direction of y, we say $f=\frac{\mathrm{dd}y}{\mathrm{d}t^2}$; we must therefore make $\frac{\mathrm{dd}y}{\mathrm{d}t^2}=\frac{\mathrm{dd}y}{\mathrm{d}x^2}gy$, or $\frac{\mathrm{d}x^2}{\mathrm{d}t^2}-gy$, and $\frac{\mathrm{d}x}{\mathrm{d}t}=\surd(gy)$: consequently if x flow with the constant velocity $v=\frac{\mathrm{d}x}{\mathrm{d}t}=\surd(gy)$, the second fluxion of y will always represent the actual acceleration of the surface of the fluid, the part of the curve corresponding to the time t always representing the actual position of the particle, as well as its motion. But $\surd(gy)$ is the velocity acquired by a body in falling through $\frac{1}{2}y$, since in general $v^2=2gs$, (232) and $v=\surd(2gs)$, or $=\surd(2g\mathrm{M})$. In this simple manner we attain a strict demonstration, on the premised supposition respecting the nature of the fluid, that the velocity of the surface will be represented by that of the surface of a wave advancing with the horizontal velocity thus determined, or in other words, that the wave will actually advance with that velocity.

But in this form the solution is limited to the case of a wave already in progress. It may, however, readily be extended to all possible cases. For since the actions of any two or more forces are always expressed by the addition or subtraction of the results produced, in any given time, by their single operations, it may easily be understood that any two or more minute impressions may be propagated in a similar manner through the canal, without impeding each other; the inclination of the surface, which is the original cause of the acting force, being the joint effect of the inclinations produced by the separate impressions, and producing singly the same force, as would have resulted from the combination of the two separate inclinations; and the elevation or depression becoming always the sum or difference of those which belong to the separate agitations. If then we suppose two similar impulses, waves, or series of waves, to meet each other in directions precisely opposite, they will still pursue their course: and at the instant when they meet in such a manner as to destroy completely each other's horizontal and vertical motions, the elevation and depression of each series will coincide and be redoubled, and the fluid will be quiescent, with an undulated surface: but in the next instant the two series will proceed uninterrupted, as before: consequently the fluid being supposed to be initially in the same state, its progressive changes will be represented by the effects of the two series of waves meeting each other, and the place of each point will be determined by the middle between the two places which it would have held by the separate effects of the two series, that is, by the mean between the elevation or depression of the two points supposed in the proposition.

Corollary 1. The points, in which the similar parts of the two opposite series of waves continue to meet, will always be free from horizontal motion; hence it follows that a solid obstacle in a vertical direction might be interposed without altering the phenomenon: and consequently that any fixed obstacle meeting the waves would produce precisely the same effect on the subsequent state of either series, as is produced by the opposition of a similar series, and would reflect it in a form similar to that of the opposite series, which would have travelled over it, if it had originated from a primitive cause of motion on the other side of the obstacle.

Scholium. It will appear, by considering the combination of the horizontal with the vertical motion, that each particle of the surface will describe an oval figure, which it will be simplest to suppose an ellipsis; the motion in the upper part of the orbit being direct with regard to the progress of the wave, and in the lower part retrograde: and the orbit will be of the same form and magnitude for each particle of the surface, when the canal is supposed to be prismatic.

379. Theorem. The divergence of a wave makes no sensible difference in the velocity of its propagation, and its height will vary as the square root of the distance from the centre.

The immediate horizontal force is the same for a diverging wave as for a prismatic canal, its measure being always

$\frac{dy}{dx}g$, as well for the parts lying without the sides of a supposed prismatic canal, as for the parts contained within it, the inclination of the surface being the same without as within those limits, and the fluxion of the height being in the same proportion to that of the length x, notwithstanding that the pressure in one direction is derived, for the extreme parts, from the surface of the collateral portion of the wave: consequently the force, as referred to the surface of the fluid, will still be expressed by $\frac{ddy}{dx^2}gy$. It will, however, be modified by the depression attending a progressive motion, necessary for preserving the continuity of the fluid, which must obviously be such that $-\delta y$ may be to δx, the progressive velocity, as y to x, and $\delta y = -\delta x \frac{y}{x}$: and the accelerative force $\frac{dy}{dx}g$, considered with regard to its effect at the surface, will be modified in the same proportion as the velocity, so that instead of $\frac{dy}{dx}g$, it will become $-\frac{dy}{dx}g\frac{y}{x} = -\frac{dy}{xdx}gy$, consequently the joint acceleration of the surface will be $\left(\frac{ddy}{dx^2} - \frac{dy}{xdx}\right)gy$. Now $\frac{ddy}{dx^2} = \frac{1}{2r}$, (194) which is the reciprocal of the diameter of the circle of curvature, and $\frac{dy}{xdx}$ is the reciprocal of $x\frac{dx}{dy}$, the height of the intersection of the vertical line passing through the centre of divergence with the perpendicular to the surface of the wave, which will be very great in comparison with the diameter of curvature, when the distance from the centre becomes considerable: and the second part of the expression will become a small disturbing force, depending on the tangent of the inclination of the surface, which represents the fluent of the curvature, or of the accelerating force, and being therefore proportional to the velocity: so that like the resistance of a pendulum proportional to the velocity, it will not sensibly affect the whole period of the alternate motion, or the propagation of the wave depending on it. We obtain the law of the diminution of the height of the waves in diverging, from the principle of the preservation of impetus (319), since the mass affected at once by the similar velocities increases directly as the distance from the centre x, when the depth is equable, consequently all the velocities concerned must decrease as the square root of x, in order that the sum of the masses, multiplied by the squares of the velocities may remain constant. There will always be a continual but insensible reflection, which will preserve the centre of gravity immoveable, though it consumes no considerable part of the impetus; except at the very origin of the wave, where there seems to be something like a vibratory motion from this reflection, for a short space, at the beginning of the motion.

SCHOLIUM. It is obvious that the surface of a wave so diminishing cannot be supposed to glide on unaltered, but the demonstration shows that the motion of each point of the surface is the same as that of a surface, affected by a series of equal waves, of the magnitude of the actual wave at the given point, which is the condition supposed in the comparison of the force with the curvature.

380. "400." THEOREM. All minute impulses are conveyed through a homogeneous

elastic medium with a uniform velocity, equal to that which a heavy body would acquire, by falling through half M, the height of the medium causing the pressure.

In this case we have to call the density y, instead of the height of an incompressible fluid in article 378, and to imagine the surface of the wave to be that of a curve representing the density by its ordinate y, which is equal to the height of a uniform column of the medium capable of producing the pressure, or in other words, to the height of the modulus of elasticity of the medium: then $\frac{dy}{dx}g$ will be the direct accelerating force, and $\frac{ddy}{dx^2}gy$ the acceleration of the ordinate of the curve of density, since here again the variation of density δy is to y, as δdx to dx: and the same conclusion is inferred, respecting the velocity with which the curve of densities must advance, in order that it may represent the instantaneous change at each point, and consequently for all the points in succession.

381. "397, Sch." THEOREM. Every small change of form is propagated along an elastic chord, with a velocity equal to that which is due to half the length M, of a portion of the chord, of which the weight is equal to the force producing the tension, and is reflected from the extremities in an opposite direction.

This proposition, though not belonging to the motions of fluids, is inserted here to complete the analogy between the height of a liquid, the modulus of elasticity of an elastic medium, and the modulus of tension of a vibrating chord. The force, impelling any small portion of the chord towards the quiescent position, or axis, is obviously expressed by the diagonal of the elementary parallelogram, formed by its extreme tangents, that is the line intercepted between the intersection of those tangents and a line equal and parallel to the second drawn from the extremity of the first, or in other words, by the second fluxion of the ordinate, when the tangent represents the first fluxion of the axis, the curve being always supposed infinitely near to the axis, and in general the force will be to the tension as the second difference $\Delta\Delta y$ to the first difference Δx: but the tension is to the weight of the element Δx as M to Δx, consequently the tension of Δx is $\frac{M}{\Delta x}g$, and the accelerative force $\frac{\Delta\Delta y}{\Delta x} \cdot \frac{M}{\Delta x}g = \frac{\Delta\Delta y}{\Delta x^2}Mg - \frac{d^2 y}{dx^2}Mg$, which we may make $= f = \frac{d^2 y}{dt^2}$, and we shall have $v = \sqrt{(gM)}$, as $v = \sqrt{(gy)}$ in article 378; and the velocity will be that which is due to half the height M.

The reflection at the extremities of the chord may be represented by delineating the initial figure, and repeating it in an inverted position below the absciss: then taking, in the absciss, each way, a distance proportional to the time; and the half sum of the corresponding ordinates will indicate the place of the point at the expiration of that time. The chord will thus represent a portion of the surface of a liquid agitated by a series of waves: and on the other hand a wave reflected backwards and forwards within a prismatic canal of its own length, abruptly terminated at each end, will exhibit a vibration precisely resembling that of an elastic chord. It may be inferred from the consideration of the motion of a chord so continued, that the point corresponding to the end of the primitive chord will always remain at rest; whence it follows that the motion of the chord, terminated by such a fixed point, must be the same as if it were continued in the manner described, the reasoning being the same as in the case of the reflection of a wave.

APPENDIX A.

OF THE COHESION OF FLUIDS.

---

382. THEOREM. If there be a series of equal particles, arranged at equal intervals in a right line, each attracting or repelling its immediate neighbour, only with a constant force f; the force VM, acting on any obstacle M at one end of the whole line u, supposing the other to be fixed, will be equal to f.

The general principle of virtual velocities is $\Sigma mS\delta s = 0$, (l, 305) or, taking any one of the forces combined with each other as the result of the rest, and in an opposite direction, $MV\delta u = \Sigma mS\delta s$: and in applying this principle, the variations may be taken in any manner capable of representing their relations to each other, without confining them to such as are likely to occur in the natural phenomena to be considered; and the motive force VM may always be found, if we can determine its equal $\frac{\Sigma mS\delta s}{\delta u}$. Now if the number of particles concerned be m, and their masses equal to unity, we shall have $\delta s = \frac{\delta u}{m}$, since we may suppose the particles to remain equally distributed throughout the line after the variation of their distances, and S being $=f$, we have $\Sigma m S \delta s = f \delta u$; consequently $VM = f$.

383. THEOREM. If an attractive or repulsive force extend to a given distance c among a series of m particles situated at equal distances in a right line, the mutual forces of any two particles being f, and their masses each unity, the tension acting on an obstacle at the end of the line u will be $\frac{mmcc}{uu} f$.

The number of particles in the line u being m, the number acting at any one point will be $2m \frac{c}{u}$; and when the length u is varied, the variation of the distance of the remotest of these particles will be $\delta u \frac{c}{u}$, while that of the particles at a smaller distance will be proportionally smaller: and the mean variation of the distances of the particles within the respective spheres of action will be half the extreme variation. For each particle, therefore, the variation $\Sigma m S \delta s$ will be $\frac{1}{2} \delta u \frac{c}{u} 2c \frac{m}{u} f = \frac{cc}{uu} m f \delta u$, and for the whole line, consisting of m particles, $m^2 \frac{cc}{uu} f \delta u$, which, divided by δu, gives $VM = \frac{mmcc}{uu} f$.

COROLLARY 1. Hence, if u be given, the tension will vary as the square of the number of particles or density m, and as the square of the extent of the sphere of action c, conjointly.

COROLLARY 2. If there be two forces, a cohesive force

C, and a repulsive force R, holding each other in equilibrium, but extending to the different distances c and r, they will balance each other, in this hypothetical case, if $c^2C=r^2R$, that is, if the primitive forces of the single pairs of particles be inversely as the squares of the minute distances, to which they extend.

SCHOLIUM. It is obvious that the length u is indifferent to the force, since m must vary as u, and $\frac{m}{u}$ must remain constant, when the density is given.

384. THEOREM. If a fluid, composed of cohesive and repulsive particles, holding each other in equilibrium, be contained between two parallel surfaces, of unlimited extent, the equal and opposite forces, acting on either of the surfaces M, will be $\frac{\pi}{3}\, d^2\, c^4\, Mf$; d being the density, and π the circumference of a circle divided by its diameter.

The number of particles in the space Mu being dMu, the number of those, which are within the limits of the sphere of action of each particle, will be $d\frac{4}{3}\pi c^3$. Supposing now the distances of the particles to be varied by a slight change of the density; it is evident that the variation of the density will be in the triplicate proportion of that of the distances, since if $d=x^3$, $\mathrm{d}d=3x^2\,\mathrm{d}x$; and the variation of the whole space Mu being $M\delta u$, that of the density $\delta d=-\delta u\,\frac{d}{u}$, and that of any linear distance c will be $\delta c = -\frac{1}{3}\,\delta d\,\frac{c}{d}=\frac{1}{3}\,\delta u\,\frac{c}{u}$, which will be the variation of the distance of the particles, at the surface of the sphere of action, from its centre. But the mean distance of each elementary pyramid from its vertex, or of the whole sphere from the centre, is $\frac{3}{4}$ of the height or the radius, since the products of the elements of the content into the distance added together and divided by the content, or $\frac{Sx.x^2 Dx}{Sx^2 Dx} = \frac{3}{4}$. The mean variation of distance for the whole fluid is therefore $\frac{1}{4} c \frac{\delta u}{u}$; and this variation, multiplied by the number of particles within the sphere of action, becomes $\frac{d}{3} \pi c^4 \frac{\delta u}{u}$; which being again multiplied by the number of centres Mud, and by the force f, and divided by δu, gives us $V = \frac{\pi}{3} d^2 c^4 Mf$, for the whole force acting on the surface M.

COROLLARY. In this case if the two forces C and R hold each other in equilibrium, we must have $c^4 C = r^4 R$, and C must be to R, for each pair of particles, as r^4 to c^4: each force still varying as the square of the density.

SCHOLIUM 1. The determination of the attractive or repulsive force of a sphere thus constituted may be illustrated and confirmed by a simpler mode of considering the joint action of the particles of each hemisphere, which is easily shown to be half as great as if they were collected into one line. For it is obvious that each particle in any spherical surface must have its action on the central point reduced in the proportion that the radius bears to its distance from the plane dividing the hemispheres, consequently the whole force will be represented by the distance of the centre of gravity of the surface, multiplied into the mass, or the number of particles contained in it. Now the centre of gravity of a spherical surface is situated in the middle of its absciss or verse sine, since the increments of the surface are proportional to those of the verse sine (183). Hence it follows, that the joint force of all the particles in each surface is half what it would be, if they were all situated in the given direction: and the proportion being the same for all the concentric surfaces, it must also remain the same for the whole hemisphere. If we had only to consider the attractions of a series of particles, situated in a circular circumference, upon a central particle, it might be shown, in a similar manner, that they would be together equal to that of a number of particles represented by the chord, supposed to be placed at the middle of the arc.

SCHOLIUM 2. If any of the elastic fluids, with which we are acquainted, be considered as thus constituted, we must suppose the fourth power of the distance r to vary inversely as the density d, since the force V is found to vary simply as the density, and $\frac{V}{d} - \frac{\pi}{3} dc^4 \mathrm{M}f$ is constant. It would have been more natural to expect, that if c were not constant, its cube c^3 would have varied inversely as the density, supposing the number of particles cooperating to be given. But in the Newtonian demonstration the elementary force f is also supposed to vary inversely as the distance, while the number of particles cooperating is invariable. In this case the number of particles in the space Mu are as dMu, and the elementary forces as $d^{\frac{1}{3}}$, the variations of the distances, for a given value of δu, being as $d^{-\frac{1}{3}}$, so that the products of these quantities remain constant, and the effective force is as the number of particles concerned, or simply as d.

385. LEMMA. If the height of a cone be a, the radius of the base b, and the oblique side c, the mean distance of the base from the vertex will be $\frac{2}{3} \cdot \frac{c^3 - a^3}{b^2}$.

For, if the fluxion of the radius of the base be $\mathrm{d}x$, the product of the elementary ring $2\pi x \Delta x$, into its distance $\surd(a^2+x^2)$, will be $2\pi x \Delta x \surd(a^2+x^2)$; and since $\mathrm{d}\left\{(a^2+x^2)^{\frac{3}{2}}\right\} = \frac{3}{2} \times 2x\mathrm{d}x \surd(a^2+x^2)$, we have $\int 2\pi x \surd(a^2+x^2)\mathrm{d}x = \frac{2\pi}{3}(a^2+x^2)^{\frac{3}{2}}$, which becomes initially $\frac{2\pi}{3} a^3$, and when $x=b$, $\frac{2\pi}{3} c^3$, and the difference, divided by πb^2, the area of the base, that is, $\frac{2}{3} \frac{c^3-a^3}{b^2}$, or $\frac{2}{3} \frac{c^3-a^3}{c^2-a^2}$, will be the mean distance of the base from the vertex.

COROLLARY. For a solid cone, the mean distance becomes $\frac{3}{4}$ of that of the base, as in the case of the sphere: and the expression becomes, in this case, $\frac{1}{2} \frac{c^3-a^3}{c^2-a^2}$

386. THEOREM. The deficiency of the mutual actions of the superficial particles of a fluid, of limited extent, deducts from the tension $\frac{1}{5}$ of the whole force of a stratum equal in thickness to the radius of the sphere of equal action.

For the interior parts of the fluid, the actions of all the particles will be the same as in a fluid of unlimited extent, that is, $\frac{\pi}{3} c^4 Mf$, calling the density unity, since its finite variations do not enter into the present question. But for the particles within the distance c of the surface, the forces will be able to act on such a number of other particles only, as are contained in a segment of the sphere, of which the verse sine is $c+x$, the distance from the surface being x, which are not only fewer than in the whole sphere, but are also at a smaller mean distance from the centre.

Each of these segments may be divided into two portions; that which is contained between the centre and the spherical circumference, and the cone, which lies between the centre and the plane surface: the variation of the mean distance of the former will be the same as for the whole sphere; but for the cone, instead of the variation belonging to that of the corresponding portion of the sphere, which will be expressed by the product of its content into $\frac{3}{4}$ of the variation of the radius, we shall have the content of the cone into the variation of its mean distance, or $\frac{\pi}{3}(c^2-x^2)\,x$ into $\frac{1}{2}\,\frac{c^3-x^3}{c^2-x^2}\cdot\frac{\delta u}{u}$ that is, $\frac{\pi}{6}(c^3-x^3)\,x\,\frac{\delta u}{3u}$, instead of $2\pi c(c-x)\,\frac{c}{3}$ into $\frac{3}{4}c\,\frac{\delta u}{3u}$, or $\frac{\pi}{2}(c^4-c^3x)\,\frac{\delta u}{3u}$, the difference being $\frac{\pi}{6}(3c^4-4c^3x+x^4)\,\frac{\delta u}{3u}$, for each particle at the distance x from the surface; and in order to find the total difference for the whole stratum, we must multiply this by the fluxion of x, and find the fluent, which will be $\frac{\pi}{6}$ $(3c^4x-2c^3x^2+\frac{1}{5}x^5)\frac{\delta u}{3u}$ or, when $x=c$, $\frac{\pi}{6}\cdot\frac{6}{5}c^5\frac{\delta u}{3u}=\frac{\pi}{15}c^5\,\frac{\delta u}{u}$, and for the length u, $\frac{\pi}{15}c^5\delta u$, while the force of the whole stratum, of the thickness c, would have been $\frac{\pi}{3}c^4c$, substituting c for M in article 385, and the deficiency is to the whole force as $\frac{1}{15}$ to $\frac{1}{3}$, or as 1 to 5).

Corollary. If the cohesive force C and the repulsive R be in equilibrium for the whole fluid considered as incomparably greater in thickness than c or r, the difference of the forces with regard to the superficial stratum on each side only, will be $\frac{1}{5}\cdot\frac{\pi}{3}(c^5C-r^5R)$: now it has been shown that $c^4C^5=r^4R$, consequently $c^5C-r^5R=c^4C(c-r)$, and the joint deficiency in the cohesive force will be $\frac{1}{5}\cdot\frac{\pi}{3}c^5C\left(1-\frac{r}{c}\right)$

Corollary 2. The deficiency being positive when c is greater than r, it follows that if the superficial cohesion prevail in a fluid so constituted, it must be because r is greater than c and the defect is greatest with regard to the repulsive force. In such cases the fluid must be slightly condensed in its interior parts, so as to produce a resistance equivalent to the excess of cohesion of the surface.

Corollary 3. These conclusions are applicable, with slight modifications only, to the case of a repulsion like that of elastic fluids, as assumed by Newton. For we have only to take r equal to the radius of the actual mean sphere of action for the fluid in any given state of compression, and the superficial deficiency of the force will be very nearly as determined by this proposition, the distance r becoming in this case somewhat smaller than the whole extent of the sphere of action. The utmost possible cohesive force would be obtained from the supposition that c is incomparably smaller than r, and this force would be $\frac{1}{5}\frac{\pi}{3}r^5R$, or $\frac{1}{5}$ of the repulsive force of a stratum of the interior part of the fluid of the thickness r; but in every case that can actually occur, the superficial force must probably be much less than this.

SCHOLIUM. On the whole, we are fully justified in concluding that, since the phenomena of capillary action necessarily lead us to infer the existence of a superficial tension, and since, without this supposition, we should be obliged to admit the possibility of a perpetual source of motion, from an unequal hydrostatic pressure, upon any floating body not homogeneous; the existence of such a cohesive tension proves that the mean sphere of action of the repulsive force is more extended than that of the cohesive: a conclusion, which, though contrary to the tendency of some other modes of viewing the subject, shows the absolute insufficiency of all theories built upon the examination of one kind of corpuscular force alone. It must also be recollected that, as far as our experiments enable us to observe, the repulsive force of solids does actually extend further than the cohesive, though, with respect to its mean intensity, we have no direct method of ascertaining the comparative extent of the spheres of action of the two forces.

APPENDIX B.

OF INTERPOLATION AND EXTERMINATION.

---

387. Theorem. The fluxions of any quantity u may be found from its finite differences, taken at equal intervals with respect to another flowing quantity x, by the theorem

$$\frac{\mathrm{d}u'}{\mathrm{d}x}h=\Delta u-\tfrac{1}{2}\Delta^2 u+\tfrac{1}{3}\Delta^3 u-\tfrac{1}{4}\Delta^4 u+\tfrac{1}{5}\Delta^5 u-\tfrac{1}{6}\Delta^6 u+\ldots$$

$$\frac{\mathrm{d}^2u'}{\mathrm{d}x^2}h^2=\Delta^2 u-\tfrac{2}{2}\Delta^3 u+\tfrac{11}{12}\Delta^4 u-\tfrac{5}{6}\Delta^5 u+\tfrac{137}{180}\Delta^6 u-\tfrac{7}{10}\Delta^7 u+\ldots$$

$$\frac{\mathrm{d}^3u'}{\mathrm{d}x^3}h^3=\Delta^3 u-\tfrac{3}{2}\Delta^4 u+\tfrac{7}{4}\Delta^5 u-\tfrac{15}{8}\Delta^6 u+\tfrac{29}{15}\Delta^7 u\ \ldots$$

$$\frac{\mathrm{d}^4u'}{\mathrm{d}x^4}h^4=\Delta^4 u-\tfrac{4}{2}\Delta^5 u+\tfrac{17}{6}\Delta^6 u-\tfrac{7}{2}\Delta^7 u+\ldots$$

$$\frac{\mathrm{d}^5u'}{\mathrm{d}x^5}h^5=\Delta^5 u-\tfrac{5}{2}\Delta^6 u+\tfrac{25}{6}\Delta^7 u-\ldots$$

$$\frac{\mathrm{d}^6u'}{\mathrm{d}x^6}h^6=\Delta^6 u-\tfrac{6}{2}\Delta^7 u+\ldots$$

$$\frac{\mathrm{d}^7u'}{\mathrm{d}x^7}h^7=\Delta^7 u-\tfrac{7}{2}\Delta^8 u+\ldots$$

We obtain, for the value of u_n, first $u+n\Delta u+n\,.\,\frac{n-1}{2}\Delta^2 u+\ldots$ (245), and secondly, putting the finite difference of $x=nh$, $u+\Delta u'=nh\frac{\mathrm{d}u'}{\mathrm{d}x}+\frac{n^2h^2}{1.2}\cdot\frac{\mathrm{d}^2u'}{\mathrm{d}x^2}+\ldots$(247). The first expression affords us, by expanding its terms, $u_n = u + \frac{n}{1}\Delta u$

$$+ \frac{n^2 - n}{1.2}\Delta^2 u + \frac{n^3 - 3n^2 + 2n}{1.2.3}\Delta^3 u + \frac{n^4 - 6n^3 + 11n^2 - 6n}{1.2.3.4}\Delta^4$$

$$u = u + \left(\frac{\Delta u}{1} - \frac{\Delta^2 u}{1.2} + \frac{2\Delta^3 u}{2.3} - \frac{6\Delta^4 u}{2..4}\ldots\right) n + \left(\frac{\Delta^2 u}{1.2} - \frac{3\Delta^3 u}{1.2.3}\right.$$

$$\left.+ \frac{11\Delta^4 u}{1..4}\ldots\right) n^2 + \left(\frac{\Delta^3 u}{1..3} - \frac{6\Delta^4 u}{1..4} + \ldots\right) n^3 + \left(\frac{\Delta^4 u}{1..4}\ldots\right) n^4; \text{ and}$$

by equating the terms containing the same powers of n (277), we have $u = u$, $nh\frac{du'}{dx} = \left(\frac{\Delta u}{1} - \frac{\Delta^2 u}{1.2}\ldots\right) n$ and $\frac{du'}{dx}h = \Delta u -$ $\frac{\Delta^2 u}{2} + \ldots$, $\frac{d^2 u'}{dx^2} = \ldots$; and the respective series may be continued to any number of terms by the actual developement of the different products.

SCHOLIUM 1. It may be observed that the coefficients of the different terms of the first series agree with those of the developement of the quantity hl $(1 + \Delta)$, and that in fact the whole may be represented to the eye by the expression $\frac{du}{dx}h = \text{hl}\,(1 + \Delta)\,u$. It was also remarked by Laplace, that the powers of this equation will afford us, with equal accuracy, the values of the higher fluxions; thus $\frac{d^2 u}{dx^2}h^2 = \left\{\text{hl}\,(1 + \Delta)\right\}^2 u$: but this mode of finding the coefficients is little more useful, in common cases, than the original computation of Euler.

SCHOLIUM 2. This theorem may very often be of use in deriving formulae from the results of observation, but it is necessary that the observations should be extremely accurate, since very minute errors will affect the higher orders of differences in a material degree.

Corollary. The fluxions, thus obtained, will enable us not only to find any intermediate values of the variable quantity u, but also the areas contained by these values as ordinates, the contents of the corresponding solids, or any other derivative quantities. If it were required, for example, to determine the magnitude of an area contained between a curve and its absciss from four equidistant ordinates, affording us four successive differences, Δu, $\Delta^2 u$, $\Delta^3 u$, and $\Delta^4 u$; we should have to deduce the four successive fluxions from the four first terms of each series, which would afford us, by substituting the values derived from the expression $\Delta^n u = u_n - n u_{n-1} + \ldots$, that is, $\Delta u = u_1 - u$, $\Delta^2 u = u_2 - 2u_1 + u$, $\Delta^3 u = u_3 - 3u_2 + 3u_1 - u$, and $\Delta^4 u = u - 4u_3 + 6u_2 - 4u_1 + u$, the equations $\frac{du'}{dx} h = -\frac{25}{12} u + 4u_1 - 3u_2 + \frac{4}{3} u_3 - \frac{1}{4} u_4$, $\frac{d^2u'}{dx^2} h^2 = \frac{35}{12} u - \frac{26}{3} u_1 + \frac{19}{2} u_2 - \frac{14}{3} u_3 + \frac{11}{12} u_4$, $\frac{d^3u'}{dx^3} h^3 = -\frac{5}{2} u + 9u_1 - 12u_2 + 7u_3 - \frac{3}{2} u$, and $\frac{d^4u'}{dx^4} h^4 = u - 4u_1 + 6u_2 - 4u_3 + u_4$. Then if we multiply each of these expressions by dh, considering h as variable, and take the fluent for the whole length $4h$, we shall have to multiply the respective coefficients by $\frac{16h}{2}$, $\frac{64h}{3}$, $\frac{256h}{4}$, and $\frac{1024h}{5}$, or if $4h = l$, by $2l$, $\frac{16}{3} l$, $16l$, and $\frac{256}{5} l$, and to divide them by 1, 2, 6, and 24, making the multipliers $2l$, $\frac{8}{3} l$, $\frac{8}{3} l$, and $\frac{32}{15} l$; the whole being equal to

$$l \left(-\frac{25}{6} u + 8u_1 - 6u_2 + \frac{8}{3} u_3 - \frac{1}{2} u_4 \right.$$

$$+ \frac{70}{9} u - \frac{208}{9} u + \frac{76}{3} u - \frac{112}{9} u + \frac{22}{9} u$$

$$-\tfrac{20}{3}u+24u_1-32u_2+\tfrac{56}{3}u_3-4u_4$$
$$+\tfrac{32}{15}u-\tfrac{128}{15}u_1+\tfrac{64}{5}u_2-\tfrac{128}{15}u_3+\tfrac{32}{15}u\ )$$
$$=l\left(-\tfrac{83}{90}u+\tfrac{16}{45}u_1+\tfrac{2}{15}u_2+\tfrac{16}{45}u_3+\tfrac{7}{90}u_4\right),$$ which is the area beyond the rectangle lu, and adding this as the correction of the fluent, we have for the true area $\frac{1}{90}l(7u+32u_1+12u_2+32u_3+7u_4)$. This interpolation is very accurate where the curve does not become extremely oblique to the absciss: but for a semicircle, or semiellipsis, it gives the area too small in the ratio of .7737 to .7854, and if great accuracy were required in a similar case, it would be proper to divide the curve into two parts, and to compute the area of each separately: or to add a little by estimation; to take, for example, $8u$ instead of $7u$, which would make the area of the semicircle .784.

SCHOLIUM 3. If the ordinates are not equidistant, it will be easiest to represent them by an equation of the form $y=a+bx+cx^2+dx^3+\ldots$ consisting of as many terms as we have values of y, and finding each of the unknown quantities a, b, c, ..., by comparing these values with each other. This process is generally a little tedious, and it is not possible to shorten it materially by any artifice, though the results may be expressed in a form which is not wholly without symmetry.

388. THEOREM. If there be any number of linear equations, involving as many unknown quantities, in the form $a_1x+b_1y+\ldots-A_1$, $a_2x+b_2y+\ldots-A_2$; we shall have $x-\dfrac{\alpha A_1-\beta A_2+\gamma A_3-\ldots}{\alpha a_1-\beta a_2+\gamma a_3-\ldots}$; the coefficients α, β, γ, be-

ing obtained from the original coefficients, by exterminating all the unknown quantities, except x, in succession.

For example, if there are two equations between x and y, we have $\alpha = b_2$ and $\beta = b_1$, and $x = \frac{b_2 A_1 - b_1 A_2}{b_2 a_1 - b_1 a_2}$: if there are three, between x, y, and z, we have $\alpha = b_2 c_3 - b_3 c_2$, $\beta = b_1 c_3 - b_3 c_1$ and $\gamma = b_1 c_2 - b_2 c_1$. It will readily appear in all cases, that at every step of the process of extermination, the quantities a and A are multiplied or divided by the same factors, so that when all the other quantities are exterminated, the factor of x, which remains, must contain all the as, with the same factors as belong to the As on the opposite side of the equation. Thus, for two equations, $a_1 x + b_1 y = A_1$, and $a_2 x + b_2 y = A_2$, multiplying the first by b_2 and the second by b_1, and taking their difference, we have $a_1 b_2 x - a_2 b_1 x = A_1 b_2 - A_2 b_1$: or dividing by b_1 and b_2 respectively, $\frac{a_1}{b_1} x - \frac{a_2}{b_2} x = \frac{A_1}{b_1} - \frac{A_2}{b_2}$, which obviously leads to the same result. For three equations

$a_1 x + b_1 y + c_1 z = A_1$;

$a_2 x + b_2 y + c_2 z = A_2$;

$a_3 x + b_3 y + c_3 z = A_3$; we have first $a_1 b_2 x + \,..\, + b_2 c_1 z - A_1 b_2$, $a_2 b_1 x + \,..\, + b_1 c_2 z = A_2 b_1$, whence $(a_1 b_2 - a_2 b_1) x + (b_2 c_1 - b_1 c_2) z = A_1 b_2 - A_2 b_1$; and in the same manner

$(a_1 b_3 - a_3 b_1)\ x + (b_3 c_1 - b_1 c_3)\ z = A_1 b_3 - A_3 b_1$; and from these two results we obtain $(a_1 b_2 - a_2 b_1)\ (b_3 c_1 - b_1 c_3)\ x - (a_1 b_3 - a_3 b_1)\ (b_2 c_1 - b_1 c_2)\ x = (A_1 b_2 - A_2 b_1)\ (b_3 c_1 - b_1 c_3) - (A_1 b_3 - A_3 b_1)\ (b_2 c_1 - b_1 c_2)$: whence, by actual multiplication, we have *abbc*, or *Abbc*, marked thus, (1,2,3,1—1,2,1,3—2,1,3,1+2,1,1,3)—(1,3,2,1—1,3,1,2—3,1,2,1+3,1,1,2), or since $a_1 b_2 b_3 c_1 = a_1 b_3 b_2 c_1$, (—1,2,1,3—2,1,3,1+2,1,1,3)—(—1,3,1,2—3,1,2,1+3,1,1,2) which is divisible by $-b_1$, and may therefore be reduced to (1,2,3+2,3,1—2,1,3—1,3,2 — 3,2,1 + 3,1,2) = 1, (2,3 —3,2)—2, (1,3—3,1)+3, (1,2—2,1). And in the case of 4 equations, the analogy leads us to the value $a = b_2 (c_3 d_4 - c_4 d_3) - b_3 (c_2 d_4 - c_4 d_2) + b_4 (c_2 d_3 - c_3 d_2)$: but in all such cases, a numerical computation has the advantage in conciseness, because the sums or differences of two numbers are as easily multiplied as the numbers themselves.

The process may also be represented in a symmetrical manner by calling the second series of equations $a'_1 x + b'_1 y + c'_1 z + \ldots = A'_1$, $a'_2 x + \ldots = A'_2$, the third series $a''_1 x + \ldots = A''_1 \ldots$, until at last x is left alone on one side: a'_1 being $\frac{a_1}{b_1} - \frac{a_2}{b_2}$, $b'_1 = 0$, $c'_1 = \frac{c_1}{b_1} - \frac{c_2}{b_2}$, $a'_2 = \frac{a_1}{b_1} - \frac{a_3}{b_3} \ldots$, and so forth.

SCHOLIUM. We may take for an example the equation $y = a + bx + cx^2 + dx^3 + ex^4$, to be determined from five values of y; u, u_1, u_2, u_3, and u_4, corresponding to the values of x, 0, 1, 2, 3, and 4; or $a=u$;

$$b+c+d+e=u_1-a=u_1-u;$$
$$2b+4c+8d+16e=u_2-u;$$
$$3b+9c+27d+81e=u_3-u; \text{ and}$$
$$4b+16c+64d+256e=u_4-u:$$ we may here get the second series of equations most easily by multiplying the first by the coefficients of e, whence

$$16b+16c+16d+16e=16u_1-16u; \text{ consequently}$$
$$14b+12c+8d=16u_1-15u-u_2;$$

and in the same manner

$$78b+72c+54d=81u_1-80u-u_3; \text{ and}$$
$$252b+240c+192d=256u_1-255u-u_4.$$

Here the coefficients of c are obviously the most manageable, and they afford us $6b-6d=15u_1-10u-6u_2+u_3$, and $28b-32d=64u_1-45u-20u_2+u_4$; then taking $\frac{3}{16}$ of the latter from the former, we have $\frac{3}{4}b=3u_1-\frac{25}{16}u-\frac{9}{4}u_2+u_3-\frac{3}{16}u_4$; and $b=4u_1-\frac{25}{12}u-3u_2+\frac{4}{3}u_3-\frac{1}{4}u_4$; which agrees with the result obtained, from the inversion of Taylor's theorem, for $\frac{du'}{dx}h$: and this method, though less elegant, has the advantage of being more readily applicable to the case of ordinates not equidistant.

HOWLETT and BRIMMER,
Printers, 10, Frith-street, Soho.

Printed by BoD™in Norderstedt, Germany

9 781333 258573